A Compendium of Partial Differential Equation Models: Method of Lines Analysis with Matlab

The mathematical modeling of physical and chemical systems is used extensively throughout science, engineering, and applied mathematics. In order to make use of mathematical models, it is necessary to have solutions to the model equations. Generally, this requires numerical methods because of the complexity and number of equations.

A Compendium of Partial Differential Equation Models presents numerical methods and associated computer codes in Matlab for the solution of a spectrum of models expressed as partial differential equations (PDEs), one of the most widely used forms of mathematics in science and engineering. The authors focus on the method of lines (MOL), a well-established numerical procedure for all major classes of PDEs in which the boundary-value partial derivatives are approximated algebraically by finite differences. This reduces the PDEs to ordinary differential equations (ODEs) and thus makes the computer code easy to understand, implement, and modify. Also, the ODEs (via MOL) can be combined with any other ODEs that are part of the model (so that MOL naturally accommodates ODE/PDE models).

This book uniquely includes a detailed, line-by-line discussion of computer code as related to the associated equations of the PDE model.

William E. Schiesser is the Emeritus R. L. McCann Professor of Chemical Engineering and a Professor of Mathematics at Lehigh University. He is also a visiting professor at the University of Pennsylvania and the coauthor of the Cambridge book *Computational Transport Phenomena.*

Graham W. Griffiths is a visiting professor in the School of Engineering and Mathematical Sciences of City University, London, and a principal consultant with Honeywell Systems. He is also a founder of Special Analysis and Simulation Technology Ltd. and has worked extensively in the field of dynamic simulation of chemical processes.

A COMPENDIUM OF PARTIAL DIFFERENTIAL EQUATION MODELS

Method of Lines Analysis with Matlab

William E. Schiesser
Lehigh University, Bethlehem, PA, USA

Graham W. Griffiths
City University, London, UK

CAMBRIDGE
UNIVERSITY PRESS

CAMBRIDGE UNIVERSITY PRESS
Cambridge, New York, Melbourne, Madrid, Cape Town, Singapore, São Paulo, Delhi

Cambridge University Press
32 Avenue of the Americas, New York, NY 10013-2473, USA

www.cambridge.org
Information on this title: www.cambridge.org/9780521519861

First published 2009

Printed in the United States of America

A catalog record for this publication is available from the British Library.

Library of Congress Cataloging in Publication data

Schiesser, W. E.
A compendium of partial differential equation models : method of lines analysis with MATLAB / William E. Schiesser, Graham W. Griffiths.
p. cm.
Includes index.
ISBN 978-0-521-51986-1 (hardback)
1. Differential equations, Partial – Mathematical models. 2. MATLAB. I. Griffiths, Graham W. II. Title
QA377.S3538 2009
515′.353–dc22 2008045816

ISBN 978-0-521-51986-1 hardback

To Dolores and Patricia for their patience and support

Contents

Preface

In the analysis and the quest for an understanding of a physical system, generally, the formulation and use of a mathematical model that is thought to describe the system is an essential step. That is, a mathematical model is formulated (as a system of equations) that is thought to quantitatively define the interrelationships between phenomena that define the characteristics of the physical system. The mathematical model is usually tested against observations of the physical system, and if the agreement is considered acceptable, the model is then taken as a representation of the physical system, at least until improvements in the observations lead to refinements and extensions of the model. Often the model serves as a guide to new observations. Ideally, this process of refinement of the observations and model leads to improvements of the model and thus enhanced understanding of the physical system.

However, this process of comparing observations with a proposed model is not possible until the model equations are solved to give a solution that is then the basis for the comparison with observations. The solution of the model equations is often a challenge. Typically in science and engineering this involves the integration of systems of ordinary and partial differential equations (ODE/PDEs). The intent of this volume is to assist scientists and engineers in the process of solving differential equation models by explaining some numerical, computer-based methods that have generally been proved to be effective for the solution of a spectrum of ODE/PDE system problems.

For PDE models, we have focused on the method of lines (MOL), a well-established numerical procedure in which the PDE spatial (boundary-value) partial derivatives are approximated algebraically, in our case, by finite differences (FDs). The resulting differential equations have only one independent variable remaining, an initial-value variable, typically time in a physical application. Thus, the MOL approximation replaces a PDE system with an initial-value ODE system. This ODE system is then integrated using a standard routine, which, for the Matlab analysis used in the example applications, is one of the Matlab library integrators. In this way, we can take advantage of the recent progress in ODE numerical integrators.

However, while we have presented our MOL solutions in terms of Matlab code, it is not our intention to provide optimized Matlab code but, rather, to provide code that will be readily understood and that can be easily converted to other computer languages. This approach has been adopted in view of our experience that there is considerable interest in numerical solutions written in other computer languages such as Fortran, C, C++, and Java. Nevertheless, discussion of specific Matlab proprietary routines is included where this is thought to be of benefit to the reader.

Important variations on the MOL are possible. For example, the PDE spatial derivatives can be approximated by finite elements, finite volumes, weighted residual methods, and spectral methods. All of these approaches have been used and are described in the numerical analysis literature. For our purposes, and to keep the discussion to a reasonable length, we have focused on FDs. Specifically, we provide library routines for FDs of orders 2–10.

Our approach to describing the numerical methods is by example. Each chapter has a common format consisting of:

- An initial statement of the concepts in mathematics and computation discussed in the chapter.
- A statement of the equations to be solved numerically. These equations are a mathematical model that can originate from the analysis of a physical system. However, we have broadened the usual definition of a mathematical model for a physical system to also include equations that test a numerical method or algorithm, and in this sense, they are a model for the algorithm.

  Parenthetically, the selected PDE applications include some of the classical (we might even say "famous") PDEs. For example, we discuss the Euler and Navier Stokes equations of fluid dynamics with the Burgers equation as a special case, the Maxwell equations of electromagnetic field theory with the wave equation as a special case, and the Korteweg–deVries equation to illustrate some basic properties of solitons (as illustrated on the cover). The versatility of the MOL analysis is illustrated by linear and nonlinear PDEs in one dimension (1D), 2D, and 3D with a variety of boundary conditions, for example, Dirichlet, Neumann, and third type.
- A listing of a complete, commented computer program or code, written in Matlab, to solve the model equations. Thus, the programming is all in one place, and therefore a back-and-forth study of the chapter and programming located elsewhere (e.g., on a CD or in a Web link) is not required (although all of the Matlab routines are available from the Web site http://www.pdecomp.net).
- A step-by-step explanation of the code, with emphasis on the associated mathematics and computational algorithms at each step.
- A discussion of the output from the code, both numerically tabulated and plotted. In particular, the details of the solution that demonstrate features of the model equations and characteristics of the numerical algorithm are highlighted. The graphical output is typically in 2D and 3D, and in some applications includes movies/animations.
- The output is also evaluated with respect to accuracy, either by comparison with an analytical (exact) solution when available or by inference from changes in the approximations used in the numerical algorithms.
- A summary at the end of the chapter to reiterate (a) the general features and limitations of the numerical algorithm, and (b) the class of problems that the numerical algorithm can address.

All of the models in this volume are based on PDEs. However, because of the use of the MOL (again, in which the PDEs are replaced by systems of approximating ODEs), both ODEs and PDEs are covered along with associated algorithms. Our expectation is that the different types of models, covering all of the major classes of PDEs (parabolic, hyperbolic, elliptic), will provide a starting point for the numerical study of the ODE/PDE system of interest. This might be a straightforward modification of a computer code or extend to the development of a new code based on the ideas presented in one or more examples.

To this end, the chapters are essentially self-contained; they do not require reading the preceding chapters. Rather, we have tried to explain all of the relevant ideas within each chapter, which in some instances requires some repetition between chapters. Also, other chapters are occasionally mentioned for additional details, but it is not necessary to read those chapters. Six appendices are also included to cover concepts that are relevant to more than a single chapter.

We hope this format of self-contained chapters, rather than a chapter-to-chapter format, will be helpful in minimizing the reading and studying required to start the solution to the ODE/PDE system of interest. We welcome your comments about this organization, and your questions about any of the concepts and details presented, as reflected in the following lists of topics. We think these lists, along with the table of contents and the concluding index, will point to the chapters and pages relevant to the problem of interest.

Topic	**Chapter**
Burgers equation	5
Characteristics of hyperbolic PDEs	8
Complex PDEs	6
Conservation principles	13
Continuation methods	11
Coordinate-free operators	5, 9, 14
Cylindrical coordinates	13
Cubic Schrodinger equation	6
D'Alembert solution	8
Differential-algebraic equations (DAEs)	12
Differential operators	5, 9
Dirichlet boundary conditions	2, 5, 11
Discontinuous solutions	5, 8
Euler equations	5
Exact (analytical) solutions	1, 2, 3, 4, 5, 6, 7, 8, 10, 11, 12
Finite-difference library routines	1, 2, 3, 5, 9, 14
Finite differences (FDs)	1, 2
Front sharpening	5
Green's function analysis	3
h- and p-refinement	5, 9, 11
Helmholtz's equation	10
Higher-order FDs	1, 4
Implicit ODEs	12
Infinite spatial domains	6, 7, 8

Topic	Chapter
Inhomogeneous PDEs	4, 11
Integral invariants	3, 7
Jacobian matrix	7, 9
Korteweg–deVries equation	7
Laplace's equation	10, 11
Linear PDEs	2
Maxwell's equations	9
Method of lines (MOL)	1
Mixed boundary conditions	11, 13
Mixed (hyperbolic–parabolic) PDEs	9, 13
Mixed partial derivatives	12
Navier Stokes equations	5
Neumann boundary conditions	2, 5, 11, 13, 14
Nonlinear PDEs	4, 5, 6, 7, 13
Numerical quadrature	3, 7
Numerical solution accuracy	1, 2, 3, 4, 5, 6, 7, 8, 10, 11, 12
PDE derivations	5, 13
PDE simplification	5, 9
PDE test problems	9
Poisson's equation	10
Robin boundary conditions	11, 13
Second-order PDEs	8, 9
Shock formation	5
Simultaneous PDEs	4, 6, 13
Singularities	13, 14
Solitons	7
Source terms	11, 13, 14
Sparse matrix integration	6, 7, 9
Spatial convergence	5, 9
Spherical coordinates	14
Stagewise differentiation	3, 5, 12
Tensors	5
Third-type boundary conditions	11, 13
Three-dimensional PDEs	11
Traveling wave solutions	5, 6, 7, 8
Two-dimensional PDEs	10, 13, 14
Units in PDEs	13
Variable-coefficient PDEs	4, 13, 14
Vector operators	5, 9
Wave equation	8, 9

This list contains primarily mathematical topics. The programming in each of the chapters is also a major topic.

The six appendices cover the following topics:

Topic	Appendix
Algebraic grid points	4
Analytical solutions	3
Anisotropic diffusion	1
Cartesian coordinates	1
Conservation principles	1
Cylindrical coordinates	1
Differential grid points	4
Differential operators	1
Dirichlet boundary conditions	4
Finite-difference order conditions	2
Finite-difference test problems	2
Finite differences (FDs)	2, 5
Library FD routines	5
Movies/animations	6
PDE derivations	1
Spherical coordinates	1
Tensors	1
Time-varying boundary conditions	4
Traveling waves	3
Truncation error	2
Vector operators	1

We have assumed a background of basic calculus and ODEs. Since the central algorithm is the MOL, we begin with a MOL introduction in Chapter 1. Then the chapters progress through example applications of increasing complexity and diversity. The preceding list serves as a guide for specific topics.

We have not included exercises at the end of the chapters since we think variations in the applications and the associated Matlab codes provide ample opportunities for exploration and further study. References are provided at the end of the chapters and appendices when we think they would provide useful additional background, but we have not attempted a comprehensive list of references on any particular topic.

Our intent for this volume is to present mathematical and computational methods that can be applied to a broad spectrum of ODE/PDE models. In particular, we are attempting to assist engineers and scientists who have an ODE/PDE problem of interest and who wish to produce an accurate numerical solution with reasonable computational effort without having to first delve into the myriad details of numerical methods and computer programming. We hope this book is of assistance toward meeting this objective.

William E. Schiesser
Bethlehem, PA, USA

Graham W. Griffiths
Nayland, Suffolk, UK
August 1, 2008

1

An Introduction to the Method of Lines[1]

The chapters in this book pertain particularly to mathematical models expressed as *partial differential equations* (PDEs). The computer-based numerical solution of the PDE models is implemented primarily through the *method of lines* (MOL). We therefore start with this chapter, which is an introduction to the MOL. Although the reader may be familiar with the MOL, we suggest reading this chapter since it describes some aspects and details of our use of the MOL that appear in the subsequent chapters. We start with some basic features of PDEs.

SOME PDE BASICS

Our physical world is most generally described in scientific and engineering terms with respect to three-dimensional (3D) space and time, which we abbreviate as *spacetime*. PDEs provide a mathematical description of physical spacetime, and they are therefore among the most widely used forms of mathematics. As a consequence, methods for the solution of PDEs, such as the MOL, are of broad interest in science and engineering.

As a basic illustrative example of a PDE, we consider

$$\frac{\partial u}{\partial t} = D\frac{\partial^2 u}{\partial x^2} \tag{1.1}$$

where

u dependent variable (dependent on x and t)
t independent variable representing time
x independent variable representing one dimension of 3D space
D constant explained next

Note that Eq. (1.1) has two independent variables, x and t, which is the reason it is classified as a PDE (any differential equation with more than one independent

[1] This chapter is based on an article that originally appeared in the online encyclopedia *Scholarpedia* [1].

variable is a PDE). A differential equation with only one independent variable is generally termed an *ordinary differential equation* (ODE); we will consider ODEs later as part of the MOL.

Equation (1.1) is termed the *diffusion equation*. When applied to heat transfer, it is *Fourier's second law*; the dependent variable u is temperature and D is the *thermal diffusivity*. When Eq. (1.1) is applied to mass diffusion, it is *Fick's second law*; u is mass concentration and D is the *coefficient of diffusion* or the *diffusivity*.

$\partial u/\partial t$ is a partial derivative of u with respect to t (x is held constant when taking this partial derivative, which is why *partial* is used to describe this derivative). Equation (1.1) is *first order in t* since the highest-order partial derivative in t is first order; it is *second order in x* since the highest-order derivative in x is second order. Equation (1.1) is *linear* or *first degree* since all of the terms are to the first power (note that *order* and *degree* can easily be confused).

INITIAL AND BOUNDARY CONDITIONS

Before we consider a solution to Eq. (1.1), we must specify some *auxiliary conditions* to complete the statement of the PDE problem. The number of required auxiliary conditions is determined by the *highest-order derivative in each independent variable*. Since Eq. (1.1) is first order in t and second order in x, it requires one auxiliary condition in t and two auxiliary conditions in x. To have a complete, *well-posed* problem, some additional conditions may have to be included, for example, that specify valid ranges for coefficients. However, this is a more advanced topic and will not be developed further here.

t is termed an *initial-value variable* and therefore requires *one initial condition* (IC). It is an initial-value variable since it starts at an initial value, t_0, and moves forward over a *finite interval* $t_0 \leq t \leq t_f$ or a *semi-infinite interval* $t_0 \leq t \leq \infty$ without any additional conditions being imposed. Typically in a PDE application, the initial-value variable is time, as in the case of Eq. (1.1).

x is termed a *boundary-value variable* and therefore requires *two boundary conditions* (BCs). It is a boundary-value variable since it varies over a *finite interval* $x_0 \leq x \leq x_f$, a *semi-infinite interval* $x_0 \leq x \leq \infty$, or a *fully infinite interval* $-\infty \leq x \leq \infty$, and at *two different values of* x, conditions are imposed on u in Eq. (1.1). Typically, the two values of x correspond to boundaries of a physical system, and hence the name *boundary conditions*.

As examples of auxiliary conditions for Eq. (1.1),

- An IC could be

$$u(x, t = 0) = u_0 \tag{1.2}$$

where u_0 is a given function of x (typically a constant) for $x_0 \leq x \leq x_f$.

- Two BCs could be

$$u(x = x_0, t) = u_b \tag{1.3a}$$

$$\frac{\partial u(x = x_f, t)}{\partial x} = 0 \tag{1.3b}$$

where u_b is a given boundary (constant) value of u for all t.

- Another common possibility is where the IC is given as earlier and the BCs are $u(x = x_0, t) = f_0(t)$ and $u_x(x = x_f, t) = f_b(t)$.

An important consideration is the possibility of *discontinuities at the boundaries*, produced, for example, by differences in ICs and BCs at the boundaries, which can cause computational difficulties, particularly for *hyperbolic PDEs* (such as the classic linear wave equation $\partial^2 u/\partial t^2 = \partial^2 u/\partial x^2$).

BCs can be of three types:

1. If the dependent variable is specified, as in BC (1.3a), the BC is termed *Dirichlet*.
2. If the derivative of the dependent variable is specified, as in BC (1.3b), the BC is termed *Neumann*.
3. If both the dependent variable and its derivative appear in the BC, it is termed a *BC of the third type* or a *Robin BC*.

TYPES OF PDE SOLUTIONS

Equations (1.1)–(1.3) constitute a complete (*well-posed*) PDE problem and we can now consider what we mean by a solution to this problem. Briefly, the solution of a PDE problem is a *function that defines the dependent variable as a function of the independent variables* – in this case, $u(x, t)$. In other words, we seek a function that when substituted in the PDE and all of its auxiliary conditions satisfies simultaneously all of these equations.

The solution can be of two types:

1. If the solution is an actual mathematical function, it is termed an *analytical solution*. While analytical solutions are the "gold standard" for PDE solutions in the sense that they are exact, they are also generally difficult to derive mathematically for all but the simplest PDE problems (in much the same way that solutions to nonlinear algebraic equations generally cannot be derived mathematically except for certain classes of nonlinear equations).
2. If the solution is in numerical form, for example, $u(x, t)$ tabulated numerically as a function of x and t, it is termed a *numerical solution*. Ideally, the numerical solution is simply a numerical evaluation of the analytical solution. But since an analytical solution is generally unavailable for realistic PDE problems in science and engineering, the *numerical solution is an approximation to the analytical solution*, and our expectation is that it represents the analytical solution with good accuracy. However, numerical solutions can be computed with modern-day computers for very complex problems, and they will generally have good accuracy (even though this cannot be established directly by comparison with the analytical solution since the latter is usually unknown).

The focus of the MOL is *the calculation of accurate numerical solutions*.

PDE SUBSCRIPT NOTATION

Before we go on to the general classes of PDEs that the MOL can handle, we briefly discuss an alternative notation for PDEs. Instead of writing the partial derivatives as in Eq. (1.1), we adopt a *subscript notation* that is easier to state and bears a closer resemblance to the associated computer coding. For example, we can write Eq. (1.1) as

$$u_t = Du_{xx} \tag{1.4}$$

where, for example, u_t is subscript notation for $\partial u/\partial t$. In other words, a partial derivative is represented as the dependent variable with a subscript that defines the independent variable. For a derivative that is of order n, the independent variable is repeated n times; for example, for Eq. (1.1), u_{xx} represents $\partial^2 u/\partial x^2$.

A GENERAL PDE SYSTEM

Using the subscript notation, we can now consider some general PDEs. For example, a general PDE first order in t can be considered:

$$\overline{u}_t = \overline{f}(\overline{x}, t, \overline{u}, \overline{u}_{\overline{x}}, \overline{u}_{\overline{xx}}, \ldots) \tag{1.5}$$

where an overbar (overline) denotes a vector. For example, $\overline{u}$ denotes a vector of n dependent variables

$$\overline{u} = (u_1, u_2, \ldots, u_n)^T$$

that is, a system of n simultaneous PDEs. Similarly, $\overline{f}$ denotes an n vector of derivative functions

$$\overline{f} = (f_1, f_2, \ldots, f_n)^T$$

where T denotes a *transpose* (here a row vector is transposed to a column vector). Note also that $\overline{x}$ is a vector of spatial coordinates, so that, for example, in *Cartesian coordinates* $\overline{x} = (x, y, z)^T$, while in *cylindrical coordinates* $\overline{x} = (r, \theta, z)^T$. Thus, Eq. (1.5) can represent PDEs in one, two, and three spatial dimensions.

Since Eq. (1.5) is first order in t, it requires one IC

$$\overline{u}(\overline{x}, t = 0) = \overline{u}_0(\overline{x}, \overline{u}, \overline{u}_{\overline{x}}, \overline{u}_{\overline{xx}}, \ldots) \tag{1.6}$$

where $\overline{u}_0$ is an n vector of IC functions

$$\overline{u}_0 = (u_{10}, u_{20}, \ldots, u_{n0})^T$$

The derivative vector $\overline{f}$ of Eq. (1.5) includes functions of various spatial derivatives, $(\overline{u}, \overline{u}_{\overline{x}}, \overline{u}_{\overline{xx}}, \ldots)$, and therefore we cannot state a priori the required number of BCs. For example, if the highest-order derivative in $\overline{x}$ in all of the derivative functions is second order, then we require $2 \times d \times n$ BCs, where d is the number of spatial dimensions. Thus, for Eq. (1.4), the number of required BCs is 2 (second order in x) $\times 1$ (one dimensional) $\times 1$ (one PDE) $= 2$.

We state the general BC requirement of Eq. (1.5) as

$$\overline{f}_b(\overline{x}_b, \overline{u}, \overline{u}_{\overline{x}}, \overline{u}_{\overline{xx}}, \ldots) = 0 \tag{1.7}$$

where the subscript b denotes *boundary*. The vector of BC functions, $\overline{f}_b$, has a length (number of functions) determined by the highest-order derivative in $\overline{x}$ in each PDE (in Eq. (1.5)), as discussed previously.

PDE GEOMETRIC CLASSIFICATION

Equations (1.5)–(1.7) constitute a general PDE system to which the MOL can be applied. Before proceeding to the details of how this might be done, we need to discuss the three basic forms of the PDEs as classified geometrically. This *geometric classification* can be done rigorously if certain mathematical forms of the functions in Eqs. (1.5)–(1.7) are assumed. However, we will adopt a somewhat more descriptive (less rigorous but more general) form of these functions for the specification of the three geometric classes.

If the derivative functions in Eq. (1.5) contain *only first-order derivatives in* $\bar{x}$, the PDEs are classified as *first-order hyperbolic*. As an example, the equation

$$u_t + vu_x = 0 \tag{1.8}$$

is generally called the *linear advection equation*; in physical applications, v is a linear or flow velocity. Although Eq. (1.8) is possibly the simplest PDE, this simplicity is deceptive in the sense that it can be very difficult to integrate numerically since it *propagates discontinuities*, a distinctive feature of first-order hyperbolic PDEs.

Equation (1.8) is termed a *conservation law* since it typically expresses conservation of *mass, energy, or momentum* under the conditions for which it is derived, that is, the *assumptions on which the equation is based*. Conservation laws are a bedrock of PDE mathematical models in science and engineering, and an extensive literature pertaining to their solution, both analytical and numerical, has evolved over many years.

An example of a *first-order hyperbolic system* (using the notation $u_1 \Rightarrow u$, $u_2 \Rightarrow v$) is

$$u_t = v_x \tag{1.9a}$$

$$v_t = u_x \tag{1.9b}$$

Equations (1.9a) and (1.9b) constitute a system of *two linear, constant-coefficient, first-order hyperbolic PDEs*.

Differentiation and algebraic substitution can occasionally be used to eliminate some dependent variables in systems of PDEs. For example, if Eq. (1.9a) is differentiated with respect to t and Eq. (1.9b) is differentiated with respect to x

$$u_{tt} = v_{xt}$$

$$v_{tx} = u_{xx}$$

we can then eliminate the mixed partial derivative between these two equations (assuming v_{xt} in the first equation equals v_{tx} in the second equation) to obtain

$$u_{tt} = u_{xx} \tag{1.10}$$

Equation (1.10) is the *second-order hyperbolic wave equation*.

If the derivative functions in Eq. (1.5) contain *only second-order derivatives in* $\bar{x}$, the PDEs are classified as *parabolic*. Equation (1.1) is an example of a parabolic PDE.

Finally, if a PDE *contains no derivatives in t* (e.g., the LHS of Eq. (1.5) is zero), it is classified as *elliptic*. As an example,

$$u_{xx} + u_{yy} = 0 \tag{1.11}$$

is *Laplace's equation*, where x and y are spatial independent variables in Cartesian coordinates. Note that with no derivatives in t, *elliptic PDEs require no ICs*; that is, they are entirely boundary-value PDEs.

PDEs with *mixed geometric characteristics* are possible and, in fact, are quite common in applications. For example, the PDE

$$u_t = -u_x + u_{xx} \tag{1.12}$$

is *hyperbolic–parabolic*. Since it frequently models convection (hyperbolic) through the term u_x and diffusion (parabolic) through the term u_{xx}, it is generally termed a *convection–diffusion equation*. If additionally, it includes a function of the dependent variable u such as

$$u_t = -u_x + u_{xx} + f(u) \tag{1.13}$$

then it might be termed a *convection–diffusion–reaction equation* since $f(u)$ typically models the rate of a chemical reaction. If the function is only for the independent variables, that is,

$$u_t = -u_x + u_{xx} + g(x, t) \tag{1.14}$$

the equation could be labeled an *inhomogeneous PDE*.

This discussion clearly indicates that PDE problems come in a very wide variety, depending, for example, on linearity, types of coefficients (constant, variable), coordinate system, geometric classification (hyperbolic, elliptic, parabolic), number of dependent variables (number of simultaneous PDEs), number of independent variables (number of dimensions), types of BCs, smoothness of the IC, and so on, so it might seem impossible to formulate numerical procedures with any generality that can address a relatively broad spectrum of PDEs. But in fact, the MOL provides a surprising degree of generality, although the success in applying it to a new PDE problem depends to some extent on the experience and inventiveness of the analyst; that is, MOL is not a single, straightforward, clearly defined approach to PDE problems, but rather is a general concept (or philosophy) that requires specification of details for each new PDE problem. We now proceed to illustrate the formulation of a MOL numerical algorithm, with the caveat that this will not be a general discussion of MOL as it is applied to any conceivable PDE problem.

ELEMENTS OF THE MOL

The basic idea of the MOL is to *replace the spatial (boundary-value) derivatives in the PDE with algebraic approximations*. Once this is done, the spatial derivatives are no longer stated explicitly in terms of the spatial independent variables. Thus, in effect, *only the initial-value variable, typically time in a physical problem, remains*. In other words, with only one remaining independent variable, we have a *system of ODEs that approximate the original PDE*. The challenge, then, is to formulate the approximating system of ODEs. Once this is done, we can apply any integration algorithm for initial-value ODEs to compute an approximate numerical solution to the PDE. Thus, one of the salient features of the MOL is the use of *existing, and generally well-established, numerical methods for ODEs.*

To illustrate this procedure, we consider the MOL solution of Eq. (1.8). First we need to replace the spatial derivative u_x with an algebraic approximation. In this case we will use a *finite difference* (FD), such as

$$u_x \approx \frac{u_i - u_{i-1}}{\Delta x} \tag{1.15}$$

where i is an *index designating a position along a grid in x and Δx is the spacing in x along the grid*. Thus, for the left-end value of x, $i = 1$, and for the right-end value of x, $i = M$; that is, the grid in x has M points. Then the MOL approximation of Eq. (1.8) is

$$\frac{du_i}{dt} = -v\frac{u_i - u_{i-1}}{\Delta x}, \quad 1 \le i \le M \tag{1.16}$$

Note that Eq. (1.16) is written as an ODE since there is now *only one independent variable, t*. Note also that Eq. (1.16) represents a system of M ODEs.

This transformation of a PDE, Eq. (1.8), to a system of ODEs, Eq. (1.16), illustrates the essence of the MOL, namely, *the replacement of the spatial derivatives, in this case u_x, so that a system of ODEs approximates the original PDE*. Then, *to compute the solution of the PDE, we compute a solution to the approximating system of ODEs*. But before considering this integration in t, we have to complete the specification of the PDE problem. Since Eq. (1.8) is first order in t and first order in x, it requires one IC and one BC. These will be taken as

$$u(x, t = 0) = f(x) \tag{1.17a}$$

$$u(x = 0, t) = g(t) \tag{1.17b}$$

Since Eq. (1.16) constitutes M first-order, initial-value ODEs, M initial conditions are required, and from Eq. (1.17a), these are

$$u(x_i, t = 0) = f(x_i), \quad 1 \le i \le M \tag{1.18a}$$

Also, application of BC (1.17b) gives for grid point $i = 1$

$$u(x_1, t) = g(t) \tag{1.18b}$$

Equations (1.16) and (1.18) now constitute the complete MOL approximation of Eq. (1.8) subject to Eqs. (1.17a) and (1.17b). The solution of this ODE system gives the M functions

$$u_1(t), u_2(t), \ldots, u_{M-1}(t), u_M(t) \tag{1.19}$$

that is, an approximation to $u(x, t)$ at the grid points $i = 1, 2, \ldots, M$.

Before we go on to consider the numerical integration of the approximating ODEs, in this case Eq. (1.16), we briefly consider further the FD approximation of Eq. (1.15), which can be written as

$$u_x \approx \frac{u_i - u_{i-1}}{\Delta x} + O(\Delta x) \tag{1.20}$$

where $O(\Delta x)$ denotes *of order* Δx; that is, the truncation error of the approximation of Eq. (1.16) is *proportional to* Δx *to the first power (varies linearly with* Δx*)*; thus, Eq. (1.20) is also termed a *first-order* FD (since Δx is to the first power in the order or truncation error term). The term *truncation error* reflects the fact that the FD of Eq. (1.15) comes from a *truncated Taylor series.*

Note that the numerator of Eq. (1.15), $u_i - u_{i-1}$, is a difference in two values of u. Also, the denominator Δx remains finite (nonzero). Hence the name *finite difference* (and it is an approximation because of the truncated Taylor series, so a more complete description is *first-order FD approximation*). In fact, in the limit $\Delta x \to 0$, the approximation of Eq. (1.15) becomes *exactly the derivative.* However, in a practical computer-based calculation, Δx remains finite, so Eq. (1.15) remains an approximation.

Also, Eq. (1.8) typically describes the flow of a physical quantity such as concentration of a chemical species or temperature, represented by u, from left to right with respect to x with velocity v. Then, using the FD approximation of Eq. (1.20) at i involves u_i and u_{i-1}. In a flowing system, u_{i-1} is to the left (in x) of u_i or is *upstream* or *upwind* of u_i (to use a nautical analogy). Thus, Eq. (1.20) is termed a *first-order upwind FD approximation.* Generally, for strongly convective systems such as that modeled by Eq. (1.8), *some form of upwinding is required in the numerical solution of the descriptive PDEs*; we will look at this requirement further in the subsequent discussion.

ODE INTEGRATION WITHIN THE MOL

We now consider briefly the numerical integration of the M ODEs of Eq. (1.16). If the derivative du_i/dt is approximated by a first-order FD

$$\frac{du_i}{dt} \approx \frac{u_i^{n+1} - u_i^n}{\Delta t} + O(\Delta t) \tag{1.21}$$

where n is an index for the variable t (t moves forward in steps denoted or indexed by n), then an FD approximation of Eq. (1.16) is

$$\frac{u_i^{n+1} - u_i^n}{\Delta t} = -v\frac{u_i^n - u_{i-1}^n}{\Delta x}$$

or solving for u_i^{n+1},

$$u_i^{n+1} = u_i^n - (v\Delta t/\Delta x)(u_i^n - u_{i-1}^n), \quad i = 1, 2, \ldots, M \tag{1.22}$$

Equation (1.22) has the important characteristic that it gives u_i^{n+1} *explicitly*; that is, we can solve for the solution at the advanced point in t, $n+1$, from the solution at the base point n. In other words, explicit numerical integration of Eq. (1.16) is by the *forward FD* of Eq. (1.21), and this procedure is generally termed the *forward Euler method*, which is the most basic form of ODE integration.

While the explicit form of Eq. (1.22) is computationally convenient, it has a possible limitation. If the time step Δt is *above a critical value, the calculation becomes unstable*, which is manifest by successive changes in the dependent variable, $\Delta u = u_i^{n+1} - u_i^n$, becoming larger and eventually unbounded as the calculation moves forward in t (for increasing n). In fact, for the solution of Eq. (1.8) by the method of Eq. (1.22) to remain stable, the dimensionless group $(v\Delta t/\Delta x)$, which is called the *Courant-Friedricks-Lewy or CFL number*, must remain below a critical value – in this case, unity. Note that this *stability limit* places an upper limit on Δt for a given v and Δx; if one attempts to increase the accuracy of Eq. (1.22) by using a smaller Δx (larger number of grid points in x by increasing M), a smaller value of Δt is required to keep the CFL number below its critical value. Thus, there is a *conflicting requirement of improving accuracy while maintaining stability.*

The stability limit of the explicit Euler method as implemented via the forward FD of Eq. (1.21) can be circumvented by using a *backward FD* for the derivative in t

$$\frac{du_i}{dt} \approx \frac{u_i^n - u_i^{n-1}}{\Delta t} + O(\Delta t) \tag{1.23}$$

so that the FD approximation of Eq. (1.16) becomes

$$\frac{u_i^n - u_i^{n-1}}{\Delta t} = -v\frac{u_i^n - u_{i-1}^n}{\Delta x}$$

or after rearrangement (with $(v\Delta t/\Delta x) = \alpha$),

$$(1+\alpha)u_i^n - \alpha u_{i-1}^n = u_i^{n-1}, \quad i = 1, 2, \ldots, M \tag{1.24}$$

Note that we cannot now solve Eq. (1.24) explicitly for the solution at the advanced point u_i^n in terms of the solution at the base point u_i^{n-1}. Rather, Eq. (1.24) is *implicit* in u_i^n because u_{i-1}^n is also unknown; that is, we must solve Eq. (1.24) written for each grid point $i = 1, 2, \ldots, M$ as a simultaneous system of *bidiagonal equations* (bidiagonal because each of Eq. (1.24) has two unknowns so that *simultaneous solution of the full set of approximating algebraic equations is required* to obtain the complete numerical solution $u_1^n, u_2^n, \ldots, u_M^n$). Thus, the solution of Eq. (1.24) is termed the *implicit Euler method.*

We could then naturally ask why use Eq. (1.24) when Eq. (1.22) is so much easier to use (explicit calculation of the solution at the next step in t of Eq. (1.22) vs. the implicit calculation of Eq. (1.24)). The answer is that the implicit calculation of Eq. (1.24) is often worthwhile because the *implicit Euler method has no stability limit* (is *unconditionally stable* in comparison with the explicit method, with the stability limit stated in terms of the CFL condition). However, there is a price to pay for the improved stability of the implicit Euler method; that is, we must solve a *system of simultaneous algebraic equations*; Eq. (1.24) is an example. Furthermore, if the original ODE system approximating the PDE is nonlinear, we have to solve a *system of nonlinear algebraic equations.* (Equation (1.24) is linear, so the solution is much easier.) The system of nonlinear equations is typically solved by a *variant of Newton's method* that can become very demanding computationally if the number of ODEs is large (due to the use of a large number of spatial grid points in the MOL approximation of the PDE, especially when we attempt the solution of 2D and 3D PDEs). If you have had some experience with Newton's method, you may appreciate that the *Jacobian matrix* of the nonlinear algebraic system can *become very large and sparse as the number of spatial grid points increases.*

Additionally, although there is no limit for Δt with regard to stability for the implicit method, there is a *limit with regard to accuracy*. In fact, the first-order upwind approximation of u_x in Eq. (1.8), Eq. (1.20), and the first-order approximation of u_t in Eq. (1.8), Eq. (1.21) or (1.23), taken together limit the accuracy of the resulting FD approximation of Eq. (1.8). One way around this accuracy limitation is to use *higher-order FD approximations for the derivatives in Eq. (1.8).*

For example, if we consider the second-order approximation of u_x at i

$$u_x \approx \frac{u_{i+1} - u_{i-1}}{2\Delta x} + O(\Delta x^2) \tag{1.25}$$

substitution in Eq. (1.8) gives the MOL approximation of Eq. (1.8)

$$\frac{du_i}{dt} = -v\frac{u_{i+1} - u_{i-1}}{2\Delta x}, \quad 1 \le i \le M \tag{1.26}$$

We could then reason that if the integration in t is performed by the explicit Euler method, that is, we use the approximation of Eq. (1.21) for $u_t = du_i/dt$, the resulting numerical solution should be more accurate than the solution from Eq. (1.22). In fact, the MOL approximation based on this idea

$$u_i^{n+1} = u_i^n - \frac{v\Delta t}{2\Delta x}(u_{i+1}^n - u_{i-1}^n), \quad i = 1, 2, \ldots, M \tag{1.27}$$

is *unconditionally unstable*; this conclusion can be demonstrated by a *von Neumann stability analysis* that we will not cover here. This surprising result demonstrates that *replacing the derivatives in PDEs with higher-order approximations does not necessarily guarantee more accurate solutions, or even stable solutions.*

NUMERICAL DIFFUSION AND OSCILLATION

Even if the implicit Euler method is used for the integration in t of Eq. (1.26) to achieve stability (or a more sophisticated explicit integrator in t is used that automatically adjusts Δt to achieve a prescribed accuracy), we would find that the solution *oscillates* unrealistically. This numerical distortion is one of two generally observed forms of numerical error. The other numerical distortion is *diffusion* that would be manifest in the solution from Eq. (1.22). Briefly, the solution would exhibit excessive smoothing or rounding at points in x where the solution changes rapidly. This overall observation that a *first-order approximation of u_x produces numerical diffusion, while higher-order approximations of u_x produce numerical oscillation* is predicted by the *Godunov order barrier theorem for the Riemann problem* [2]. To explain briefly, the order barrier is first order and *any linear FD approximation above first order will be oscillatory*. Equation (1.8) is an example of the Riemann problem [2] if IC Eq. (1.17a) is discontinuous; for example, $u(x, t = 0) = h(t)$, where $h(t)$ is the *Heaviside unit step function*. The (exact) analytical solution is the IC function $f(x)$ of Eq. (1.17a) moving left to right with velocity v (from Eq. (1.8)) and without distortion, that is, $u(x, t) = f(x - vt)$; however, the numerical solution will oscillate if u_x in Eq. (1.8) is replaced with a linear approximation of second or higher order.

We should also mention one point of terminology for FD approximations. The RHS of Eq. (1.25) is an example of a *centered approximation* since the two points at $i + 1$ and $i - 1$ are centered around the point i. Equation (1.20) is an example of a *noncentered, one-sided*, or *upwind approximation* since the points i and $i - 1$ are not centered with respect to i.

Finally, to conclude the discussion of first-order hyperbolic PDEs such as Eq. (1.8), since the Godunov theorem indicates that FD approximations above first order will produce numerical oscillations in the solution, the question remains if there are approximations above first order that are nonoscillatory. To answer this question we note first that *the Godunov theorem applies to linear approximations*; for example, Eq. (1.25) is a linear approximation since u on the RHS is to the first power. If, however, we consider nonlinear approximations of u_x, we can in fact develop approximations that are nonoscillatory. The details of such nonlinear approximations are beyond the scope of this discussion, so we will merely mention that they are termed *high-resolution* methods that seek a *total variation diminishing solution*. Such methods, which include *flux limiter* and *weighted essentially nonoscillatory* methods, seek to avoid nonreal oscillations when shocks or discontinuities occur in the solution (such as in the Riemann problem) [3].

So far we have considered only the MOL solution of first-order PDEs, for example, Eq. (1.8). We conclude this discussion of the MOL by considering a second-order PDE, the parabolic Eq. (1.1). To begin, we need an approximation for the second derivative u_{xx}. A commonly used *second-order, central approximation* is (again, derived from the Taylor series, so the term $O(\Delta x^2)$ represents the truncation error)

$$u_{xx} \approx \frac{u_{i+1} - 2u_i + u_{i-1}}{\Delta x^2} + O(\Delta x^2) \tag{1.28}$$

Substitution of Eq. (1.28) in Eq. (1.8) gives a system of approximating ODEs

$$\frac{du_i}{dt} = D\frac{u_{i+1} - 2u_i + u_{i-1}}{\Delta x^2}, \quad i = 1, 2, \ldots, M \tag{1.29}$$

Equation (1.29) is then integrated subject to IC (1.2) and BCs (1.3a) and (1.3b). This integration in t can be by the explicit Euler method, the implicit Euler method, or any other higher-order integrator for initial-value ODEs. Generally stability is not as much of a concern as with the previous hyperbolic PDEs (a characteristic of parabolic PDEs that tend to smooth solutions rather than hyperbolic PDEs that tend to propagate nonsmooth conditions). However, stability constraints do exist for explicit methods. For example, for the explicit Euler method with a step Δt in t, the stability constraint is $D\Delta t/\Delta x^2 <$ constant (so that as Δx is reduced to achieve better spatial accuracy in x, Δt must also be reduced to maintain stability).

Before proceeding with the integration of Eq. (1.29), we must include BCs (1.3a) and (1.3b). The Dirichlet BC at $x = x_0$, Eq. (1.3a), is merely

$$u_1 = u_b \tag{1.30}$$

and therefore the ODE of Eq. (1.29) for $i = 1$ is not required and the ODE for $i = 2$ becomes

$$\frac{du_2}{dt} = D\frac{u_3 - 2u_2 + u_b}{\Delta x^2} \tag{1.31}$$

DIFFERENTIAL ALGEBRAIC EQUATIONS

Equation (1.30) is *algebraic*, and therefore in combination with the ODEs of Eq. (1.29), we have a *differential algebraic (DAE) system*.

At $i = M$, we have Eq. (1.29)

$$\frac{du_M}{dt} = D\frac{u_{M+1} - 2u_M + u_{M-1}}{\Delta x^2} \tag{1.32}$$

Note that u_{M+1} is outside the grid in x; that is, $M + 1$ is a *fictitious point*. But we must assign a value to u_{M+1} if Eq. (1.32) is to be integrated. Since this requirement occurs at the boundary point $i = M$, we obtain this value by approximating BC (1.3b) using the centered FD approximation of Eq. (1.25)

$$u_x \approx \frac{u_{M+1} - u_{M-1}}{2\Delta x} = 0$$

or

$$u_{M+1} = u_{M-1} \tag{1.33}$$

We can add Eq. (1.33) as an *algebraic equation to our system of equations*, that is, continue to use the DAE format, or we can substitute u_{M+1} from Eq. (1.33) into the Eq. (1.32)

$$\frac{du_M}{dt} = D\frac{u_{M-1} - 2u_M + u_{M-1}}{\Delta x^2} \tag{1.34}$$

and arrive at an ODE system (Eq. (1.29) for $i = 3, \ldots, M - 1$, Eq. (1.31) for $i = 2$, and Eq. (1.34) for $i = M$). Both approaches, either an ODE system or a DAE system, have been used in MOL studies. Either way, we now have a complete formulation of the MOL ODE or DAE system, including the BCs at $i = 1, M$ in Eq. (1.29). The integration of these equations then gives the numerical solution $u_1(t), u_2(t), \ldots, u_M(t)$. The preceding discussion is based on a relatively basic DAE system, but it indicates that integrators designed for DAE systems can play an important role in MOL analysis.

If the implicit Euler method is applied to Eq. (1.29), we have

$$\frac{u_i^{n+1} - u_i^n}{\Delta t} = D\frac{u_{i+1}^{n+1} - 2u_i^{n+1} + u_{i-1}^{n+1}}{\Delta x^2}, \quad i = 1, 2, \ldots, M$$

or (with $\alpha = D\Delta t/\Delta x^2$),

$$u_{i+1}^{n+1} - (1/\alpha + 2)u_i^{n+1} + u_{i-1}^{n+1} = (1/\alpha)u_i^n, \quad i = 1, 2, \ldots, M$$

which is a *tridiagonal* system of algebraic equations (three unknowns in each equation). Since such *banded* systems (the nonzero elements are banded around the main diagonal) are common in the numerical solution of PDE systems, special algorithms have been developed to take advantage of the banded structure, typically by not storing and using the zero elements outside the band. These special algorithms that take advantage of the *structure of the problem equations* can result in major savings in computation time. In the case of tridiagonal equations, the special algorithm is generally called *Thomas' method*. If the coefficient matrix of the algebraic system does not have a well-defined structure, such as bidiagonal or tridiagonal, but consists of mostly zeros with a relatively small number of nonzero elements, which is often the case in the numerical solution of PDEs, the coefficient matrix is said to be *sparse*; special algorithms and associated software for sparse systems have been developed that can result in very substantial savings in the storage and numerical manipulation of sparse matrices.

Generally when applying the MOL, the integration of the approximating ODE/DAEs (e.g., Eqs. (1.21) and (1.29)) is accomplished by using *library routines for initial-value* ODE/DAEs. In other words, the explicit programming of the ODE/DAE integration (such as the explicit or implicit Euler method) is avoided; rather, an established integrator is used. This has the advantage that (1) the detailed programming of the integration can be avoided, particularly the linear algebra (solution of simultaneous equations) required by an implicit integrator, so that the MOL analysis is substantially simplified, and (2) library routines (usually written by experts) include features that make these routines especially effective (robust) and efficient such as automatic integration step size adjustment and the use of higher-order integration methods (beyond the first-order accuracy of the Euler methods); also, generally, they have been thoroughly tested. Thus, the use of quality ODE/DAE library routines is usually an essential part of MOL analysis. We therefore list at the end of this chapter some public domain sources of such library routines.

HIGHER DIMENSIONS AND DIFFERENT COORDINATE SYSTEMS

To conclude this discussion of the MOL solution of PDEs, we cover two additional points. First, we have considered PDEs in only Cartesian coordinates, and in fact, just one Cartesian coordinate, x. But MOL analysis can in principle be carried out in any coordinate system. Thus, Eq. (1.1) can be generalized to

$$\frac{\partial u}{\partial t} = D\nabla^2 u \tag{1.35}$$

where ∇^2 is the *coordinate independent Laplacian operator* that can then be expressed in terms of a particular coordinate system. For example, in cylindrical coordinates, Eq. (1.35) is

$$\frac{\partial u}{\partial t} = D\left(\frac{\partial^2 u}{\partial r^2} + \frac{1}{r}\frac{\partial u}{\partial r} + \frac{1}{r^2}\frac{\partial^2 u}{\partial \theta^2} + \frac{\partial^2 u}{\partial z^2}\right) \tag{1.36}$$

and in spherical coordinates, Eq. (1.35) is

$$\frac{\partial u}{\partial t} = D\left[\frac{\partial^2 u}{\partial r^2} + \frac{2}{r}\frac{\partial u}{\partial r} + \frac{1}{r^2}\left(\frac{\partial^2 u}{\partial \theta^2} + \frac{\cos\theta}{\sin\theta}\frac{\partial u}{\partial \theta}\right) + \frac{1}{r^2\sin^2\theta}\frac{\partial^2 u}{\partial \phi^2}\right] \tag{1.37}$$

The challenge then in applying the MOL to PDEs such as Eqs. (1.36) and (1.37) is the algebraic approximation of the RHS ($\nabla^2 u$) using, for example, FDs, *finite elements* or *finite volumes*; all of these approximations have been used in MOL analysis, as well as *Galerkin*, *least squares*, *spectral*, and other methods. A particularly demanding step is *regularization of singularities* such as at $r = 0$ (note the number of divisions by r in the RHS of Eqs. (1.36) and (1.37)) and at $\theta = 0, \pi/2$ (note the divisions by $\sin(\theta)$ in Eq. (1.37)). Thus the application of the MOL typically requires analysis based on the experience and creativity of the analyst (i.e., it is generally not a mechanical procedure from beginning to end).

The complexity of the numerical solution of higher-dimensional PDEs in various coordinate systems prompts the question of why a particular coordinate system would be selected over others. The mathematical answer is that the judicious choice of a coordinate system *facilitates the implementation of the BCs in the numerical solution*. The answer based on physical considerations is that the *coordinate system is selected to reflect the geometry of the problem system*. For example, if the physical system has the shape of a cylinder, cylindrical coordinates would be used. This choice then facilitates the implementation of the BC at the exterior surface of the physical system (exterior surface of the cylinder). However, this can also lead to complications such as the $r = 0$ singularities in Eq. (1.36) (due to the variable $1/r$ and $1/r^2$ coefficients). The resolution of these complications is generally worth the effort rather than the use of a coordinate system that does not naturally conform to the geometry of the physical system. If the physical system is not shaped in accordance with a particular coordinate system, that is, has an *irregular geometry*, then an approximation to the physical geometry is used, generally termed *body-fitted coordinates*.

h- AND *p*-REFINEMENT

Increasing or decreasing the grid spacing over parts or all of the problem domain is termed *h-refinement*. The name comes from the common convention in the numerical analysis literature of using h as the symbol for the grid spacing.

Modifying the order of the derivative approximation is termed *p-refinement*. The name comes from the convention of using the symbol p for the order of the approximation (e.g., $O(\Delta x)^2$ for a second-order approximation with $p = 2$).

h-refinement seeks to refine the grid spacing using *local truncation error* estimates or other refinement parameters in order to improve the accuracy of the solution. Similarly, p-refinement seeks to refine the order of derivative approximations (using the same refinement parameters). Generally there is no unique, general, best combination of h- and p-refinement and the solution of large problems usually requires some trial and error for a trade-off between accuracy and computational effort; *the goal is to reach a prespecified bound on the global error with a minimal amount of work* [4]. We will not discuss this aspect further here, but refer to [4–6] for further discussion.

ORIGIN OF THE NAME "METHOD OF LINES"

As a concluding point, we might consider the origin of the name *method of lines*. If we consider Eq. (1.29), integration of this ODE system produces the solution $u_2(t), u_3(t), \ldots, u_M(t)$. (*Note:* $u_1(t) = u_b$, a constant, from BC (1.30).) We could then plot these functions in an x–$u(x, t)$ plane as a vertical line at each x ($i = 2, 3, \ldots, M$), with the height of each line equal to $u(x_i, t)$. In other words, the plot of the solution would be a set of vertical parallel lines suggesting the name *method of lines* [7].

To illustrate this interpretation, consider the diffusion equation problem of Eq. (1.1) with

- Diffusion coefficient $D = 1$
- Dirichlet BC at the left end, $u(x = -5, t) = 0$

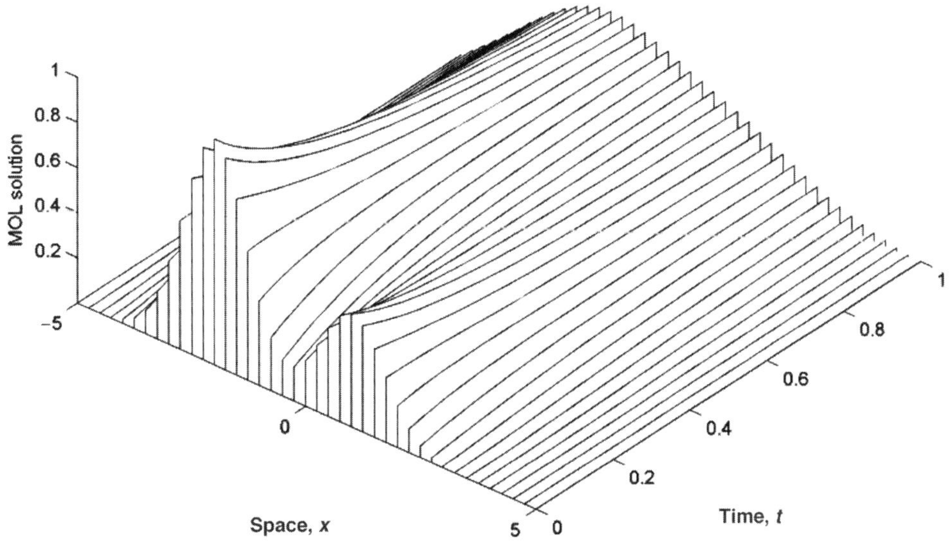

Figure 1.1. MOL solution of Eq. (1.1) illustrating the origin of the *method of lines*

- Neumann BC at the right end, $u_x(x = 5, t) = 0$ (spatial domain $-5 \leq x \leq 5$)
- Time domain $0 \leq t \leq 1$
- Initial condition $u(x, t = 0) = (1/2)e^{-(x-1)^2} + e^{-(x+2)^2}$

The MOL solution for the problem is shown in Figure (1.1). This numerical solution was obtained using Matlab and the MOL library routine `dss044` [7] with the number of grid points $M = 41$ (so that the grid spacing is $[5 - (-5)]/(41 - 1) = 0.25$).

The result of Figure 1.1 matches very well the *infinite-domain analytical solution*

$$u(x, t) = \frac{1}{2\sqrt{4Dt + 1}}\left(e^{\frac{3(2x+1)}{4Dt+1}} + 2\right)e^{-\frac{(x+2)^2}{4Dt+1}} \tag{1.38}$$

This agreement is illustrated in Figure 1.2 where the analytical result has been superimposed on the MOL solution. This comparison illustrates an important distinction between the analytical and numerical (MOL) solutions. The analytical solution is for an infinite domain, $-\infty \leq x \leq \infty$, while the MOL solution is computed on a finite domain (as required by a computer), $-5 \leq x \leq 5$ [1]. The agreement between the analytical and numerical solutions reflects the property that both solutions remain at essentially zero for $u(x = -5, t)$ and $u(x = 5, t)$ for $t \leq 1$ as indicated in Figure 1.2.[2]

[2] The exact analytical solution for the finite-domain problem is considerably more complicated than Eq. (1.38) but could be derived by a finite Fourier sine transform ([8], pp. 405–415) or a Green's function ([9], pp. 48, 58).

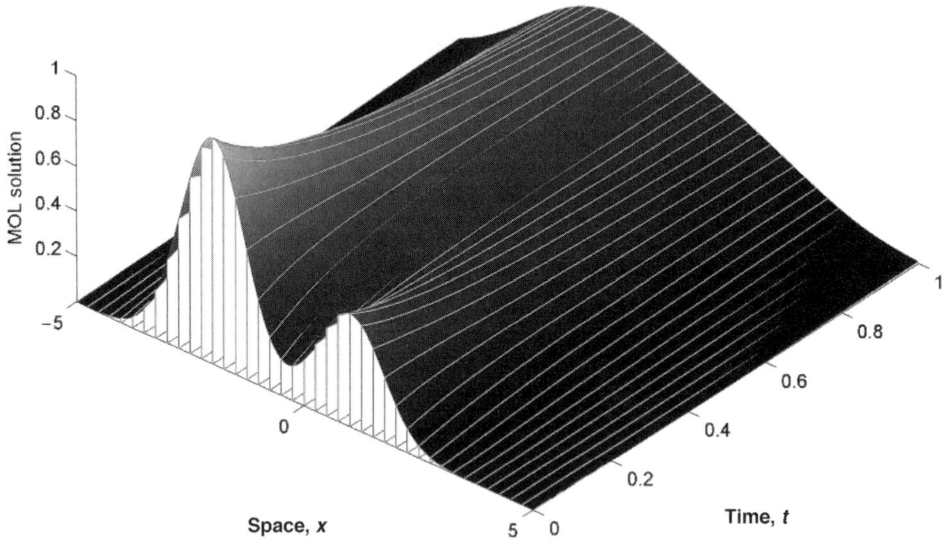

Figure 1.2. Superposition of the MOL solution of Eq. (1.1) and the analytical solution of Eq. (1.38)

SOURCES OF ODE/DAE INTEGRATORS

One of the very useful aspects of MOL is that it enables tried-and-tested ODE/DAE numerical routines to be used, many of which are in the public domain. The following sources are a good starting point for these routines. For example, the LSODE and VODE series of ODE/DAE integrators [2s], DASSL for DAEs [2s], and the SUNDIALS library [5s] are widely used in MOL analysis; test problems for ODE/DAE routines are also available [6s]. Additionally, routines that can be called from MOL codes are available to perform a variety of complementary computations (e.g., functional approximation by interpolation, evaluation of integrals, maximization and minimization in optimization associated with the solution of PDEs) [1s, 3s, 4s].

[1s] http://www.netlib.org/
[2s] http://www.netlib.org/ode/index.html (emphasis on ODE/DAE software that can be used in MOL analysis)
[3s] http://gams.nist.gov/
[4s] http://www.acm.org/toms/
[5s] http://www.llnl.gov/CASC/software.html
[6s] http://www.dm.uniba.it/~testset/

REFERENCES

[1] Hamdi, S., W. E. Schiesser, and G. W. Griffiths (2007), Method of Lines, *Scholarpedia*, **2**(7):2859; available online at http://www.scholarpedia.org/article/method_of_lines
[2] Wesseling, P. (2001), *Principles of Computational Fluid Dynamics*, Springer, Berlin
[3] Shu, C.-W. (1998), Essentially Non-Oscillatory and Weighted Essential Non-Oscillatory Schemes for Hyperbolic Conservation Laws, In: B. Cockburn, C. Johnson, C.-W. Shu,

and E. Tadmor (Eds.), *Advanced Numerical Approximation of Nonlinear Hyperbolic Equations*, Lecture Notes in Mathematics, vol. 1697, Springer, Berlin, pp. 325–432

[4] de Sterck, H., T. A. Manteuffel, S. F. McCormick, J. Nolting, J. Ruge, and L. Tang (2008), Efficiency-Based *h*- and *hp*-Refinement Strategies for Finite Element Methods, *Num. Linear Algebr. Appl.*, **15**: 89–114.

[5] Aftosmis, M. J. and M. J. Berger (2002), Multilevel Error Estimation and Adaptive *h*-Refinement for Cartesian Meshes with Embedded Boundaries, In: *AIAA Paper 2002-0863, 40th AIAA Aerospace Sciences Meeting and Exhibit*, January 14–17, 2002, Reno, NV

[6] Dong, S. and G. E. Karniadakis (May 9, 2003), p-Refinement and p-Threads, *Comput. Methods Appl. Mech. Eng.*, **192**(19): 2191–2201

[7] Schiesser, W. E. (1991), *Numerical Method of Lines Integration of Partial Differential Equations*, Academic Press, San Diego, CA

[8] Schiesser, W. E. (1994), *Computational Mathematics in Engineering and Applied Science: ODEs, DAEs, and PDEs*, CRC Press, Boca Raton, FL

[9] Polyanin, A. (2002), *Handbook of Linear Partial Differential Equations for Engineers and Scientists*, Chapman & Hall/CRC, Boca Raton, FL

2

A One-Dimensional, Linear Partial Differential Equation

This partial differential equation (PDE) problem is considered for the following reasons:

1. The PDE has an exact solution that can be used to assess the accuracy of the numerical method of lines (MOL) solution.
2. Both Dirichlet and Neumann boundary conditions (BCs) are included in the analysis.
3. The use of library routines for the finite-difference (FD) approximation of the spatial (boundary-value) derivative is illustrated.
4. The explicit programming of the FD approximations is included for comparison with the use of the library routines.
5. Some basic methods for assessing the accuracy of the MOL solution are presented.

The PDE is the *one-dimensional (1D) heat conduction equation in Cartesian coordinates:*

$$u_t = u_{xx} \tag{2.1}$$

Here we have used subscript notation for partial derivatives, so

$$u_t \leftrightarrow \frac{\partial u}{\partial t}$$

$$u_{xx} \leftrightarrow \frac{\partial^2 u}{\partial x^2}$$

The initial condition (IC) is

$$u(x, t = 0) = \sin(\pi x/2) \tag{2.2}$$

A *Dirichlet* BC is specified at $x = 0$,

$$u(x = 0, t) = 0 \tag{2.3}$$

and a *Neumann* BC is specified at $x = 1$,

$$u_x(x = 1, t) = 0 \tag{2.4}$$

The analytical solution to Eqs. (2.1)–(2.4) is

$$u(x, t) = e^{-(\pi^2/4)t} \sin(\pi x/2) \tag{2.5}$$

A main program in Matlab for the MOL solution of Eqs. (2.1)–(2.4) with the analytical solution, Eq. (2.5), included for comparison with the MOL solution, is given in Listing 2.1.

```
%
% Clear previous files
  clear all
  clc
%
% Parameters shared with the ODE routine
  global ncall ndss
%
% Initial condition
  n=21;
  for i=1:n
     u0(i)=sin((pi/2.0)*(i-1)/(n-1));
  end
%
% Independent variable for ODE integration
  t0=0.0;
  tf=2.5;
  tout=linspace(t0,tf,n);
  nout=n;
  ncall=0;
%
% ODE integration
  mf=1;
  reltol=1.0e-04; abstol=1.0e-04;
  options=odeset('RelTol',reltol,'AbsTol',abstol);
  if(mf==1) % explicit FDs
    [t,u]=ode15s(@pde_1,tout,u0,options); end
  if(mf==2) ndss=4; % ndss = 2, 4, 6, 8 or 10 required
    [t,u]=ode15s(@pde_2,tout,u0,options); end
  if(mf==3) ndss=44; % ndss = 42, 44, 46, 48 or 50 required
    [t,u]=ode15s(@pde_3,tout,u0,options); end
%
% Store numerical and analytical solutions, errors at x = 1/2
  n2=(n-1)/2.0+1;
  sine=sin(pi/2.0*0.5);
  for i=1:nout
    u_plot(i)=u(i,n2);
    u_anal(i)=exp(-pi^2/4.0*t(i))*sine;
    err_plot(i)=u_plot(i)-u_anal(i);
  end
```

```
%
% Display selected output
  fprintf('\n mf = %2d  abstol = %8.1e  reltol = %8.1e\n', ...
          mf,abstol,reltol);
  fprintf('\n     t        u(0.5,t)  u_anal(0.5,t)
          err u(0.5,t)\n');
  for i=1:5:nout
    fprintf('%6.3f%15.6f%15.6f%15.7f\n', ...
            t(i),u_plot(i),u_anal(i),err_plot(i));
  end
  fprintf('\n ncall = %4d\n',ncall);
%
% Plot numerical solution and errors at x = 1/2
  figure(1);
  subplot(1,2,1)
  plot(t,u_plot); axis tight
  title('u(0.5,t) vs t'); xlabel('t'); ylabel('u(0.5,t)')
  subplot(1,2,2)
  plot(t,err_plot); axis tight
  title('Err u(0.5,t) vs t'); xlabel('t');
        ylabel('Err u(0.5,t)');
  print -deps pde.eps; print -dps pde.ps; print -dpng pde.png
%
% Plot numerical solution in 3D perspective
  figure(2);
  colormap('Gray');
  C=ones(n);
  g=linspace(0,1,n); % For distance x
  h1=waterfall(t,g,u',C);
  axis('tight');
  grid off
  xlabel('t, time')
  ylabel('x, distance')
  zlabel('u(x,t)')
  s1=sprintf('Diffusion Equation - MOL Solution');
  sTmp=sprintf('u(x,0) = sin(\\pi x/2 )');
  s2=sprintf('Initial condition: %s',sTmp);
  title([{s1}, {s2}],'fontsize',12);
  v=[0.8616   -0.5076    0.0000   -0.1770
     0.3712    0.6301    0.6820   -0.8417
     0.3462    0.5876   -0.7313    8.5590
          0         0         0    1.0000];
  view(v);
  rotate3d on;
```

Listing 2.1. Main program pde_1_main

We can note the following points about the main program given in Listing 2.1:

1. After declaring some parameters `global` so that they can be shared with other routines called via this main program, IC (2.2) is computed over a 21-point grid in x.

```
%
% Clear previous files
  clear all
  clc
%
% Parameters shared with the ODE routine
  global ncall ndss
%
% Initial condition
  n=21;
  for i=1:n
    u0(i)=sin((pi/2.0)*(i-1)/(n-1));
  end
```

2. The independent variable `t` is defined over the interval $0 \leq t \leq 2.5$; again, a 21-point grid is used.

```
%
% Independent variable for ODE integration
  t0=0.0;
  tf=2.5;
  tout=linspace(t0,tf,n);
  nout=n;
  ncall=0;
```

3. The 21 ordinary differential equations (ODEs) are then integrated by a call to the Matlab integrator `ode15s`.

```
%
% ODE integration
  mf=1;
  reltol=1.0e-04; abstol=1.0e-04;
  options=odeset('RelTol',reltol,'AbsTol',abstol);
  if(mf==1) % explicit FDs
    [t,u]=ode15s(@pde_1,tout,u0,options); end
```

```
  if(mf==2) ndss=4; % ndss = 2, 4, 6, 8 or 10 required
    [t,u]=ode15s(@pde_2,tout,u0,options); end
  if(mf==3) ndss=44; % ndss = 42, 44, 46, 48 or 50 required
    [t,u]=ode15s(@pde_3,tout,u0,options); end
```

Three cases are programmed corresponding to `mf=1,2,3`, for which three different ODE routines, pde_1, pde_2, and pde_3, are called (these routines are discussed subsequently). The variable `ndss` refers to a library of differentiation routines for use in the MOL solution of PDEs; the use of `ndss` is illustrated in the subsequent discussion. Note that a stiff integrator, `ode15s`, was selected because the 21 ODEs are sufficiently stiff that a nonstiff integrator results in a large number of calls to the ODE routine.

4. Selected numerical results are stored for subsequent tabular and plotted output.

```
%
% Store numerical and analytical solutions, errors at x = 1/2
  n2=(n-1)/2.0+1;
  sine=sin(pi/2.0*0.5);
  for i=1:nout
    u_plot(i)=u(i,n2);
    u_anal(i)=exp(-pi^2/4.0*t(i))*sine;
    err_plot(i)=u_plot(i)-u_anal(i);
  end
```

5. Selected tabular numerical output is displayed.

```
%
% Display selected output
  fprintf('\n mf = %2d   abstol = %8.1e  reltol = %8.1e\n',...
          mf,abstol,reltol);
  fprintf('\n   t    u(0.5,t)  u_anal(0.5,t)  err u(0.5,t)\n');
  for i=1:5:nout
    fprintf('%6.3f%15.6f%15.6f%15.7f\n',...
            t(i),u_plot(i),u_anal(i),err_plot(i));
  end
  fprintf('\n ncall = %4d\n',ncall);
```

The output from this code is given in Table 2.1.

Table 2.1. Output for `mf=1` from `pde_1_main` and `pde_1`

```
mf =  1   abstol = 1.0e-004   reltol = 1.0e-004

     t       u(0.5,t)  u_anal(0.5,t)   err u(0.5,t)
0.000        0.707107       0.707107      0.0000000
0.625        0.151387       0.151268      0.0001182
1.250        0.032370       0.032360      0.0000093
1.875        0.006894       0.006923     -0.0000283
2.500        0.001472       0.001481     -0.0000091

ncall =   85
```

The output displayed in Table 2.1 indicates that the MOL solution agrees with the analytical solution to at least three significant figures. Also, `ode15s` calls the derivative routine only 85 times (in contrast with the nonstiff integrator `ode45`, which requires approximately 5,000–10,000 calls, clearly indicating the advantage of a stiff integrator for this problem).

6. The MOL solution and its error (computed from the analytical solution) are plotted.

```
%
% Plot numerical solution and errors at x = 1/2
  figure(1);
  subplot(1,2,1)
  plot(t,u_plot); axis tight
  title('u(0.5,t) vs t'); xlabel('t'); ylabel('u(0.5,t)')
  subplot(1,2,2)
  plot(t,err_plot); axis tight
  title('Err u(0.5,t) vs t'); xlabel('t'); ...
        ylabel('Err u(0.5,t)')
  print -deps pde.eps; print -dps pde.ps;  print -dpng pde.png
```

The plotted error output shown in Figure 2.1 indicates that the error in the MOL solution varied between approximately -3×10^{-5} and 16×10^{-5}, which is not quite within the error range specified in the program

```
reltol=1.0e-04; abstol=1.0e-04;
```

The fact that the error tolerances illustrated in Figure 2.1 were not satisfied does not necessarily mean that `ode15s` failed to adjust the integration

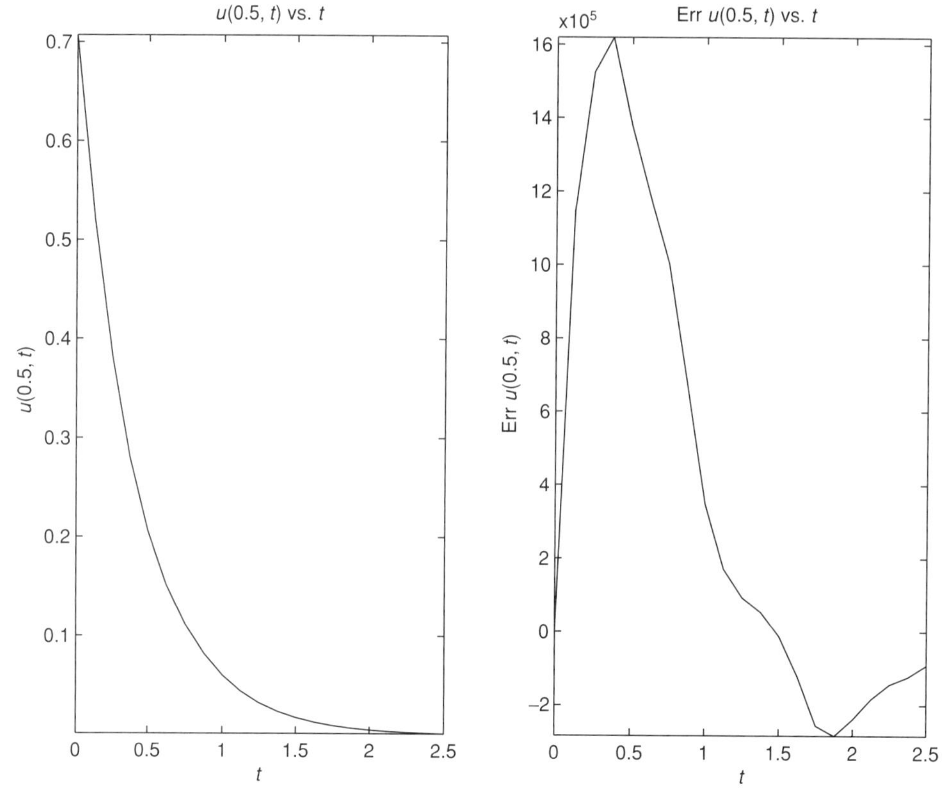

Figure 2.1. Two-dimensional graphical output from `pde_1_main`; `mf=1`

interval to meet these error tolerances. Rather, the error of approximately 1.6×10^{-4} is due to the limited accuracy of the second-order FD approximation of $\partial^2 u/\partial x^2$ programmed in `pde_1`. This conclusion is confirmed when the main program calls `pde_2` (for `mf=2`) or `pde_3` (for `mf=3`), as discussed subsequently; these two routines have FD approximations that are more accurate than in `pde_1`, so the errors fall below the specified tolerances.

This analysis indicates that two sources of errors result from the MOL solution of PDEs such as Eq. (2.1): (1) errors due to the integration in t (by `ode15s`) and (b) errors due to the approximation of the spatial derivatives such as $\partial^2 u/\partial x^2$ programmed in the derivative routine such as `pde_1`. In other words, we have to be attentive to integration errors in the *initial-* and *boundary-value independent variables*.

In summary, a comparison of the numerical and analytical solutions indicates that 21 grid points in x were not sufficient when using the second-order FDs in `pde_1`. However, in general, we will not have an analytical solution such as Eq. (2.5) to determine if the number of spatial grid points is adequate. In this case, some experimentation with the number of grid points, and the observation of the resulting solutions to infer the degree of accuracy or *spatial convergence*, may be required.

7. A 3D plot is also produced.

```
%
% Plot numerical solution in 3D perspective
  figure(2);
  colormap('Gray');
  C=ones(n);
  g=linspace(0,1,n); % For distance x
  h1=waterfall(t,g,u',C);
  axis('tight');
  grid off
  xlabel('t, time')
  ylabel('x, distance')
  zlabel('u(x,t)')
  s1=sprintf('Diffusion Equation - MOL Solution');
  sTmp=sprintf('u(x,0) = sin(\\pi x/2 )');
  s2=sprintf('Initial condition: %s',sTmp);
  title([{s1}, {s2}],'fontsize',12);
  v=[0.8616   -0.5076    0.0000   -0.1770
      0.3712    0.6301    0.6820   -0.8417
      0.3462    0.5876   -0.7313    8.5590
           0         0         0    1.0000];
  view(v);
  rotate3d on;
```

The plotted output shown in Figure 2.2 clearly indicates the origin of the *lines* in the *method of lines* (also discussed in Chapter 1).

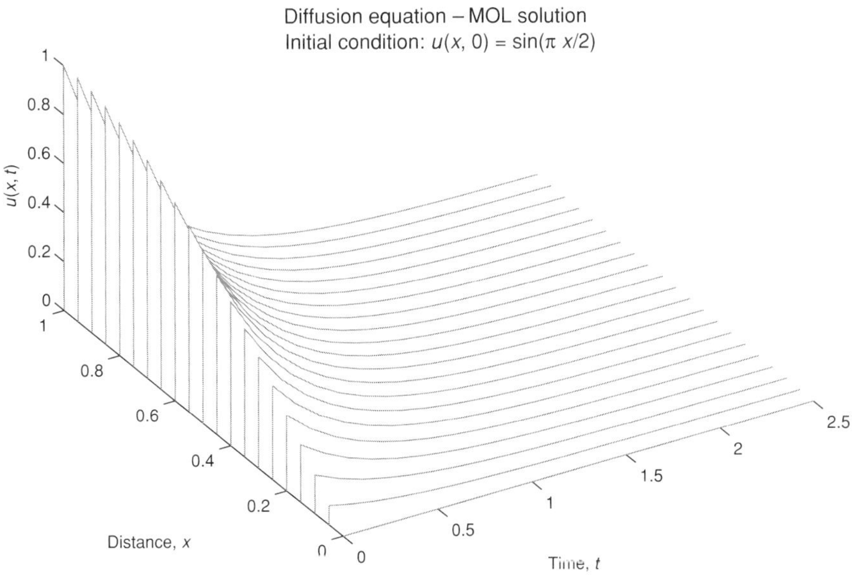

Figure 2.2. Three-dimensional graphical output from `pde_1_main`; `mf=1`

The programming of the approximating MOL/ODEs is in one of the three routines called by ode15s. We now consider each of these routines. For mf=1, ode15s calls function pde_1 (see Listing 2.2).

```
  function ut=pde_1(t,u)
%
% Problem parameters
  global ncall
  xl=0.0;
  xu=1.0;
%
% PDE
  n=length(u);
  dx2=((xu-xl)/(n-1))^2;
  for i=1:n
    if(i==1)      ut(i)=0.0;
    elseif(i==n)  ut(i)=2.0*(u(i-1)-u(i))/dx2;
    else          ut(i)=(u(i+1)-2.0*u(i)+u(i-1))/dx2;
    end
  end
  ut=ut';
%
% Increment calls to pde_1
  ncall=ncall+1;
```

Listing 2.2. Routine pde_1

We can note the following points about pde_1:

1. After the call definition of the function, some problem parameters are defined.

```
  function ut=pde_1(t,u)
%
% Problem parameters
  global ncall
  xl=0.0;
  xu=1.0;
```

xl and xu could have also been set in the main program and passed to pde_1 as global variables. The defining statement at the beginning of pde_1 indicates that the independent variable t and dependent variable vector u are inputs to pde_1, while the output is the vector of t derivatives, ut; in other words, all of the n ODE derivatives in t must be defined in pde_1.

2. The FD approximation of Eq. (2.1) is then programmed.

```
%
% PDE
  n=length(u);
  dx2=((xu-xl)/(n-1))^2;
  for i=1:n
    if(i==1)      ut(i)=0.0;
    elseif(i==n) ut(i)=2.0*(u(i-1)-u(i))/dx2;
    else          ut(i)=(u(i+1)-2.0*u(i)+u(i-1))/dx2;
  end
  end
  ut=ut';
```

The number of ODEs (21) is determined by the length command n=length(u); so that the programming is general (the number of ODEs can easily be changed in the main program). The square of the FD interval, dx2, is then computed.

3. The MOL programming of the 21 ODEs is done in the for loop. For BC (2.3), the coding is

```
    if(i==1) ut(i)=0.0;
```

since the value of $u(x = 0, t) = 0$ does not change after being set as an IC in the main program (and therefore its time derivative is zero).

4. For BC (2.4), the coding is

```
    elseif(i==n) ut(i)=2.0*(u(i-1)-u(i))/dx2;
```

which follows directly from the FD approximation of BC (2.4),

$$u_x \approx \frac{u(i+1) - u(i-1)}{\Delta x} = 0$$

or with $i = n$,

$$u(n+1) = u(n-1)$$

Note that the *fictitious value* $u(n+1)$ can then be replaced in the ODE at $i = n$ by $u(n-1)$.

5. For the remaining interior points, the programming is

```
    else ut(i)=(u(i+1)-2.0*u(i)+u(i-1))/dx2;
```

which follows from the FD approximation of the second derivative

$$u_{xx} \approx \frac{(u(i+1) - 2u(i) + u(i-1))}{\Delta x^2}$$

6. Since the Matlab ODE integrators require a column vector of derivatives, a final transpose of ut is required.

```
  ut=ut';
%
% Increment calls to pde_1
  ncall=ncall+1;
```

Finally, the number of calls to pde_1 is incremented so that at the end of the solution, the value of ncall displayed by the main program gives an indication of the computational effort required to produce the entire solution. The numerical and graphical output for this case (mf=1) was discussed previously.

For mf=2, function pde_2 is called by ode15s (see Listing 2.3).

```
function ut=pde_2(t,u)
%
% Problem parameters
  global ncall ndss
  xl=0.0;
  xu=1.0;
%
% BC at x = 0 (Dirichlet)
  u(1)=0.0;
%
% Calculate ux
  n=length(u);
  if     (ndss== 2) ux=dss002(xl,xu,n,u); % second order
  elseif(ndss== 4) ux=dss004(xl,xu,n,u); % fourth order
```

```
  elseif(ndss== 6) ux=dss006(xl,xu,n,u); % sixth order
  elseif(ndss== 8) ux=dss008(xl,xu,n,u); % eighth order
  elseif(ndss==10) ux=dss010(xl,xu,n,u); % tenth order
  end
%
% BC at x = 1 (Neumann)
  ux(n)=0.0;
%
% Calculate uxx
  if     (ndss== 2) uxx=dss002(xl,xu,n,ux); % second order
  elseif(ndss== 4) uxx=dss004(xl,xu,n,ux); % fourth order
  elseif(ndss== 6) uxx=dss006(xl,xu,n,ux); % sixth order
  elseif(ndss== 8) uxx=dss008(xl,xu,n,ux); % eighth order
  elseif(ndss==10) uxx=dss010(xl,xu,n,ux); % tenth order
  end
%
% PDE
  ut=uxx';
  ut(1)=0.0;
%
% Increment calls to pde_2
  ncall=ncall+1;
```

Listing 2.3. Routine pde_2

We can note the following points about pde_2:

1. The initial statements are the same as in pde_1. Then the Dirichlet BC at $x = 0$ is programmed.

```
%
% BC at x = 0 (Dirichlet)
  u(1)=0.0;
```

   Actually, the statement u(1)=0.0; has no effect since the dependent variables can only be changed through their derivatives, that is, ut(1), in the ODE derivative routine. This code was included just to serve as a reminder of the BC at $x = 0$, which is programmed subsequently.
2. The first-order spatial derivative $\partial u/\partial x = u_x$ is then computed.

```
%
% Calculate ux
  n=length(u);
```

```
if (ndss==2)      ux=dss002(xl,xu,n,u); % second order
elseif(ndss== 4) ux=dss004(xl,xu,n,u); % fourth order
elseif(ndss== 6) ux=dss006(xl,xu,n,u); % sixth order
elseif(ndss== 8) ux=dss008(xl,xu,n,u); % eighth order
elseif(ndss==10) ux=dss010(xl,xu,n,u); % tenth order
end
```

Five library routines, dss002 to dss010, are programmed that use second-order to tenth-order FD approximations, respectively. Since ndss=4 is specified in the main program, dss004 is used in the calculation of ux.

3. BC (2.4) is then applied, followed by the calculation of the second-order spatial derivative from the first-order spatial derivative.

```
%
% BC at x = 1 (Neumann)
  ux(n)=0.0;
%
% Calculate uxx
  if     (ndss== 2) uxx=dss002(xl,xu,n,ux); % second order
  elseif(ndss== 4) uxx=dss004(xl,xu,n,ux); % fourth order
  elseif(ndss== 6) uxx=dss006(xl,xu,n,ux); % sixth order
  elseif(ndss== 8) uxx=dss008(xl,xu,n,ux); % eighth order
  elseif(ndss==10) uxx=dss010(xl,xu,n,ux); % tenth order
  end
```

Again, dss004 is called, which is the usual procedure (the order of the FD approximation is generally not changed in computing higher-order derivatives from lower-order derivatives, a process termed *stagewise differentiation*).

4. Finally, Eq. (2.1) is programmed and the Dirichlet BC at $x = 0$ (Eq. (2.3)) is applied.

```
%
% PDE
  ut=uxx';
  ut(1)=0.0;
%
% Increment calls to pde_2
  ncall=ncall+1;
```

Note the similarity of the code to the PDE (Eq. (2.1)), and also the transpose required by ode15s.

Table 2.2. Output for `mf=2` from `pde_1_main` and `pde_2`

```
mf =  2    abstol = 1.0e-004    reltol = 1.0e-004

    t         u(0.5,t)  u_anal(0.5,t)   err u(0.5,t)
0.000         0.707107        0.707107      0.0000000
0.625         0.151267        0.151268     -0.0000013
1.250         0.032318        0.032360     -0.0000418
1.875         0.006878        0.006923     -0.0000446
2.500         0.001467        0.001481     -0.0000138

ncall =    62
```

The numerical output for this case (`mf=2`) is provided in Table 2.2. The plotted error output given in Figure 2.3 indicates that the error in the MOL solution varied between approximately -5×10^{-5} and 3.2×10^{-5}, which is within the error range specified in the program

```
reltol=1.0e-04; abstol=1.0e-04;
```

Thus, switching from the second-order FDs in `pde_1` to fourth-order FDs in `pde_2` reduced the *spatial truncation error* so that the MOL solution met the specified error tolerances.

For `mf=3`, function `pde_3` is called by `ode15s`, as given in Listing 2.4.

```
  function ut=pde_3(t,u)
%
% Problem parameters
  global ncall ndss
  xl=0.0;
  xu=1.0;
%
% BC at x = 0
  u(1)=0.0;
%
% BC at x = 1
  n=length(u);
  ux(n)=0.0;
%
% Calculate uxx
  nl=1; % Dirichlet
  nu=2; % Neumann
```

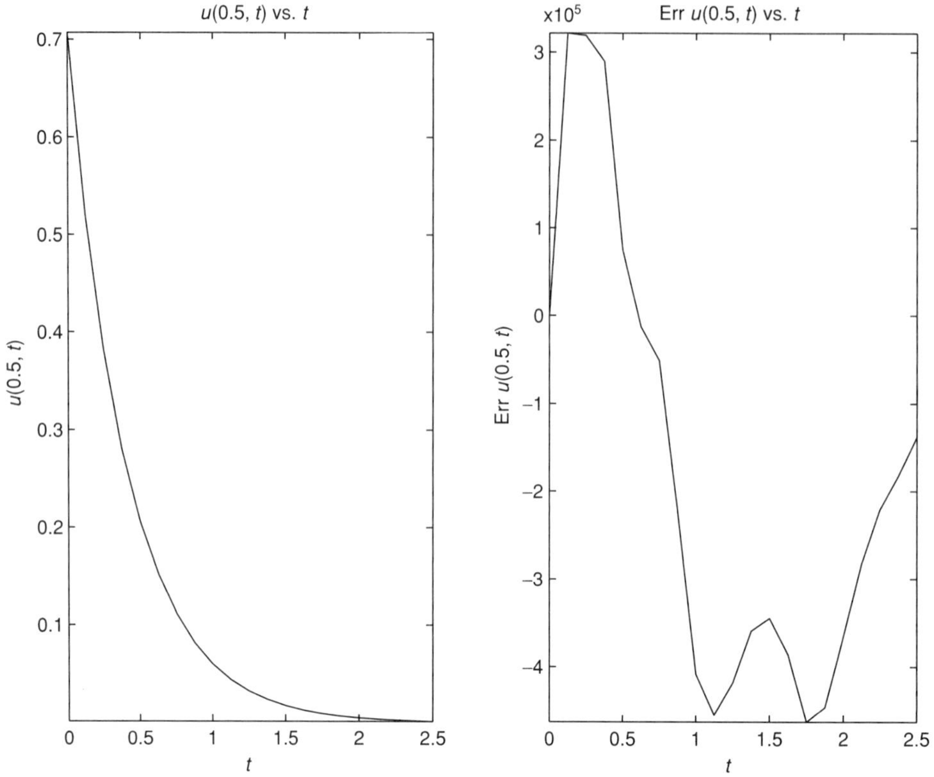

Figure 2.3. Two-dimensional graphical output from `pde_1_main`; `mf=2`

```
    if     (ndss==42) uxx=dss042(xl,xu,n,u,ux,nl,nu);
           % second order
    elseif(ndss==44) uxx=dss044(xl,xu,n,u,ux,nl,nu);
           % fourth order
    elseif(ndss==46) uxx=dss046(xl,xu,n,u,ux,nl,nu);
           % sixth order
    elseif(ndss==48) uxx=dss048(xl,xu,n,u,ux,nl,nu);
           % eighth order
    elseif(ndss==50) uxx=dss050(xl,xu,n,u,ux,nl,nu);
           % tenth order
    end
  %
  % PDE
    ut=uxx';
    ut(1)=0.0;
  %
  % Increment calls to pde_3
    ncall=ncall+1;
```

Listing 2.4. Routine pde_3

We can note the following points about pde_3:

1. The initial statements are the same as in pde_1. Then the Dirichlet BC at $x = 0$ and the Neumann BC at $x = 1$ are programmed.

```
  function ut=pde_3(t,u)
%
% Problem parameters
  global ncall ndss
  xl=0.0;
  xu=1.0;
%
% BC at x = 0
  u(1)=0.0;
%
% BC at x = 1
  n=length(u);
  ux(n)=0.0;
```

   Again, the statement u(1)=0.0; has no effect (since the dependent variables can only be changed through their derivatives, i.e., ut(1), in the ODE derivative routine). This code was included just to serve as a reminder of the BC at $x = 0$, which is programmed subsequently.
2. The second-order spatial derivative $\partial^2 u/\partial x^2 = u_{xx}$ is then computed.

```
%
% Calculate uxx
  nl=1; % Dirichlet
  nu=2; % Neumann
  if (ndss==42)     uxx=dss042(xl,xu,n,u,ux,nl,nu); % second order
  elseif(ndss==44) uxx=dss044(xl,xu,n,u,ux,nl,nu); % fourth order
  elseif(ndss==46) uxx=dss046(xl,xu,n,u,ux,nl,nu); % sixth order
  elseif(ndss==48) uxx=dss048(xl,xu,n,u,ux,nl,nu); % eighth order
  elseif(ndss==50) uxx=dss050(xl,xu,n,u,ux,nl,nu); % tenth order
  end
```

   Five library routines, dss042 to dss050, are programmed that use second-order to tenth-order FD approximations, respectively, for a second derivative. Since ndss=44 is specified in the main program, dss044 is used in the calculation of uxx. Also, these differentiation routines have two parameters that specify the type of BCs: (a) nl=1 or 2 specifies a Dirichlet or a

Table 2.3. Output for `mf=3` from `pde_1_main` and `pde_3`

```
mf =  3   abstol = 1.0e-004   reltol = 1.0e-004

     t        u(0.5,t)  u_anal(0.5,t)   err u(0.5,t)
0.000         0.707107       0.707107      0.0000000
0.625         0.151267       0.151268     -0.0000017
1.250         0.032318       0.032360     -0.0000420
1.875         0.006878       0.006923     -0.0000447
2.500         0.001467       0.001481     -0.0000138

ncall =   62
```

Neumann BC, respectively, at the lower boundary value of $x = xl(= 0)$; in this case, BC (2.3) is Dirichlet, so `nl=1`; and (b) `nu=1` or `2` specifies a Dirichlet or a Neumann BC, respectively, at the upper boundary value of $x = xu(= 1)$; in this case, BC (2.4) is Neumann, so `nu=2`.

3. Finally, Eq. (2.1) is programmed and the Dirichlet BC at $x = 0$ (Eq. (2.3)) is applied.

```
%
% PDE
  ut=uxx';
  ut(1)=0.0;
%
% Increment calls to pde_3
  ncall=ncall+1;
```

Again, the transpose is required by `ode15s`.

The numerical output for this case (`mf=3`) is given in Table 2.3. The plotted error output shown in Figure 2.4 indicates that the error in the MOL solution varied between approximately -4.8×10^{-5} and 3.2×10^{-5}, which is within the error range specified in the program

```
    reltol=1.0e-04; abstol=1.0e-04;
```

We conclude the example given in Figure 2.4 with the following observation: As the solution approaches steady state, $t \to \infty$, $u_t \to 0$, and from Eq. (2.1), $u_{xx} \to 0$.

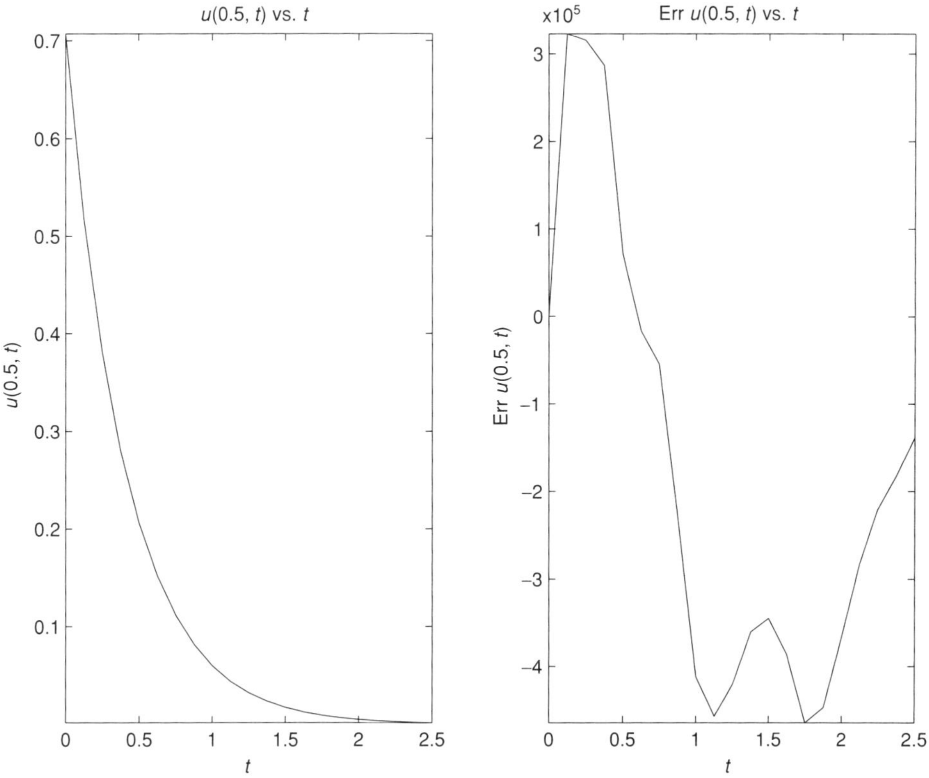

Figure 2.4. Two-dimensional graphical output from `pde_1_main`; `mf=3`

As the second derivative vanishes, the solution becomes

$$u_{xx} = 0$$

$$u_x = c_1$$

$$u = c_1 x + c_2$$

Thus, the steady-state solution is linear in x, which can serve as another check on the numerical solution (for BCs (2.3) and (2.4), $c_1 = c_2 = 0$ and thus at steady state, $u = 0$, which also follows from the analytical solution, Eq. (2.5)). This type of special case analysis is often useful in checking a numerical solution. In addition to mathematical conditions such as the linear dependency on x, physical conditions can frequently be used to check solutions, for example, conservation of mass, momentum, and energy.

Through this example application we have attempted to illustrate the basic steps of MOL/PDE analysis to arrive at a numerical solution of acceptable accuracy. We have also presented some basic ideas for assessing accuracy with respect to time and space (e.g., t and x). More advanced applications (e.g., problems expressed as systems of nonlinear PDEs) are considered in subsequent chapters.

3

Green's Function Analysis

This partial differential equation (PDE) application introduces the following mathematical concepts and computational methods:

1. The PDE has an exact solution that can be used to assess the accuracy of the numerical method of lines (MOL) solution.
2. The use of library routines for the finite-difference (FD) approximation of the spatial (boundary-value) derivative is illustrated.
3. The explicit programming of the FD approximations is included for comparison with the use of the library routines.
4. The failure of stagewise differentiation when applied to this PDE problem with a possible explanation for the failure.
5. The analytical solution, which is compared with the numerical solution to assess the accuracy of the latter, is a Green's function.
6. Computation of an invariant for the Green's function to evaluate the accuracy of the numerical solution.
7. The use of the Green's function for the derivation of other analytical solutions.

The PDE is the *one-dimensional (1D) diffusion equation in Cartesian coordinates*

$$\frac{\partial u}{\partial t} = D\frac{\partial^2 u}{\partial x^2}$$

or in subscript notation,

$$u_t = Du_{xx} \tag{3.1}$$

D is the *thermal diffusivity*, a positive constant.

The initial condition (IC) is

$$u(x, t = 0) = \delta(x) \tag{3.2}$$

where $\delta(x)$ is the *Dirac delta function* or *unit impulse function*. $\delta(x)$ has the following mathematical properties:

$$\delta(x) = 0, \quad x \neq 0 \tag{3.3a}$$

$$\int_{-\infty}^{\infty} \delta(x)\,dx = 1 \tag{3.3b}$$

$$\int_{-\infty}^{\infty} f(x)\delta(x)\,dx = f(0) \tag{3.3c}$$

which will be discussed subsequently when applied to the numerical solution.

Since Eq. (3.1) is second order in x, it requires two boundary conditions (BCs). For this problem the spatial domain in x is $-\infty \leq x \leq \infty$. But for a computer analysis, we must choose a *finite domain* (because computers work with finite numbers). Thus, we select finite boundary values for x, which are in effect at $x = \pm\infty$; that is, they are *large enough to accurately represent the infinite spatial domain*. This selection of the boundary values of x is based on a knowledge of the PDE solution, or if this is not possible, they are selected by trial and error (these ideas are illustrated by the subsequent analysis).

Additionally, we choose BCs that are *consistent with the IC* (Eq. 3.2). In this way, we ensure a smooth solution for $t > 0$; that is, we do not introduce discontinuities between the IC and the BCs (which could lead to computational problems in addition to violating the smoothness properties of the PDE solution).

In the case of IC (3.2), we can use the *homogeneous (zero) Dirichlet BCs* $u(x = x_l, t) = u(x = x_u, t) = 0$, where x_l and $x_u (x_l < x_u)$ are the finite boundary values of x we have selected so that the solution does not depart from zero at the boundaries (and therefore homogeneous BCs apply). Alternatively, we can use the *homogeneous Neumann BCs* $u_x(x = x_l, t) = u_x(x = x_u, t) = 0$, again, because the slope of the solution at $x = x_l, x_u$ does not depart from zero for the values of t considered. In the subsequent programming, we use the homogeneous Dirichlet BCs.

The analytical solution to Eq. (3.1) with the IC function $u(x, t = 0) = f(x)$ is [1]

$$u(x,t) = \frac{1}{2\sqrt{\pi D t}} \int_{-\infty}^{\infty} f(x) e^{-(x-\xi)^2/4Dt}\, d\xi \tag{3.4a}$$

For the special case of the IC function of Eq. (3.2), Eq. (3.4a)reduces to

$$u(x,t) = \frac{1}{2\sqrt{\pi D t}} \int_{-\infty}^{\infty} \delta(x) e^{-(x-\xi)^2/4Dt}\, d\xi = \frac{1}{2\sqrt{\pi D t}} e^{-x^2/4Dt} \tag{3.4b}$$

Equation (3.4b) follows from the property of the $\delta(x)$ function (Eq. 3.3c). The verification of Eq. (3.4b) as a solution of Eq. (3.1) is given in an appendix at the end of this chapter.

Equation (3.4a) can be written as

$$u(x,t) = \int_{-\infty}^{\infty} f(x) g(x, \xi, t) d\xi \tag{3.4c}$$

where

$$g(x, \xi, t) = \frac{1}{2\sqrt{\pi D t}} e^{-(x-\xi)^2/4Dt} \tag{3.4d}$$

$g(x, \xi, t)$ in Eq. (3.4d) is the *Green's function* of Eq. (3.1) for the infinite domain $-\infty \le x \le \infty$. Equation (3.4c) indicates that the Green's function can be used to derive analytical solutions to the diffusion equation for IC functions $f(x)$ that damp to zero sufficiently fast as $|x| \to \infty$ ([1], p. 95). Also, Eq. (3.4b) indicates that the Green's function can be considered as the *response of the diffusion equation to a unit impulse at* $x = \xi$ (compare Eqs. (3.4b) and (3.4d)).

Equation (3.4a) can be interpreted as the *superposition of a train of unit impulse solutions of Eq. (3.1)* throughout the spatial domain $-\infty \le x \le \infty$ (superposition achieved through integration) to produce the solution to Eq. (3.1) for the IC $u(x, t = 0) = f(x)$. Also, the solution of Eq. (3.4b) has the integral property

$$I(t) = \int_{-\infty}^{\infty} u(x, t)dx = \frac{1}{2\sqrt{\pi D t}} \int_{-\infty}^{\infty} e^{-x^2/4Dt}\, dx = 1 \tag{3.5}$$

Since the integral $I(t)$ is a function of t, it has the counterintuitive property that it is actually a constant. This *invariant* is used subsequently as a check of the numerical solution of Eq. (3.1).

A main program in Matlab for the MOL solution of Eqs. (3.1) and (3.2) is given in Listing 3.1.

```
%
% Clear previous files
  clear all
  clc
%
% Parameters shared with the ODE routine
  global ncall ndss n xl xu x dx
%
% Spatial grid
  n=101;
  xl=-10.0;
  xu= 10.0;
  dx=(xu-xl)/(n-1);
  for i=1:n
    x(i)=xl+dx*(i-1);
  end
  D=1.0;
%
% Initial condition
  n2=(n-1)/2+1;
  for i=1:n
    if(i==n2)u0(i)=25.0*dx;
    else u0(i)=0.0;
    end
  end
%
```

```
% Independent variable for ODE integration
  t0=0.0;
  tf=2.0;
  tout=(t0:0.5:tf);
  nout=5;
  ncall=0;
%
% ODE integration
  reltol=1.0e-04; abstol=1.0e-04;
  options=odeset('RelTol',reltol,'AbsTol',abstol);
  mf=3;
  if(mf==1) % explicit FDs
    [t,u]=ode15s(@pde_1,tout,u0,options); end
  if(mf==2) ndss=4; % ndss = 2, 4, 6, 8 or 10 required
    [t,u]=ode15s(@pde_2,tout,u0,options); end
  if(mf==3) ndss=44; % ndss = 42, 44, 46, 48 or 50 required
    [t,u]=ode15s(@pde_3,tout,u0,options); end
%
% Store analytical solution and the difference between the
% numerical and analytical solutions
  for it=2:nout
  for i=1:n
    u_anal(it,i)=1.0/(2.0*sqrt(D*pi*t(it)))*...
                 exp(-x(i)^2/(4.0*D*t(it)));
    err(it,i)=u(it,i)-u_anal(it,i);
  end
  end
%
% Display selected output
  fprintf('\n mf = %2d   abstol = %8.1e   reltol = %8.1e\n',...
          mf,abstol,reltol);
%
% t = 0 not included
  for it=2:nout
  fprintf('\n    t    x      u(num)     u(anal)         err\n');
  for i=1:n
    fprintf('%6.2f%6.1f%15.6f%15.6f%15.6f\n',...
           t(it),x(i),u(it,i),u_anal(it,i),err(it,i));
  end
%
%   Calculate and display the integral of the solution
    ui=u(it,:);
    uint=simp(xl,xu,n,ui);
    fprintf('\n Integral of u(x,t=%4.2f) = %7.4f\n',t(it),uint);
  end
  fprintf('\n ncall = %4d\n',ncall);
%
```

```
% Plot numerical solution
  figure(1);
  plot(x,u,'-');
  xlabel('x')
  ylabel('u(x,t)')
  title('Green's function; t = 0, 0.5, 1, 1.5, 2; numerical')
  hold on
%
% Plot numerical and analytical solutions
  figure(2);
  plot(x,u(2:nout,:),'o',x,u_anal,'-');
  xlabel('x')
  ylabel('u(x,t)')
  title('Green's function; t = 0.5, 1, 1.5, 2;
         o - numerical; solid - analytical')

% Plot numerical solution in 3D perspective
  if(mf==3)
  figure(3);
  view([220 10]);
  hold on
  h1=surfl(x,t,u);
  shading interp;
  colormap gray;
  hold off
  axis('tight');
  grid on
  zlim([0 0.5])
  ylabel('t, time')
  xlabel('x, distance')
  zlabel('u')
  s1=sprintf('Green's Function - MOL Solution');
  title(s1, 'fontsize', 12);
  rotate3d on;
  end
% print -deps -r300 pde1.eps; print -dps -r300 pde1.ps;
  print -dpng -r300 pde1.png
```

Listing 3.1. Main program pde_1_main.m

We can note the following points about this main program:

1. After declaring some parameters *global*, so that they can be shared with other routines called via this main program, a spatial grid is defined over 101 points, extending over the interval $-10 \le x \le 10$.

```
%
% Clear previous files
  clear all
  clc
%
% Parameters shared with the ODE routine
  global ncall ndss n xl xu x dx
%
% Spatial grid
  n=101;
  xl=-10.0;
  xu= 10.0;
  dx=(xu-xl)/(n-1);
  for i=1:n
    x(i)=xl+dx*(i-1);
  end
  D=1.0;
```

The computation of a numerical solution of Eqs. (3.1) and (3.2) indicates that the infinite domain $-\infty \leq x \leq \infty$ can be replaced with the finite domain $-10 \leq x \leq 10$. The justification for this is given when the numerical solution of Eqs. (3.1) and (3.2) is discussed subsequently.

2. Equation (3.2) is defined over the 101-point spatial grid. This IC presents a difficulty in the numerical representation of $\delta(x)$. Note from Eq. (3.3a) that this function is zero everywhere along the spatial grid where $x \neq 0$. This is approximated by the `for` loop where `u0(i)=0.0` except at `i=n2=51` corresponding to $x = 0$. At this point, `u0(n2)=25.0*dx`, where `dx` is the grid spacing. The scaling of 25.0 was selected so that the numerical integral of the numerical solution $u(x, t)$ (= `u(it,i)`) equaled 1 according to Eq. (3.5); this scaling also ensured that the peak values of the analytical and numerical solutions at $x = 0$, $t \neq 0$ are equal (as reflected in the numerical output discussed subsequently).

```
%
% Initial condition
  n2=(n-1)/2+1;
  for i=1:n
    if(i==n2)u0(i)=25.0*dx;
    else u0(i)=0.0;
    end
  end
```

In other words, the `for` loop used to define `u0(i)` is an attempt at a numerical approximation of $\delta(x)$ that satisfies the two basic properties of Eqs. (3.3a) and (3.3b). For the integral conditions of Eqs. (3.3b) and (3.5), two cases can be considered:

(a) For $t = 0$, the approximation of the IC can be considered as two adjacent right triangles, each with a height of `25.0*dx` and a base of `dx`. Thus the total area of the two triangles is (with $dx = (10 - (-10))/(101 - 1) = 0.2$)

$$2[(\text{height})(\text{base})/2] = 2(25\,dx)dx/2 = 2(25)(0.2)(0.2)/2 = 1$$

in agreement with Eq. (3.3b). In other words, the IC is a triangular pulse approximation of $\delta(x)$ spanning the three points $x = -dx, 0, dx$ with the correct "strength" of one, and this is achieved with the scale factor of 25. If the grid spacing is changed through a change of the number of grid points (other than 101), perhaps to achieve better spatial resolution of the numerical solution $u(x, t)$ (by increasing n), the scale factor can be changed accordingly by application of the preceding analysis to maintain the integral property of Eq. (3.3b). In the present case, the agreement between the numerical and analytical solutions is quite acceptable and therefore an increase in the number of grid points would not produce a significantly better numerical solution. This choice of $n = 101$ therefore meets the usual goal of using a grid that produces acceptable accuracy without excessive computation.

(b) For $t > 0$, the numerical integral of the numerical solution equals one (unity) to five figures, in accordance with Eq. (3.5), as demonstrated in the numerical output to follow.

3. t is then defined over the interval $0 \le t \le 2$ with the interval for displaying the numerical solution set to 0.5 so that the solution is displayed five times. The counter for the number of calls to the ordinary differential equation (ODE) routine is also initialized.

```
%
% Independent variable for ODE integration
  t0=0.0;
  tf=2.0;
  tout=(t0:0.5:tf);
  nout=5;
  ncall=0;
```

4. The 101 ODEs are then integrated by a call to the Matlab integrator `ode15s`.

```
%
% ODE integration
  reltol=1.0e-04; abstol=1.0e-04;
  options=odeset('RelTol',reltol,'AbsTol',abstol);
  mf=3;
```

```
  if(mf==1) % explicit FDs
    [t,u]=ode15s(@pde_1,tout,u0,options); end
  if(mf==2) ndss=4; % ndss = 2, 4, 6, 8 or 10 required
    [t,u]=ode15s(@pde_2,tout,u0,options); end
  if(mf==3) ndss=44; % ndss = 42, 44, 46, 48 or 50 required
    [t,u]=ode15s(@pde_3,tout,u0,options); end
```

Three cases are programmed corresponding to `mf=1,2,3`, for which three different ODE routines, `pde_1`, `pde_2`, and `pde_3`, are called (these routines are discussed subsequently). The variable `ndss` refers to a library of differentiation routines for use in the MOL solution of PDEs; the use of `ndss` is illustrated in the subsequent discussion. Note that a stiff integrator, `ode15s`, was selected because the 101 ODEs are sufficiently stiff that a nonstiff integrator results in a large number of calls to the ODE routine.

5. The analytical solution, Eq. (3.4b), is computed over the spatial grid, and the difference between the numerical and analytical solution is also computed.

```
%
% Store analytical solution and the difference between the
% numerical and analytical solutions
  for it=2:nout
  fprintf('\n   t   x       u(num)      u(anal)          err\n');
  for i=1:n
    u_anal(it,i)=1.0/(2.0*sqrt(D*pi*t(it)))*...
                 exp(-x(i)^2/(4.0*D*t(it)));
    err(it,i)=u(it,i)-u_anal(it,i);
  end
  end
```

Note that the analytical solution is not stored at $t = 0$ (corresponding to `it=1`) because of the two terms with a division by t in Eq. (3.4b), that is,

$$\frac{1}{2\sqrt{\pi D t}} \tag{3.6a}$$

$$e^{-x^2/4Dt} \tag{3.6b}$$

These two terms have some interesting properties:

(a) For $t \to 0$, term (3.6a) becomes arbitrarily large.
(b) For $t \to 0$, term (3.6b) becomes arbitrarily small (for $x \neq 0$).
(c) Since terms (3.6a) and (3.6b) multiply in Eq. (3.4b), the product approaches $\infty \bullet 0$ at $t \to 0$, but the exponential of term (3.6b) dominates, so the product approaches zero (as expected from property (3.3a)).
(d) For $x = 0$, term (3.6b) is unity, so the product of terms (3.6a) and (3.6b) becomes arbitrarily large for $t \to 0$. This arbitrarily large value

demonstrates the difficulty of representing $\delta(x)$ numerically on the spatial grid, and is approximated as `if(i==n2)u0(i)=25.0*dx;` in the programming of IC (3.2). But Eqs. (3.3a) and (3.3b) are the two essential requirements for approximating δx numerically, and this has been done through the programming of Eq. (3.2) described earlier.

6. Selected tabular numerical output is displayed.

```
%
% Display selected output
  fprintf('\n mf = %2d   abstol = %8.1e
           reltol = %8.1e\n', ... mf,abstol,reltol);
%
% t = 0 not included
  for it=2:nout
  fprintf('\n    t    x       u(num)      u(anal)        err\n');
  for i=1:n
    fprintf('%6.2f%6.1f%15.6f%15.6f%15.6f\n',...
             t(it),x(i),u(it,i),u_anal(it,i),err(it,i));
  end
```

Again, the numerical and analytical solutions at $t = 0$ are not displayed.

7. The invariant of Eq. (3.5) is computed by a call to `simp` that implements *Simpson's rule for numerical quadrature (integration)*; `simp` is discussed in an appendix to this chapter.

```
%
%   Calculate and display the integral of the solution
    ui=u(it,:);
    uint=simp(xl,xu,n,ui);
    fprintf('\n Integral of u(x,t=%4.2f) = %7.4f\n',...
             t(it),uint);
  end
  fprintf('\n ncall = %4d\n',ncall);
```

The counter for the calls to the ODE routines is displayed at the end of the numerical solution.

8. `pde_1_main` concludes with plotting (1) the numerical solution including $t = 0$ since it is finite at $x = 0$ (i.e., the coding is `if(i==n2)u0(i)=25.0*dx`, as discussed previously) and (2) the numerical and analytical solutions for $t \neq 0$.

```
%
% Plot numerical solution
  figure(1);
  plot(x,u,'-');
  xlabel('x')
  ylabel('u(x,t)')
  title('Green's function; t = 0, 0.5, 1, 1.5, 2; numerical')
  print -deps -r300 pde.eps; print -dps -r300 pde.ps;...
  print -dpng -r300 pde.png
  hold on
%
% Plot numerical and analytical solutions
  figure(2);
  plot(x,u(2:nout,:),'o',x,u_anal,'-');
  xlabel('x')
  ylabel('u(x,t)')
  title('Green's function; t = 0.5, 1, 1.5, 2;...
        o - numerical; solid - analytical')
```

9. For mf=3, a 3D plot of the solution is also produced.

```
% Plot numerical solution in 3D perspective
  if(mf==3)
  figure(3);
  view([220 10]);
  hold on
  h1=surfl(x,t,u);
  shading interp;
  colormap gray;
  hold off
  axis('tight');
  grid on
  zlim([0 0.5])
  ylabel('t, time')
  xlabel('x, distance')
  zlabel('u')
  s1=sprintf('Green's Function - MOL Solution');
  title(s1, 'fontsize', 12);
  rotate3d on;
  end
% print -deps -r300 pde1.eps; print -dps -r300 pde1.ps;
% print -dpng -r300 pde1.png
```

The output from pde_1_main is given in Table 3.1.

Table 3.1. Output for `mf=1` from `pde_1_main` and `pde_1`

mf = 1 abstol = 1.0e-004 reltol = 1.0e-004

t	x	u(num)	u(anal)	err
0.50	-10.0	0.000000	0.000000	-0.000000
0.50	-9.8	0.000000	0.000000	0.000000
0.50	-9.6	0.000000	0.000000	0.000000
0.50	-9.4	0.000000	0.000000	0.000000
0.50	-9.2	0.000000	0.000000	0.000000
0.50	-9.0	0.000000	0.000000	0.000000
	.			.
	.			.
	.			.
0.50	-6.0	0.000000	0.000000	0.000000
0.50	-5.8	0.000000	0.000000	0.000000
0.50	-5.6	0.000000	0.000000	0.000000
0.50	-5.4	0.000000	0.000000	0.000000
0.50	-5.2	0.000001	0.000001	0.000001
0.50	-5.0	0.000003	0.000001	0.000001
0.50	-4.8	0.000007	0.000004	0.000003
0.50	-4.6	0.000016	0.000010	0.000006
0.50	-4.4	0.000036	0.000025	0.000011
0.50	-4.2	0.000080	0.000059	0.000021
0.50	-4.0	0.000170	0.000134	0.000036
0.50	-3.8	0.000352	0.000292	0.000060
0.50	-3.6	0.000705	0.000612	0.000093
0.50	-3.4	0.001367	0.001232	0.000135
0.50	-3.2	0.002565	0.002384	0.000181
0.50	-3.0	0.004650	0.004432	0.000219
0.50	-2.8	0.008145	0.007915	0.000230
0.50	-2.6	0.013771	0.013583	0.000188
0.50	-2.4	0.022464	0.022395	0.000070
0.50	-2.2	0.035333	0.035475	-0.000142
0.50	-2.0	0.053554	0.053991	-0.000437
0.50	-1.8	0.078176	0.078950	-0.000774
0.50	-1.6	0.109845	0.110921	-0.001075
0.50	-1.4	0.148486	0.149727	-0.001242
0.50	-1.2	0.193008	0.194186	-0.001178
0.50	-1.0	0.241138	0.241971	-0.000833
0.50	-0.8	0.289465	0.289692	-0.000226
0.50	-0.6	0.333761	0.333225	0.000536
0.50	-0.4	0.369552	0.368270	0.001282
0.50	-0.2	0.392867	0.391043	0.001825
0.50	0.0	0.400966	0.398942	0.002023
0.50	0.2	0.392867	0.391043	0.001825
0.50	0.4	0.369552	0.368270	0.001282
0.50	0.6	0.333761	0.333225	0.000536

```
0.50   0.8      0.289465      0.289692     -0.000226
0.50   1.0      0.241138      0.241971     -0.000833
0.50   1.2      0.193008      0.194186     -0.001178
0.50   1.4      0.148486      0.149727     -0.001242
0.50   1.6      0.109845      0.110921     -0.001075
0.50   1.8      0.078176      0.078950     -0.000774
0.50   2.0      0.053554      0.053991     -0.000437
0.50   2.2      0.035333      0.035475     -0.000142
0.50   2.4      0.022464      0.022395      0.000070
0.50   2.6      0.013771      0.013583      0.000188
0.50   2.8      0.008145      0.007915      0.000230
0.50   3.0      0.004650      0.004432      0.000219
0.50   3.2      0.002565      0.002384      0.000181
0.50   3.4      0.001367      0.001232      0.000135
0.50   3.6      0.000705      0.000612      0.000093
0.50   3.8      0.000352      0.000292      0.000060
0.50   4.0      0.000170      0.000134      0.000036
0.50   4.2      0.000080      0.000059      0.000021
0.50   4.4      0.000036      0.000025      0.000011
0.50   4.6      0.000016      0.000010      0.000006
0.50   4.8      0.000007      0.000004      0.000003
0.50   5.0      0.000003      0.000001      0.000001
0.50   5.2      0.000001      0.000001      0.000001
0.50   5.4      0.000000      0.000000      0.000000
0.50   5.6      0.000000      0.000000      0.000000
0.50   5.8      0.000000      0.000000      0.000000
0.50   6.0      0.000000      0.000000      0.000000
        .                                      .
        .                                      .
        .                                      .
0.50   9.0      0.000000      0.000000      0.000000
0.50   9.2      0.000000      0.000000      0.000000
0.50   9.4      0.000000      0.000000      0.000000
0.50   9.6      0.000000      0.000000      0.000000
0.50   9.8      0.000000      0.000000      0.000000
0.50  10.0      0.000000      0.000000     -0.000000

Integral of u(x,t=0.50) =  1.0000

          .                                      .
          .                                      .
          .                                      .
                Output for t = 1, 1.5 removed

          .                                      .
          .                                      .
          .                                      .
```

(continued)

Table 3.1 (*continued*)

t	x	u(num)	u(anal)	err
2.00	-10.0	-0.000000	0.000001	-0.000001
2.00	-9.8	0.000001	0.000001	-0.000000
2.00	-9.6	0.000002	0.000002	0.000000
2.00	-9.4	0.000003	0.000003	0.000000
2.00	-9.2	0.000006	0.000005	0.000001
2.00	-9.0	0.000009	0.000008	0.000001
2.00	-8.8	0.000014	0.000012	0.000001
2.00	-8.6	0.000021	0.000019	0.000002
2.00	-8.4	0.000032	0.000029	0.000002
2.00	-8.2	0.000048	0.000045	0.000003
2.00	-8.0	0.000071	0.000067	0.000004
2.00	-7.8	0.000105	0.000099	0.000006
2.00	-7.6	0.000153	0.000146	0.000007
2.00	-7.4	0.000222	0.000212	0.000009
2.00	-7.2	0.000318	0.000306	0.000012
2.00	-7.0	0.000451	0.000436	0.000014
2.00	-6.8	0.000633	0.000616	0.000017
2.00	-6.6	0.000882	0.000861	0.000020
2.00	-6.4	0.001215	0.001192	0.000023
2.00	-6.2	0.001660	0.001633	0.000026
2.00	-6.0	0.002245	0.002216	0.000029
2.00	-5.8	0.003007	0.002976	0.000030
2.00	-5.6	0.003988	0.003958	0.000031
2.00	-5.4	0.005239	0.005210	0.000029
2.00	-5.2	0.006816	0.006791	0.000025
2.00	-5.0	0.008782	0.008764	0.000018
2.00	-4.8	0.011206	0.011197	0.000008
2.00	-4.6	0.014159	0.014164	-0.000005
2.00	-4.4	0.017716	0.017737	-0.000021
2.00	-4.2	0.021952	0.021992	-0.000040
2.00	-4.0	0.026935	0.026995	-0.000061
2.00	-3.8	0.032725	0.032808	-0.000083
2.00	-3.6	0.039371	0.039475	-0.000104
2.00	-3.4	0.046901	0.047025	-0.000124
2.00	-3.2	0.055321	0.055460	-0.000139
2.00	-3.0	0.064609	0.064759	-0.000150
2.00	-2.8	0.074710	0.074864	-0.000154
2.00	-2.6	0.085535	0.085684	-0.000150
2.00	-2.4	0.096955	0.097093	-0.000138
2.00	-2.2	0.108809	0.108926	-0.000117
2.00	-2.0	0.120896	0.120985	-0.000090
2.00	-1.8	0.132987	0.133043	-0.000055
2.00	-1.6	0.144830	0.144846	-0.000016
2.00	-1.4	0.156154	0.156127	0.000027

2.00	-1.2	0.166684	0.166612	0.000071
2.00	-1.0	0.176147	0.176033	0.000115
2.00	-0.8	0.184290	0.184135	0.000155
2.00	-0.6	0.190883	0.190694	0.000189
2.00	-0.4	0.195736	0.195521	0.000215
2.00	-0.2	0.198708	0.198476	0.000231
2.00	0.0	0.199708	0.199471	0.000237
2.00	0.2	0.198708	0.198476	0.000231
2.00	0.4	0.195736	0.195521	0.000215
2.00	0.6	0.190883	0.190694	0.000189
2.00	0.8	0.184290	0.184135	0.000155
2.00	1.0	0.176147	0.176033	0.000115
2.00	1.2	0.166684	0.166612	0.000071
2.00	1.4	0.156154	0.156127	0.000027
2.00	1.6	0.144830	0.144846	-0.000016
2.00	1.8	0.132987	0.133043	-0.000055
2.00	2.0	0.120896	0.120985	-0.000090
2.00	2.2	0.108809	0.108926	-0.000117
2.00	2.4	0.096955	0.097093	-0.000138
2.00	2.6	0.085535	0.085684	-0.000150
2.00	2.8	0.074710	0.074864	-0.000154
2.00	3.0	0.064609	0.064759	-0.000150
2.00	3.2	0.055321	0.055460	-0.000139
2.00	3.4	0.046901	0.047025	-0.000124
2.00	3.6	0.039371	0.039475	-0.000104
2.00	3.8	0.032725	0.032808	-0.000083
2.00	4.0	0.026935	0.026995	-0.000061
2.00	4.2	0.021952	0.021992	-0.000040
2.00	4.4	0.017716	0.017737	-0.000021
2.00	4.6	0.014159	0.014164	-0.000005
2.00	4.8	0.011206	0.011197	0.000008
2.00	5.0	0.008782	0.008764	0.000018
2.00	5.2	0.006816	0.006791	0.000025
2.00	5.4	0.005239	0.005210	0.000029
2.00	5.6	0.003988	0.003958	0.000031
2.00	5.8	0.003007	0.002976	0.000030
2.00	6.0	0.002245	0.002216	0.000029
2.00	6.2	0.001660	0.001633	0.000026
2.00	6.4	0.001215	0.001192	0.000023
2.00	6.6	0.000882	0.000861	0.000020
2.00	6.8	0.000633	0.000616	0.000017
2.00	7.0	0.000451	0.000436	0.000014
2.00	7.2	0.000318	0.000306	0.000012
2.00	7.4	0.000222	0.000212	0.000009
2.00	7.6	0.000153	0.000146	0.000007
2.00	7.8	0.000105	0.000099	0.000006

(continued)

Table 3.1 (*continued*)

```
2.00   8.0       0.000071       0.000067       0.000004
2.00   8.2       0.000048       0.000045       0.000003
2.00   8.4       0.000032       0.000029       0.000002
2.00   8.6       0.000021       0.000019       0.000002
2.00   8.8       0.000014       0.000012       0.000001
2.00   9.0       0.000009       0.000008       0.000001
2.00   9.2       0.000006       0.000005       0.000001
2.00   9.4       0.000003       0.000003       0.000000
2.00   9.6       0.000002       0.000002       0.000000
2.00   9.8       0.000001       0.000001      -0.000000
2.00  10.0       0.000000       0.000001      -0.000001

Integral of u(x,t=2.00) =  1.0000

ncall =  195
```

The plotted output from pde_1_main is shown in Figures 3.1 and 3.2. We can note the following details of this output:

1. The agreement between the analytical and numerical solutions is quite acceptable (from Table 3.1 and Figure 3.2); note, in particular, this agreement

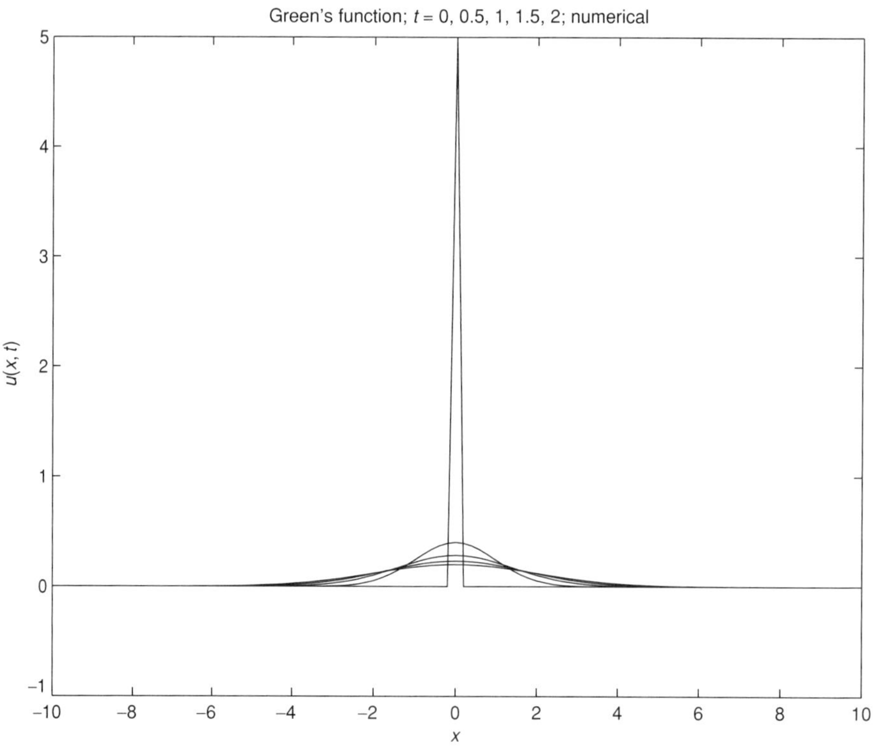

Figure 3.1. Numerical solution from pde_1_main including $t = 0$ (mf=1)

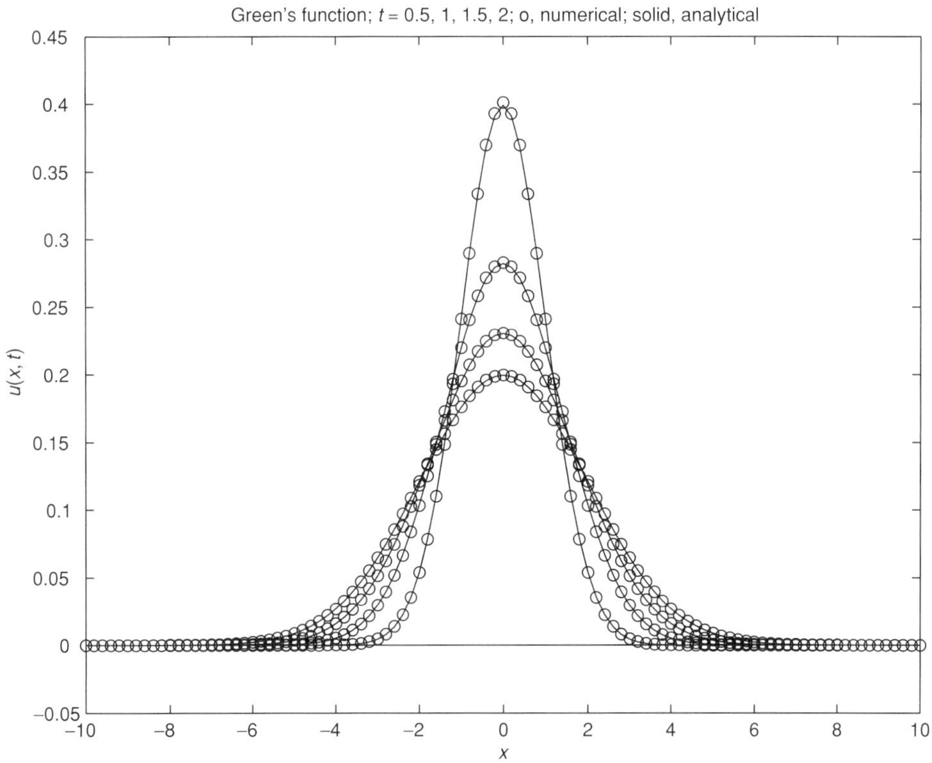

Figure 3.2. Analytical and numerical solutions from pde_1_main for $t \neq 0$; (mf=1)

in Table 3.1 at $x = 0$ for $t = 0.5$ and $t = 2$ (similar agreement occurred at $t = 1, 1.5$).

2. The integral constraint of Eq. (3.5) is closely approximated (e.g., `Integral of u(x,t=2.00) = 1.0000`), which is perhaps better than expected considering the approximation of $\delta(x)$ of Eq. (3.2) on the spatial grid of 101 points.
3. The height of the numerical approximation of $\delta(x)$ in Figure 3.1, $u(x = 0, t = 0) = 5$, is as expected from the statement `if(i==n2)u0(i)=25.0*dx;` with `dx = 0.2`.
4. The computational effort is quite modest with `ncall = 195`.

The programming of the approximating MOL/ODEs is in one of the three routines called by ode15s. We now consider each of these routines. For mf=1, ode15s calls function pde_1 (see Listing 3.2).

```
  function ut=pde_1(t,u)
%
% Problem parameters
  global ncall n dx
%
% PDE
  dx2=dx^2;
```

```
    for i=1:n
      if(i==1)       ut(i)=0.0;
      elseif(i==n) ut(i)=0.0;
      else           ut(i)=(u(i+1)-2.0*u(i)+u(i-1))/dx2;
      end
    end
    ut=ut';
%
% Increment calls to pde_1
  ncall=ncall+1;
```

Listing 3.2. Routine pde_1

We can note the following points about pde_1:

1. After the function call is defined, some problem parameters are declared as *global.*

```
  function ut=pde_1(t,u)
%
% Problem parameters
  global ncall n dx
```

2. The finite-difference (FD) approximation of Eq. (3.1) is then programmed.

```
%
% PDE
  dx2=dx^2;
  for i=1:n
    if(i==1)       ut(i)=0.0;
    elseif(i==n) ut(i)=0.0;
    else           ut(i)=(u(i+1)-2.0*u(i)+u(i-1))/dx2;
    end
  end
  ut=ut';
%
% Increment calls to pde_1
  ncall=ncall+1;
```

The MOL programming of the 101 ODEs is done in the `for` loop:

(a) For the homogeneous Dirichlet BCs $u(x = xl, t) = u(x = xu, t) = 0$, where $xl = -10$ and $xu = 10$, the coding is

```
  if(i==1)      ut(i)=0.0;
  elseif(i==n) ut(i)=0.0;
```

since the boundary values do not change after being set as part of the IC (Eq. (3.2)) in the main program (and therefore their time derivatives are zero).

(b) The FD approximation of Eq. (3.1) at the interior points is programmed as

```
  else ut(i)=(u(i+1)-2.0*u(i)+u(i-1))/dx2;
```

which follows directly from the FD approximation for u_{xx} in Eq. (3.1):

$$u_{xx} \approx \frac{(u(i+1) - 2u(i) + u(i-1))}{\Delta x^2} \tag{3.7}$$

3. Since `ode15s` requires a column vector of derivatives, a final transpose of `ut` is made.

```
  ut=ut';
%
% Increment calls to pde_1
  ncall=ncall+1;
```

Also, the number of calls to `pde_1` is incremented so that at the end of the solution, the value of `ncall` displayed by the main program gives an indication of the computational effort required to produce the entire solution (i.e., `ncall = 195`).

For `mf=2`, function `pde_2` is called by `ode15s`, as given in Listing 3.3.

```
  function ut=pde_2(t,u)
%
% Problem parameters
  global ncall ndss n xl xu
%
% BCs
  u(1)=0.0;
  u(n)=0.0;
%
```

```
% Calculate ux
  if     (ndss== 2) ux=dss002(xl,xu,n,u); % second order
  elseif(ndss== 4) ux=dss004(xl,xu,n,u); % fourth order
  elseif(ndss== 6) ux=dss006(xl,xu,n,u); % sixth order
  elseif(ndss== 8) ux=dss008(xl,xu,n,u); % eighth order
  elseif(ndss==10) ux=dss010(xl,xu,n,u); % tenth order
  end
%
% Calculate uxx
  if     (ndss== 2) uxx=dss002(xl,xu,n,ux); % second order
  elseif(ndss== 4) uxx=dss004(xl,xu,n,ux); % fourth order
  elseif(ndss== 6) uxx=dss006(xl,xu,n,ux); % sixth order
  elseif(ndss== 8) uxx=dss008(xl,xu,n,ux); % eighth order
  elseif(ndss==10) uxx=dss010(xl,xu,n,ux); % tenth order
  end
%
% PDE
  ut=uxx';
  ut(1)=0.0;
  ut(n)=0.0;
%
% Increment calls to pde_2
  ncall=ncall+1;
```

Listing 3.3. Routine pde_2.m

We can note the following points about pde_2:

1. The initial statements are the same as in pde_1. Then the Dirichlet BCs at $x = -10, 10$ are programmed.

```
  function ut=pde_2(t,u)
%
% Problem parameters
  global ncall ndss n xl xu
%
% BCs
  u(1)=0.0;
  u(n)=0.0;
```

Actually, the statements u(1)=0.0;, u(n)=0.0; have no effect since the dependent variables can only be changed through their derivatives (i.e., ut(1), ut(n)) in the ODE derivative routine (a requirement of the ODE integrator ode15s). This code was included just to serve as a reminder of the BCs at $x = -10$, $x = 10$ that are programmed subsequently.

2. The first-order spatial derivative $\partial u/\partial x = u_x$ is then computed.

```
%
% Calculate ux
  if     (ndss== 2) ux=dss002(xl,xu,n,u); % second order
  elseif(ndss== 4) ux=dss004(xl,xu,n,u); % fourth order
  elseif(ndss== 6) ux=dss006(xl,xu,n,u); % sixth order
  elseif(ndss== 8) ux=dss008(xl,xu,n,u); % eighth order
  elseif(ndss==10) ux=dss010(xl,xu,n,u); % tenth order
  end
```

Five library routines, dss002 to dss010, are programmed that use second-order to tenth-order FD approximations, respectively. Since ndss=4 is specified in the main program, dss004 is used in the calculation of ux.

3. The second-order spatial derivative is computed from the first-order spatial derivative.

```
%
% Calculate uxx
  if     (ndss== 2) uxx=dss002(xl,xu,n,ux); % second order
  elseif(ndss== 4) uxx=dss004(xl,xu,n,ux); % fourth order
  elseif(ndss== 6) uxx=dss006(xl,xu,n,ux); % sixth order
  elseif(ndss== 8) uxx=dss008(xl,xu,n,ux); % eighth order
  elseif(ndss==10) uxx=dss010(xl,xu,n,ux); % tenth order
  end
```

Again, dss004 is called, which is the usual procedure (the order of the FD approximation is generally not changed in computing higher-order derivatives from lower-order derivatives, a process termed *stagewise differentiation*).

4. Finally, Eq. (3.1) is programmed and the Dirichlet BCs at $x = -10, x = 10$ are applied.

```
%
% PDE
  ut=uxx';
  ut(1)=0.0;
  ut(n)=0.0;
%
% Increment calls to pde_2
  ncall=ncall+1;
```

Note the similarity of the code to the PDE (Eq. (3.1)), and also the transpose required by ode15s.

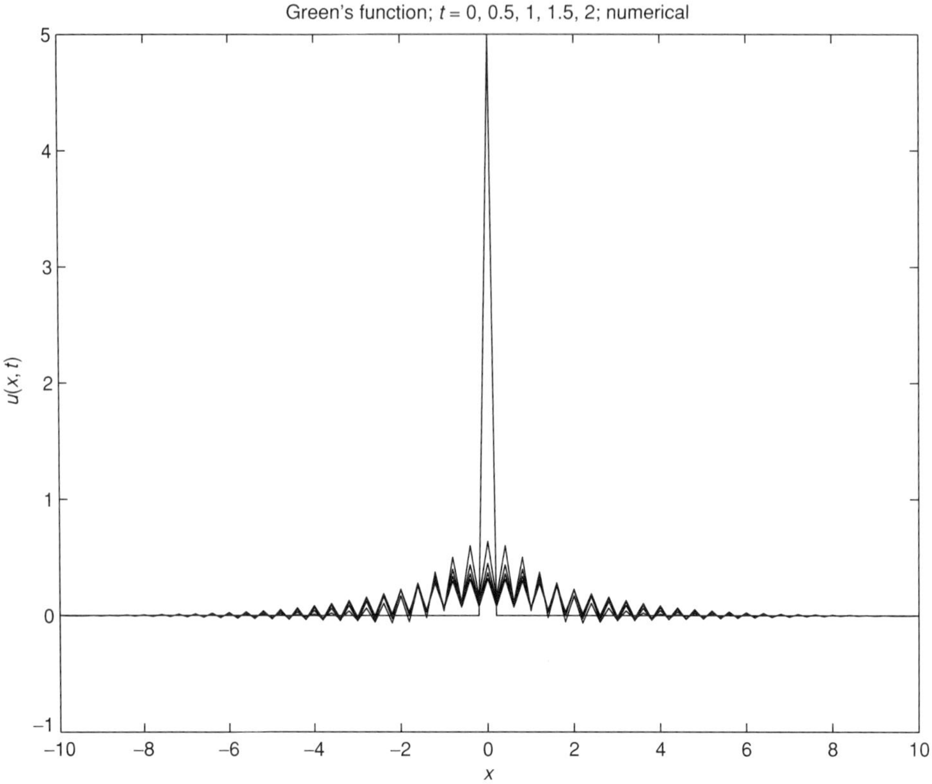

Figure 3.3. Numerical solution from pde_1_main including $t = 0$ (mf=2)

The numerical output for this case (mf=2) would logically be presented and discussed at this point. However, the agreement between the analytical and numerical solutions is not good and is actually difficult to visualize from the numerical output. Therefore, we go directly to the plotted output shown in Figures 3.3 and 3.4 that makes clearer the discrepancies between the analytical and numerical solutions.

We can note the following details of this output:

1. The agreement between the analytical and numerical solutions is unacceptable.
2. The integral constraint of Eq. (3.5) is also substantially in error (from the output of pde_1_main).

```
Integral of u(x,t=0.50) =  0.6667

Integral of u(x,t=1.00) =  0.6666

Integral of u(x,t=1.50) =  0.6659

Integral of u(x,t=2.00) =  0.6644
```

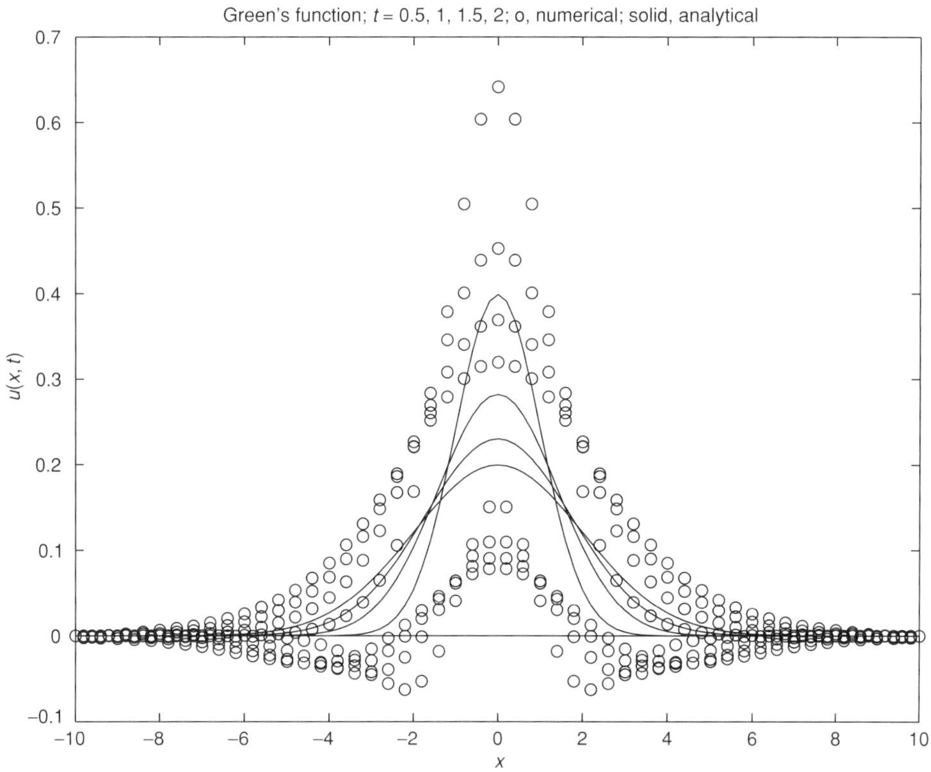

Figure 3.4. Analytical and numerical solutions from `pde_1_main` for $t \neq 0$; (`mf=2`)

Thus, we conclude that the stagewise differentiation of `pde_2` does not work correctly. One possible explanation is that the calculation of the first derivative u_x (e.g., using `dss004`) when applied to the approximation of IC (3.2), as plotted in Figure 3.1, is inaccurate because of the rapid changes in u_x at $x = 0$ for $t \approx 0$. With inaccurate values of u_x, the accurate calculation of u_{xx} (through the stagewise differentiation in `pde_2`) is not possible, and therefore the accurate MOL solution of Eq. (3.1) is not possible.

This explanation is not completely satisfactory since the direct calculation of u_{xx} through Eq. (3.7) produced good agreement between the analytical and numerical solutions (for `mf=1`). With the rapid changes in the solution at $x = 0$ displayed in Figure 3.1, we might also expect that the direct calculation of u_{xx} via Eq. (3.7) would not work well either (in fact, we might expect that u_{xx} calculated numerically would be even more prone to error than u_x). This does not, however, appear to be the case, and thus about all we can conclude from these results is that for this problem (Eqs. (3.1) and (3.2)), direct differentiation worked when stagewise differentiation did not. Clearly, some additional explanation of these results is required, but we will not explore this other than to consider one more approach to the MOL solution of Eqs. (3.1) and (3.2) using `pde_3` (`mf=3`).

This discussion also suggests another point. If we had only the numerical solution for `mf=2` (and not the analytical solution), how would we know that the numerical solution is substantially in error (since we could not compare it to an analytical solution). This is an important consideration since generally *we will not have an analytical solution to a PDE problem* (the point of computing a numerical solution is usually that we have no recourse other than a numerical solution). In other words, we are generally faced with the requirement of having to *evaluate a numerical solution to try to determine in some fashion whether it is accurate.*

We do not have a general approach to the resolution of this requirement to offer; rather each PDE problem must be considered with respect to the numerical solution, starting with the question of whether the numerical solution makes sense and is reasonable – matters of judgment rather than rigorous analysis. A practical approach to this evaluation is to observe how the numerical solution changes during h- and p-refinement; we would like to have a situation where further reductions in the grid spacing, h, and changes in the order of the approximations, p, do not change the numerical solution beyond a level that is considered acceptable, typically in an application, four significant figures.

For `mf=3`, function `pde_3` called by `ode15s` is given in Listing 3.4.

```
  function ut=pde_3(t,u)
%
% Problem parameters
  global ncall ndss n xl xu
%
% Calculate uxx
  nl=1; % Dirichlet
  nu=1; % Dirichlet
  ux=zeros(1,n);
  if     (ndss==42) uxx=dss042(xl,xu,n,u,ux,nl,nu); % second order
  elseif(ndss==44) uxx=dss044(xl,xu,n,u,ux,nl,nu); % fourth order
  elseif(ndss==46) uxx=dss046(xl,xu,n,u,ux,nl,nu); % sixth order
  elseif(ndss==48) uxx=dss048(xl,xu,n,u,ux,nl,nu); % eighth order
  elseif(ndss==50) uxx=dss050(xl,xu,n,u,ux,nl,nu); % tenth order
  end
%
% PDE
  ut=uxx';
  ut(1)=0.0;
  ut(n)=0.0;
%
% Increment calls to pde_3
  ncall=ncall+1;
```

Listing 3.4. Routine `pde_3.m`

We can note the following points about `pde_3`:

1. The initial statements are the same as in `pde_1`. Then the Dirichlet BCs at $x = -10, x = 10$ are programmed.

```
  function ut=pde_3(t,u)
%
% Problem parameters
  global ncall ndss n xl xu
%
% Calculate uxx
  nl=1; % Dirichlet
  nu=1; % Dirichlet
```

2. The second-order spatial derivative $\partial^2 u/\partial x^2 = u_{xx}$ is computed.

```
ux=zeros(1,n);
if     (ndss==42) uxx=dss042(xl,xu,n,u,ux,nl,nu); % second order
elseif(ndss==44) uxx=dss044(xl,xu,n,u,ux,nl,nu); % fourth order
elseif(ndss==46) uxx=dss046(xl,xu,n,u,ux,nl,nu); % sixth order
elseif(ndss==48) uxx=dss048(xl,xu,n,u,ux,nl,nu); % eighth order
elseif(ndss==50) uxx=dss050(xl,xu,n,u,ux,nl,nu); % tenth order
end
```

Five library routines, dss042 to dss050, are programmed that use second-order to tenth-order FD approximations, respectively, for a second derivative. Since ndss=44 is specified in the main program, dss044 is used in the calculation of uxx. Also, these differentiation routines have two parameters that specify the type of BCs: (a) nl=1 or 2 specifies a Dirichlet or a Neumann BC, respectively, at the lower boundary value of $x = xl(= -10)$, and (b) nu=1 or 2 specifies a Dirichlet or a Neumann BC, respectively, at the upper boundary value of $x = xu(= 10)$.

For the present analysis of Eq. (3.1), Dirichlet BCs at $x = -10, 10$ are specified as nl=nu=1. For these Dirichlet BCs, u_x (= ux) is not used by dss044. However, it is an input argument to dss044 and Matlab requires that input arguments to a function have at least one assigned value. This is accomplished with the statement ux=zeros(1,n);, but again, this has no effect on the calculation of uxx.

3. Finally, Eq. (3.1) is programmed and the Dirichlet BCs at $x = -10, 10$ are applied.

```
%
% PDE
  ut=uxx';
  ut(1)=0.0;
  ut(n)=0.0;
%
```

```
% Increment calls to pde_3
  ncall=ncall+1;
```

Again, the transpose is required by ode15s.

The numerical and plotted output for this case (mf=3) is given in Table 3.2.

Table 3.2. Output for mf=3 from pde_1_main and pde_3

mf = 3 abstol = 1.0e-004 reltol = 1.0e-004

t	x	u(num)	u(anal)	err
0.50	-10.0	0.000000	0.000000	-0.000000
0.50	-9.8	0.000000	0.000000	0.000000
0.50	-9.6	0.000000	0.000000	0.000000
0.50	-9.4	0.000000	0.000000	0.000000
0.50	-9.2	0.000000	0.000000	0.000000
0.50	-9.0	0.000000	0.000000	0.000000
	.			.
	.			.
	.			.
0.50	-6.0	0.000000	0.000000	-0.000000
0.50	-5.8	0.000000	0.000000	-0.000000
0.50	-5.6	0.000000	0.000000	-0.000000
0.50	-5.4	0.000000	0.000000	-0.000000
0.50	-5.2	0.000001	0.000001	-0.000000
0.50	-5.0	0.000001	0.000001	-0.000000
0.50	-4.8	0.000004	0.000004	-0.000000
0.50	-4.6	0.000010	0.000010	-0.000000
0.50	-4.4	0.000024	0.000025	-0.000001
0.50	-4.2	0.000058	0.000059	-0.000001
0.50	-4.0	0.000132	0.000134	-0.000001
0.50	-3.8	0.000290	0.000292	-0.000002
0.50	-3.6	0.000610	0.000612	-0.000002
0.50	-3.4	0.001232	0.001232	-0.000001
0.50	-3.2	0.002386	0.002384	0.000002
0.50	-3.0	0.004438	0.004432	0.000006
0.50	-2.8	0.007926	0.007915	0.000011
0.50	-2.6	0.013597	0.013583	0.000014
0.50	-2.4	0.022408	0.022395	0.000014
0.50	-2.2	0.035483	0.035475	0.000008
0.50	-2.0	0.053987	0.053991	-0.000004
0.50	-1.8	0.078932	0.078950	-0.000018
0.50	-1.6	0.110891	0.110921	-0.000030

0.50	-1.4	0.149694	0.149727	-0.000033
0.50	-1.2	0.194161	0.194186	-0.000025
0.50	-1.0	0.241961	0.241971	-0.000009
0.50	-0.8	0.289698	0.289692	0.000007
0.50	-0.6	0.333242	0.333225	0.000017
0.50	-0.4	0.368290	0.368270	0.000020
0.50	-0.2	0.391062	0.391043	0.000020
0.50	0.0	0.398961	0.398942	0.000019
0.50	0.2	0.391062	0.391043	0.000020
0.50	0.4	0.368290	0.368270	0.000020
0.50	0.6	0.333242	0.333225	0.000017
0.50	0.8	0.289698	0.289692	0.000007
0.50	1.0	0.241961	0.241971	-0.000009
0.50	1.2	0.194161	0.194186	-0.000025
0.50	1.4	0.149694	0.149727	-0.000033
0.50	1.6	0.110891	0.110921	-0.000030
0.50	1.8	0.078932	0.078950	-0.000018
0.50	2.0	0.053987	0.053991	-0.000004
0.50	2.2	0.035483	0.035475	0.000008
0.50	2.4	0.022408	0.022395	0.000014
0.50	2.6	0.013597	0.013583	0.000014
0.50	2.8	0.007926	0.007915	0.000011
0.50	3.0	0.004438	0.004432	0.000006
0.50	3.2	0.002386	0.002384	0.000002
0.50	3.4	0.001232	0.001232	-0.000001
0.50	3.6	0.000610	0.000612	-0.000002
0.50	3.8	0.000290	0.000292	-0.000002
0.50	4.0	0.000132	0.000134	-0.000001
0.50	4.2	0.000058	0.000059	-0.000001
0.50	4.4	0.000024	0.000025	-0.000001
0.50	4.6	0.000010	0.000010	-0.000000
0.50	4.8	0.000004	0.000004	-0.000000
0.50	5.0	0.000001	0.000001	-0.000000
0.50	5.2	0.000001	0.000001	-0.000000
0.50	5.4	0.000000	0.000000	-0.000000
0.50	5.6	0.000000	0.000000	-0.000000
0.50	5.8	0.000000	0.000000	-0.000000
0.50	6.0	0.000000	0.000000	-0.000000
	.			.
	.			.
	.			.
0.50	9.0	0.000000	0.000000	0.000000
0.50	9.2	0.000000	0.000000	0.000000
0.50	9.4	0.000000	0.000000	0.000000

(continued)

Table 3.2 (*continued*)

```
0.50   9.6        0.000000        0.000000        0.000000
0.50   9.8        0.000000        0.000000        0.000000
0.50  10.0        0.000000        0.000000       -0.000000

Integral of u(x,t=0.50) =  1.0000

          .                                          .
          .                                          .
          .                                          .

               Output for t = 1, 1.5 removed

          .                                          .
          .                                          .
          .                                          .

   t      x          u(num)          u(anal)             err
2.00 -10.0       -0.000000        0.000001       -0.000001
2.00  -9.8        0.000001        0.000001       -0.000000
2.00  -9.6        0.000002        0.000002       -0.000000
2.00  -9.4        0.000003        0.000003       -0.000000
2.00  -9.2        0.000005        0.000005       -0.000000
2.00  -9.0        0.000008        0.000008       -0.000000
2.00  -8.8        0.000012        0.000012       -0.000000
2.00  -8.6        0.000019        0.000019       -0.000000
2.00  -8.4        0.000029        0.000029       -0.000000
2.00  -8.2        0.000044        0.000045       -0.000000
2.00  -8.0        0.000067        0.000067       -0.000000
2.00  -7.8        0.000099        0.000099       -0.000000
2.00  -7.6        0.000146        0.000146       -0.000000
2.00  -7.4        0.000212        0.000212       -0.000000
2.00  -7.2        0.000306        0.000306       -0.000000
2.00  -7.0        0.000436        0.000436       -0.000000
2.00  -6.8        0.000616        0.000616        0.000000
2.00  -6.6        0.000862        0.000861        0.000000
2.00  -6.4        0.001193        0.001192        0.000001
2.00  -6.2        0.001634        0.001633        0.000001
2.00  -6.0        0.002217        0.002216        0.000001
2.00  -5.8        0.002978        0.002976        0.000002
2.00  -5.6        0.003959        0.003958        0.000002
2.00  -5.4        0.005212        0.005210        0.000002
2.00  -5.2        0.006793        0.006791        0.000002
2.00  -5.0        0.008765        0.008764        0.000001
2.00  -4.8        0.011198        0.011197        0.000001
2.00  -4.6        0.014163        0.014164       -0.000000
```

2.00	-4.4	0.017736	0.017737	-0.000001
2.00	-4.2	0.021989	0.021992	-0.000002
2.00	-4.0	0.026992	0.026995	-0.000003
2.00	-3.8	0.032804	0.032808	-0.000004
2.00	-3.6	0.039470	0.039475	-0.000005
2.00	-3.4	0.047020	0.047025	-0.000005
2.00	-3.2	0.055456	0.055460	-0.000005
2.00	-3.0	0.064755	0.064759	-0.000004
2.00	-2.8	0.074861	0.074864	-0.000002
2.00	-2.6	0.085684	0.085684	-0.000000
2.00	-2.4	0.097095	0.097093	0.000002
2.00	-2.2	0.108931	0.108926	0.000005
2.00	-2.0	0.120994	0.120985	0.000008
2.00	-1.8	0.133053	0.133043	0.000010
2.00	-1.6	0.144857	0.144846	0.000012
2.00	-1.4	0.156138	0.156127	0.000011
2.00	-1.2	0.166621	0.166612	0.000009
2.00	-1.0	0.176037	0.176033	0.000004
2.00	-0.8	0.184134	0.184135	-0.000001
2.00	-0.6	0.190688	0.190694	-0.000006
2.00	-0.4	0.195510	0.195521	-0.000011
2.00	-0.2	0.198462	0.198476	-0.000014
2.00	0.0	0.199456	0.199471	-0.000015
2.00	0.2	0.198462	0.198476	-0.000014
2.00	0.4	0.195510	0.195521	-0.000011
2.00	0.6	0.190688	0.190694	-0.000006
2.00	0.8	0.184134	0.184135	-0.000001
2.00	1.0	0.176037	0.176033	0.000004
2.00	1.2	0.166621	0.166612	0.000009
2.00	1.4	0.156138	0.156127	0.000011
2.00	1.6	0.144857	0.144846	0.000012
2.00	1.8	0.133053	0.133043	0.000010
2.00	2.0	0.120994	0.120985	0.000008
2.00	2.2	0.108931	0.108926	0.000005
2.00	2.4	0.097095	0.097093	0.000002
2.00	2.6	0.085684	0.085684	-0.000000
2.00	2.8	0.074861	0.074864	-0.000002
2.00	3.0	0.064755	0.064759	-0.000004
2.00	3.2	0.055456	0.055460	-0.000005
2.00	3.4	0.047020	0.047025	-0.000005
2.00	3.6	0.039470	0.039475	-0.000005
2.00	3.8	0.032804	0.032808	-0.000004
2.00	4.0	0.026992	0.026995	-0.000003
2.00	4.2	0.021989	0.021992	-0.000002
2.00	4.4	0.017736	0.017737	-0.000001

(continued)

Table 3.2 (*continued*)

```
2.00   4.6     0.014163     0.014164    -0.000000
2.00   4.8     0.011198     0.011197     0.000001
2.00   5.0     0.008765     0.008764     0.000001
2.00   5.2     0.006793     0.006791     0.000002
2.00   5.4     0.005212     0.005210     0.000002
2.00   5.6     0.003959     0.003958     0.000002
2.00   5.8     0.002978     0.002976     0.000002
2.00   6.0     0.002217     0.002216     0.000001
2.00   6.2     0.001634     0.001633     0.000001
2.00   6.4     0.001193     0.001192     0.000001
2.00   6.6     0.000862     0.000861     0.000000
2.00   6.8     0.000616     0.000616     0.000000
2.00   7.0     0.000436     0.000436    -0.000000
2.00   7.2     0.000306     0.000306    -0.000000
2.00   7.4     0.000212     0.000212    -0.000000
2.00   7.6     0.000146     0.000146    -0.000000
2.00   7.8     0.000099     0.000099    -0.000000
2.00   8.0     0.000067     0.000067    -0.000000
2.00   8.2     0.000044     0.000045    -0.000000
2.00   8.4     0.000029     0.000029    -0.000000
2.00   8.6     0.000019     0.000019    -0.000000
2.00   8.8     0.000012     0.000012    -0.000000
2.00   9.0     0.000008     0.000008    -0.000000
2.00   9.2     0.000005     0.000005    -0.000000
2.00   9.4     0.000003     0.000003    -0.000000
2.00   9.6     0.000002     0.000002    -0.000000
2.00   9.8     0.000001     0.000001    -0.000000
2.00  10.0     0.000000     0.000001    -0.000001

Integral of u(x,t=2.00) =  1.0000

ncall =  199
```

The plotted output from `pde_1_main` is shown in Figures 3.5 and 3.6. The 3D plot from `pde_1_main` is shown in Figure 3.7. Note in particular the numerical approximation of the IC, Eq. (3.2), for small t and how this sharp change in $u(x, t)$ diffuses away for later t.

We can note the following details of this output:

1. The agreement between the analytical and numerical solutions is quite acceptable (from Table 3.2 and Figure 3.6) and is, in fact, better than that from `pde_1` (from Table 3.1 and Figure 3.2).
2. The integral constraint of Eq. (3.5) is closely approximated (e.g., `Integral of u(x,t=2.00) = 1.0000`).
3. The computational effort is quite modest with `ncall = 199`.

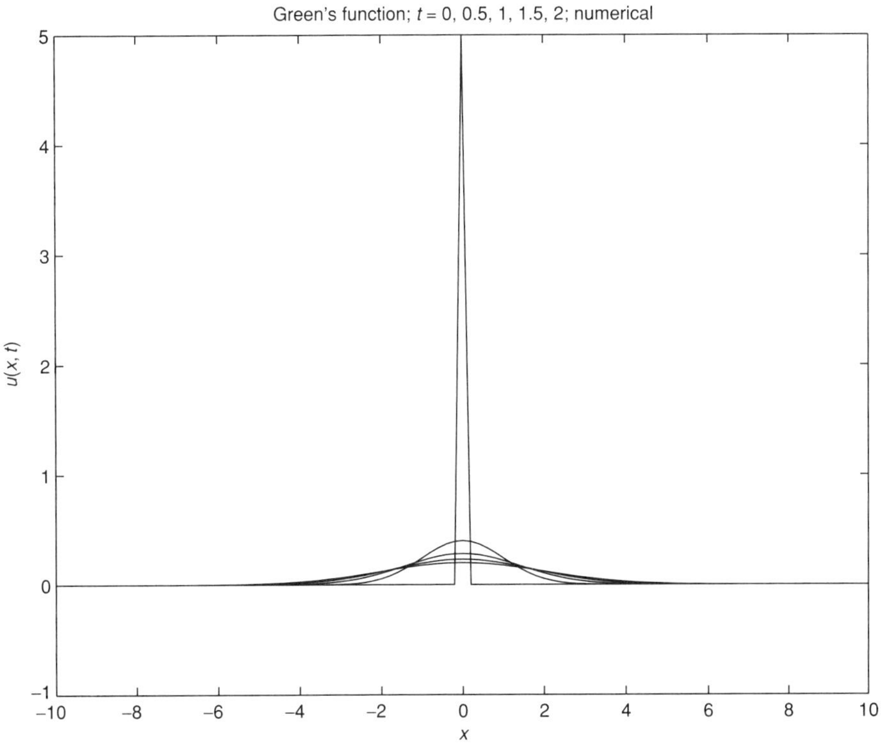

Figure 3.5. Numerical solution from `pde_1_main` including $t = 0$ (`mf=3`)

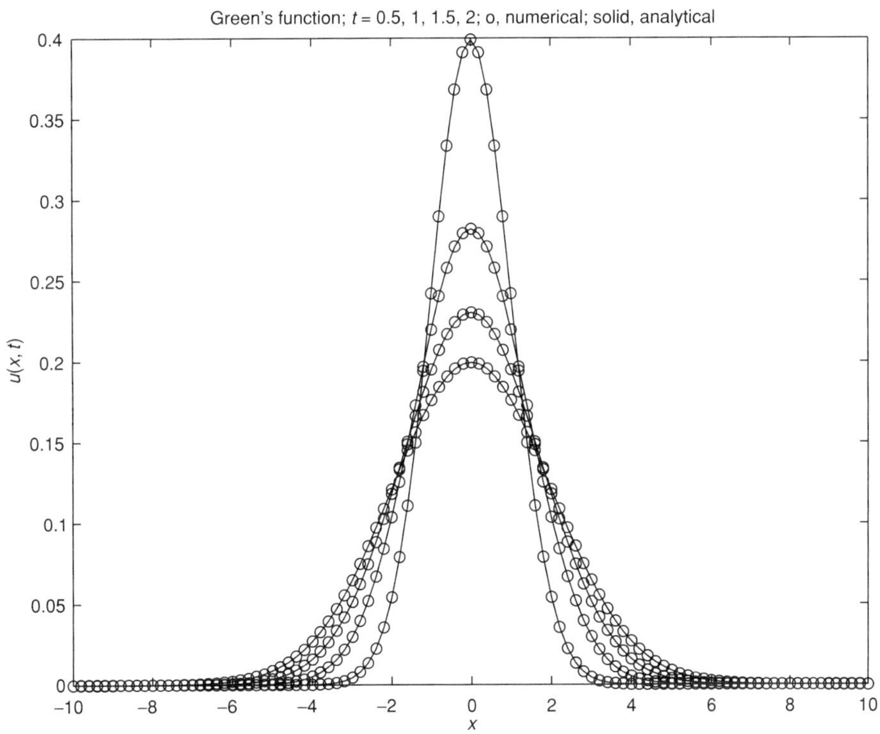

Figure 3.6. Analytical and numerical solutions from `pde_1_main` for $t \neq 0$; (`mf=3`)

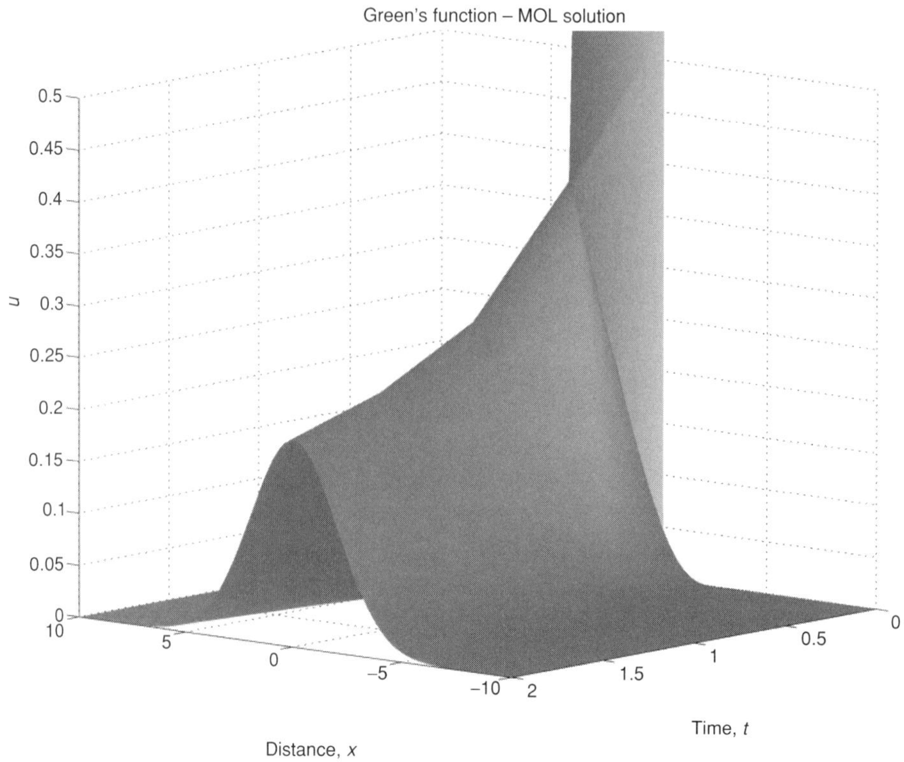

Figure 3.7. Numerical solution from `pde_1_main` including $t = 0$ (`mf=3`) in 3D

Thus, we can again conclude that direct calculation of u_{xx} (via Eq. (3.7)) works well at least for this problem (Eqs. (3.1) and (3.2)).

We conclude this chapter with some additional brief discussion:

1. The Green's function, Eq. (3.4d), is an analytical solution to the diffusion equation (3.1) for the delta function $\delta(x)$ IC of Eq. (3.2). In particular, this Green's function is for an infinite spatial domain. Solutions of Eq. (3.1) for other IC functions represented as $f(x)$ are available through the superposition of an integral (Eq. 3.4a).
2. These properties of the Green's function are applicable to a broad spectrum of (linear) PDE problems. Here are some considerations when attempting a Green's function solution of a PDE:
   (a) Spatial domain, which can be infinite (e.g., $-\infty \leq x \leq \infty$), semi-infinite (e.g., $0 \leq x \leq \infty$), or finite (e.g., $x_l \leq x \leq x_u$). The extent of the spatial domain has a major effect on the form of the Green's function. For example, for an infinite domain, the Green's function tends to zero for large $|x|$ (as with the exponential of Eq. (3.4d)). For a finite domain, the Green's function is generally an infinite series of eigenfunctions.
   (b) Also, for the semi-infinite and finite domains, the form of the Green's function depends on the type of boundary conditions, that is, Dirichlet, Neumann, or a third-type (Robin) BC.

(c) The form of the Green's function is also determined by the coordinate system (Cartesian coordinates in the case of Eq. (3.1)). For example, the Green's function in cylindrical and spherical coordinates is composed of functions specific to these coordinate systems.
(d) If the PDE is 1D, 2D, or 3D, the corresponding Green's functions is made up of parts for each coordinate dimension. For example, for a 3D PDE specified on a finite domain in each coordinate, the Green's function would be a triple infinite series.
(e) As with most analytical methods of PDE solution, the Green's function is quite limited in the class of problems to which it can be applied. For example, it generally applies only to linear PDEs, and practically to a single linear PDE.
(f) But since the Green's function provides analytical PDE solutions, it can be used to test numerical solutions as in the preceding application.
(g) Extensive collections of Green's functions are available for the variety of PDE characteristics mentioned earlier (spatial domain, coordinate system, boundary conditions, number of dimensions), notably those by Polyanin [2].
(h) Finally, the preceding example demonstrates that the Green's function is a convenient analytical method for the solution of PDEs, primarily because (a) a delta function IC facilitates the evaluation of integrals (such as Eq. (3.4b)) and (b) solutions for other ICs (than a delta function) can be readily constructed (using the idea of superposition as in Eq. (3.4d)). However, the delta function is not easy to handle numerically and requires some form of numerical approximation (such as was used in the `for` loop for IC (3.2) of `pde_1_main`).

3. The success of direct numerical differentiation (via Eq. (3.7)) and the failure of stagewise differentiation (in `pde_2`) indicates that numerical procedures that seem reasonable (such as stagewise differentiation) do not always work. Thus, PDE/MOL analysis is to some extent a trial-and-error process to arrive at an accurate numerical solution (with reasonable computing effort).
4. As a related matter, you could ask why stagewise differentiation would ever be used. The answer, generally, is that it facilitates PDE/MOL solutions when direct differentiation cannot be applied as easily. For example, the nonlinear differential group

$$\frac{\partial\left(u\dfrac{\partial u}{\partial x}\right)}{\partial x} = (uu_x)_x \tag{3.8}$$

could occur when analyzing the nonlinear diffusion equation

$$u_t = (D(u)u_x)_x \tag{3.9}$$

with $D(u) = u$.

Stagewise differentiation could be applied conveniently to term (3.8) by first calculating u_x, then multiplying this derivative by u to produce uu_x, and then differentiating this product to produce $(uu_x)_x$. Direct calculation of term (3.8) would probably first require expansion as $uu_{xx} + (u_x)^2$. Both approaches would be worth trying (and hopefully at least one would work, and ideally both would give essentially the same numerical solution). A third approach would be to program an FD approximation specifically for the term of interest; for example,

$$(uu_x)_x \approx \frac{u_{i+1}\left(\dfrac{u_{i+2}-u_i}{2\Delta x}\right) - u_{i-1}\left(\dfrac{u_i - u_{i-2}}{2\Delta x}\right)}{2\Delta x} \tag{3.10}$$

Care is required in this approach for developing an approximation to a term in a PDE to be sure, for example, that it converges to the term of interest as the grid spacing becomes arbitrarily small. One way to check such an approximation is to assume a functional form for $u(x,t)$, for example, a polynomial, which can be substituted into the approximation and then evaluated numerically. Also, the exact value of the term can be computed (using the assumed functional form for $u(x,t)$) and compared with the numerical value.

APPENDIX A

A.1. Verification of Eq. (3.4b) as the Solution to Eq. (3.1)

Equation (3.4b) can be verified as a solution to Eq. (3.1) by direct substitution. First, the derivative u_t is

$$u_t(x,t) = \frac{1}{2\sqrt{\pi D t}} e^{-x^2/4Dt}(-x^2/4D)(-1/t^2) + \frac{1}{2\sqrt{\pi D}}(-1/2)t^{-3/2}\, e^{-x^2/4Dt}$$

$$= \left[\frac{1}{2\sqrt{\pi D t}}(-x^2/4D)(-1/t^2) + \frac{1}{2\sqrt{\pi D}}(-1/2)t^{-3/2}\right] e^{-x^2/4Dt}$$

$$= \frac{1}{2\sqrt{\pi D} t^{3/2}}\left[\frac{1}{t}(x^2/4D) - (1/2)\right] e^{-x^2/4Dt} \tag{3.11a}$$

u_x is

$$u_x(x,t) = \frac{1}{2\sqrt{\pi D t}} e^{-x^2/4Dt}(-2x/4Dt)$$

Differentiation of this result with respect to x gives u_{xx}

$$Du_{xx}(x,t) = \frac{D}{2\sqrt{\pi D t}}\left[e^{-x^2/4Dt}(-2/4Dt) + e^{-x^2/4Dt}(-2x/4Dt)^2\right]$$

$$= \frac{1}{2\sqrt{\pi D} t^{3/2}}\left[(-1/2) + (x^2/4Dt)\right] e^{-x^2/4Dt} \tag{3.11b}$$

We see then that Eqs. (3.11a) and (3.11b) satisfy Eq. (3.1).

A.2. The Function simp

The integral of Eq. (3.5) is evaluated numerically by application of Simpson's rule

$$
\begin{aligned}
u_1(t) &= \int_{x_l}^{x_u} u(x,t)dx \\
&\approx \frac{h}{3}\left[u(1) + \sum_{i=2}^{n-2} 4u(i) + 2u(i+1) + u(n)\right] \qquad (3.12)
\end{aligned}
$$

where x_u and x_l are the upper and lower limits of the integral. The integration interval is therefore $h = (x_u - x_l)/(n-1)$, with n the number of quadrature (spatial grid) points in x (must be odd). For Eq. (3.5), $x_l = -\infty$, $x_u = \infty$, $n = 101$.

Function simp is a straightforward implementation of Eq. 3.12 (see Listing 3.5).

```
  function uint=simp(xl,xu,n,u)
%
% Function simp computes the integral of the numerical
% solution by Simpson's rule
%
    h=(xu-xl)/(n-1);
    uint=u(1)-u(n);
      for i=3:2:n
        uint=uint+4.0*u(i-1)+2.0*u(i);
      end
      uint=h/3.0*uint;
    end
```

Listing 3.5. Numerical quadrature routine simp applied to Eq. (3.5)

REFERENCES

[1] Farlow, S. J. (1993), The D'Alembert Solution of the Wave Equation, *Partial Differential Equations for Scientists and Engineers*, Dover Publications, New York, Chapter 17, pp. 129–136

[2] Polyanin, A. D. (2002), *Handbook of Linear Partial Differential Equations for Scientists and Engineers*, Chapman and Hall/CRC, Boca Raton, FL

4

Two Nonlinear, Variable-Coefficient, Inhomogeneous Partial Differential Equations

This 2×2 partial differential equation (PDE) problem (two PDEs in two unknowns) introduces the following mathematical concepts and computational methods:

1. The two PDEs have a variety of mathematical properties; that is, they are
   - simultaneous,
   - nonlinear,
   - variable coefficient, and
   - inhomogeneous.

   These features are briefly explained in the discussion that follows. The subsequent programming illustrates how these features can be included in a computer code.
2. The two PDEs have an exact solution that can be used to assess the accuracy of the method of lines (MOL) numerical solution.
3. The effectiveness of higher-order finite differences (FDs) in improving the accuracy of numerical MOL solutions to PDEs.

The PDEs are [1]:

$$u_t = ((v-1)u_x)_x + (16xt - 2t - 16(v-1)(u-1))(u-1) + 10x\,e^{-4x} \tag{4.1a}$$

$$v_t = v_{xx} + u_x + 4u - 4 + x^2 - 2t - 10t\,e^{-4x} \tag{4.1b}$$

Here we have used the subscript notation of partial derivatives; for example,

$$u_t \leftrightarrow \frac{\partial u}{\partial t}, \quad ((v-1)u_x)_x \leftrightarrow \frac{\partial}{\partial x}\left[(v-1)\frac{\partial u}{\partial x}\right]$$

Also, we consider t to be an *initial-value variable* and x to be a *boundary-value variable* (the distinction is explained subsequently). Since each PDE is first order in t, each requires one initial condition (IC); these ICs are taken as

$$u(x, t=0) = 1, \qquad v(x, t=0) = 1 \tag{4.2a,b}$$

Equations (4.2a) and (4.2b) are termed *inhomogeneous* ICs since they are not zero on the RHS.

Equations (4.1a) and (4.1b) are second order in x and therefore each equation requires two boundary conditions (BCs); these BCs are taken as

$$u(x=0,t) = v(x=0,t) = 1 \tag{4.3a,b}$$

$$3u(x=1,t) + u_x(x=1,t) = 3; \qquad 5v_x(x=1,t) = e^4(u(x=1,t) - 1) \tag{4.3c,d}$$

Note that Eqs. (4.3a) and (4.3b) are applied at $x = 0$, while Eqs. (4.3c) and (4.3d) are applied at $x = 1$, that is two different values of x which is why they are called boundary conditions; generally, BCs represent conditions at different boundaries of a physical system.

Equations (4.3a) and (4.3b) are termed *BCs of the first type* or *Dirichlet BCs* because they involve just the dependent variables, $u(x, t)$ and $v(x, t)$ at $x = 0$. Equations (4.3c) and (4.3d) are termed *BCs of the third type* or *Robin BCs* because they involve the dependent variables $u(x, t)$ and $v(x, t)$ and their first spatial derivatives $u_x(x, t)$ and $v_x(x, t)$ at $x = 1$. To complete the picture, BCs that involve only the derivatives $u_x(x, t)$ and $v_x(x, t)$ are termed *BCs of the second type* or *Neumann BCs*; examples would be the homogeneous Neumann BCs $u_x(x = 1, t) = 0$ and $v_x(x = 0, t)$. All of the BCs as stated are *linear*; that is, the dependent variables and their derivatives are to the first power. BCs can also be nonlinear; for example, the third-type BCs $u_x(x = 1, t) = h(u_a^4 - u(x = 1, t)^4)$ and $v_x(x = 1, t) = h(v_a^4 - v(x = 1, t)^4)$ are nonlinear, where h and u_a and v_a are given positive constants.

To continue this discussion of terminology a bit further, we return to the PDEs, Eqs. (4.1a) and (4.1b). Some of the terminology is listed in item 1. Equations (4.1a) and (4.1b) are

- *Simultaneous* since the dependent variables $u(x, t)$ and $v(x, t)$ appear in both equations so that the PDEs must be solved together.
- *Nonlinear* since the dependent variables $u(x, t)$ and $v(x, t)$ and their derivatives are to other than the first power, for example, in Eq. (4.1a) $((v - 1)u_x)_x$ and $(v - 1)(u - 1)$; note however that Eq. (4.1b) is linear.
- *Variable coefficient* since the dependent variables $u(x, t)$ and $v(x, t)$ are multiplied by functions of the *independent variables*, for example, in Eq. (4.1a) $(16xt - 2t)(u - 1)$.
- *Inhomogeneous* since terms dependent on only the *independent variables* appear in the equations, for example, $10x\,e^{-4x}$ in both equations; these terms are also termed *nonhomogenenous*.
- *Mixed type*. To explain, since Eqs. (4.1a) and (4.1b) contain first-order derivatives in t and second-order derivatives in x, they can be termed *parabolic*. But Eqs. (4.1a) and (4.1b) also contain first-order derivatives in x so that they can be termed *first-order hyperbolic*. Thus, Eqs. (4.1a) and (4.1b) can be termed *hyperbolic–parabolic*, or in more physical terms, *convective–diffusive*.

The subsequent programming demonstrates that all of these variations in mathematical features can be accommodated by the MOL approach.

Equations (4.1)–(4.3) have an analytical solution

$$u(x,t) = 1 + 10xt\,e^{-4x}, \qquad v(x,t) = 1 + x^2t \tag{4.4a,b}$$

This analytical solution is used to assess the accuracy of the MOL numerical solution in the subsequent programming. Also, Eqs. (4.4a) and (4.4b) suggest that this problem is artificial (fabricated or contrived) in the sense that it does not originate from a physical application. However, it requires coding of all of the mathematical features discussed previously and therefore we view it as a very useful and comprehensive test problem.

A program to implement Eqs. (4.1) and (4.4) follows. First the main program (pde_1_main) is given in Listing 4.1.

```
%
% Clear previous files
  clear all
  clc
%
% Parameters shared with other routines
  global ncall   ncase...
              n       x    ndss
%
% Two cases: ncase = 1, second order finite differences
%
%            ncase = 2, fourth order finite difference
%
  for ncase=1:2
%
%   Initial condition
    n=41;
    t0=0.0;
    y0=inital_1(t0);
%
%   Independent variable for ODE integration
    tf=1.0;
    nout=21;
    tout=linspace(t0,tf,nout);
    ncall=0;
%
%   ODE integration ;
    reltol=1.0e-04; abstol=1.0e-04;
    options=odeset('RelTol',reltol,'AbsTol',abstol);
    if(ncase==1) ndss=2; % ndss = 2, 4, 6, 8 or 10 required
      [t,y]=ode15s(@pde_1,tout,y0,options); end
    if(ncase==2) ndss=4; % ndss = 2, 4, 6, 8 or 10 required
      [t,y]=ode15s(@pde_1,tout,y0,options); end
%
%   One vector to two vectors
    for it=1:nout
```

```
    for i=1:n
      u(it,i)=y(it,i);
      v(it,i)=y(it,i+n);
    end
    end
%
%   Display a heading for the numerical and analytical
%   solutions
    fprintf('\n\n  ncase = %2d\n',ncase);
    fprintf('\n  u =  numerical u(x,t)');
    fprintf('\n ua = analytical u(x,t)');
    fprintf('\n diff u = u - ua');
    fprintf('\n  v =  numerical v(x,t)');
    fprintf('\n va = analytical v(x,t)');
    fprintf('\n diff v = v - uv\n');
%
%   Analytical solutions, differences between these solutions
%   at selected x
    for it=1:nout
      fprintf('\n    t       x        u      ua     diff u');
      fprintf('\n                     v      va     diff v');
    for i=1:10:n
      ua(i)=1.0+10.0*x(i)*t(it)*exp(-4.0*x(i));
      du(i)=u(it,i)-ua(i);
      va(i)=1.0+(x(i)^2)*t(it);
      dv(i)=v(it,i)-va(i);
%
%   Display the numerical solutions, analytical solutions,
%   differences between the solutions
    fprintf('\n %4.2f%9.3f%9.4f%9.4f%12.2e\n
            %23.4f%9.4f%12.2e\n',t(it),x(i),u(it,i),ua(i), ...
            du(i),v(it,i),va(i),dv(i));
    end
    end
    fprintf('  ncall = %5d\n',ncall);
%
% Plot solutions
  if(ncase==1)
%
%   Parametric plots
    figure(1);
    subplot(1,2,1)
    plot(x,u,'-k'); axis tight
    title('u(x,t) vs x, t = 0, 0.05, 0.1, ..., 1.0');
          xlabel('x'); ylabel('u(x,t)')
    subplot(1,2,2)
    plot(x,v,'-k'); axis tight
```

```
      title('v(x,t) vs x, t = 0, 0.05, 0.1, ..., 1.0');
            xlabel('x'); ylabel('v(x,t)')
%
%     Surface plots
      figure(2);
      surf(u); axis tight
      xlabel('x grid number');
      ylabel('t grid number');
      zlabel('u(x,t)');
      title('Two nonlinear PDEs');
      view([-115 36]);
      colormap gray
      rotate3d on;
      figure(3);
      surf(v); axis tight
      xlabel('x grid number');
      ylabel('t grid number');
      zlabel('v(x,t)');
      title('Two nonlinear PDEs');
      view([170 24]);
      colormap gray
      rotate3d on;
%     print -r300 -deps pde.eps; print -r300 -dps pde.ps;
%     print -r300 -dpng pde.png
    end
%
% Next case
    end
```

Listing 4.1. Main program pde_1_main

We can note the following points about this program:

1. After placing some of the problem parameters and variables in global, two cases are programmed with a for loop for the calculation of the spatial derivatives by second- and fourth-piorder FDs.

```
%
% Clear previous files
    clear all
    clc
%
% Parameters shared with other routines
    global ncall   ncase...
              n       x    ndss
%
```

```
% Two cases: ncase = 1, second order finite differences
%
%            ncase = 2, fourth order finite difference
%
  for ncase=1:2
```

2. ICs (4.2a) and (4.2b) are defined on a 41-point spatial grid (through a call to `inital_1` discussed subsequently) and a 21-point time grid (for $0 \le t \le 1$).

```
%
%   Initial condition
    n=41;
    t0=0.0;
    y0=inital_1(t0);
%
%   Independent variable for ODE integration
    tf=1.0;
    nout=21;
    tout=linspace(t0,tf,nout);
    ncall=0;
```

3. The ODE integration within the MOL PDE solution is performed by `ode15s`; note that the IC vector `y0` is of length `2n` ($n = 41$), with the first `n` values assigned to `u` and the second `n` values assigned to `v`.

```
%
%   ODE integration ;
    reltol=1.0e-04; abstol=1.0e-04;
    options=odeset('RelTol',reltol,'AbsTol',abstol);
    if(ncase==1) ndss=2; % ndss = 2, 4, 6, 8 or 10 required
      [t,y]=ode15s(@pde_1,tout,y0,options); end
    if(ncase==2) ndss=4; % ndss = 2, 4, 6, 8 or 10 required
      [t,y]=ode15s(@pde_1,tout,y0,options); end
%
%   One vector to two vectors
    for it=1:nout
    for i=1:n
      u(it,i)=y(it,i);
      v(it,i)=y(it,i+n);
    end
    end
```

After the ODE numerical solution is returned in y, the solution of Eqs. (4.1a) and (4.1b) is put into arrays u and v.

4. The numerical solution is then displayed along with the analytical solution (computed according to Eq. (4.4)).

```
%
%   Display a heading for the numerical and
%   analytical solutions
    fprintf('\n\n  ncase = %2d\n',ncase);
    fprintf('\n  u =  numerical u(x,t)');
    fprintf('\n ua = analytical u(x,t)');
    fprintf('\n diff u = u - ua');
    fprintf('\n  v =  numerical v(x,t)');
    fprintf('\n va = analytical v(x,t)');
    fprintf('\n diff v = v - uv\n');
%
%   Analytical solutions, differences between these solutions
%   at selected x
    for it=1:nout
      fprintf('\n    t       x        u      ua      diff u');
      fprintf('\n                     v      va      diff v');
    for i=1:10:n
      ua(i)=1.0+10.0*x(i)*t(it)*exp(-4.0*x(i));
      du(i)=u(it,i)-ua(i);
      va(i)=1.0+(x(i)^2)*t(it);
      dv(i)=v(it,i)-va(i);
%
%   Display the numerical solutions, analytical solutions,
%   differences between the solutions
    fprintf('\n %4.2f%9.3f%9.4f%9.4f%12.2e\n
            %23.4f%9.4f%12.2e\n',t(it),x(i),u(it,i),ua(i), ...
            du(i),v(it,i),va(i),dv(i));
    end
    end
    fprintf('  ncall = %5d\n',ncall);
```

5. The numerical solutions are plotted as a function of x with t as a parameter.

```
%
% Plot solutions
  if(ncase==1)
```

```
%
%    Parametric plots
     figure(1);
     subplot(1,2,1)
     plot(x,u,'-k'); axis tight
     title('u(x,t) vs x, t = 0, 0.05, 0.1, ..., 1.0');
          xlabel('x'); ylabel('u(x,t)')
     subplot(1,2,2)
     plot(x,v,'-k'); axis tight
     title('v(x,t) vs x, t = 0, 0.05, 0.1, ..., 1.0');
          xlabel('x'); ylabel('v(x,t)')
```

6. Finally, the numerical solutions are plotted in three-dimensional (3D) perspective.

```
%
%    Surface plots
     figure(2);
     surf(u); axis tight
     xlabel('x grid number');
     ylabel('t grid number');
     zlabel('u(x,t)');
     title('Two nonlinear PDEs');
     view([-115 36]);
     colormap gray
     rotate3d on;
     figure(3);
     surf(v); axis tight
     xlabel('x grid number');
     ylabel('t grid number');
     zlabel('v(x,t)');
     title('Two nonlinear PDEs');
     view([170 24]);
     colormap gray
     rotate3d on;
%    print -r300 -deps pde.eps; print -r300 -dps pde.ps;
%    print -r300 -dpng pde.png
   end
%
% Next case
   end
```

The initialization routine inital_1 is given in Listing 4.2.

```
  function y=inital_1(t0)
%
% Function inital_1 is called by the main program to define
% the initial conditions in the MOL solution of two nonlinear
% PDEs
%
  global ncall   ncase...
             u       v...
             n       x       e4    ndss
%
% Spatial increment
  dx=1.0/(n-1);
%
% Values of x along the spatial grid
  x=[0.0:dx:1.0];
%
% Initial conditions
  for i=1:n
    u(i)=1.0;
    v(i)=1.0;
    y(i)  =u(i);
    y(i+n)=v(i);
  end
%
% Constant e^4 used in boundary condition
  e4=exp(1.0)^4;
%
% Initialize calls to pde_1
  ncall=0;
```

Listing 4.2. Initialization routine inital_1

The following points can be noted:

1. After defining the function and the global parameters and variables, the spatial grid in x is defined for $0 \leq x \leq 1$.

```
  function y=inital_1(t0)
%
% Function inital_1 is called by the main program to define
% the initial conditions in the MOL solution of two
% nonlinear PDEs
%
```

```
  global ncall  ncase...
             u       v...
             n       x       e4     ndss
%
% Spatial increment
  dx=1.0/(n-1);
%
% Values of x along the spatial grid
  x=[0.0:dx:1.0];
```

2. ICs (4.2a) and (4.2b) are then set for $u(x, t)$ and $v(x, t)$ and subsequently put into one dependent-variable array y.

```
%
% Initial conditions
  for i=1:n
    u(i)=1.0;
    v(i)=1.0;
    y(i)  =u(i);
    y(i+n)=v(i);
  end
%
% Constant e^4 used in boundary condition
  e4=exp(1.0)^4;
%
% Initialize calls to pde_1
  ncall=0;
```

Finally, e^4 is computed here to avoid repeated calculation in the ODE routine pde_1 (listed next) and the counter for the calls to pde_1 is initialized.

The ODE routine pde_1 called by ode15s in pde_1_main is given in Listing 4.3.

```
  function yt=pde_1(t,y)
%
% Function pde_1 defines the ODEs in the MOL solution of two
% nonlinear PDEs
%
  global ncall  ncase...
             u       v...
             ut      vt...
             ux      uxx      vx      vxx...
```

```
              n        x       e4    ndss
%
% One vector to two vectors
  for i=1:n
    u(i)=y(i);
    v(i)=y(i+n);
  end
%
% Boundary conditions at x = 0
  u(1)=1.0;
  ut(1)=0.0;
  v(1)=1.0;
  vt(1)=0.0;
%
% First order spatial derivatives
  xl=x(1);
  xu=x(n);
%
% Three point centered differences
  if(ncase==1)ux=dss002(xl,xu,n,u); end
  if(ncase==1)vx=dss002(xl,xu,n,v); end
%
% Five point centered differences
  if(ncase==2)ux=dss004(xl,xu,n,u); end
  if(ncase==2)vx=dss004(xl,xu,n,v); end
%
% Boundary conditions at x = 1
  ux(n)=3.0-3.0*u(n);
  vx(n)=e4*(u(n)-1.0)/5.0;
%
% Second order spatial derivatives
%
% Three point centered  differences
  if(ncase==1)vxx=dss002(xl,xu,n,vx); end
%
% Five point centered differences
  if(ncase==2)vxx=dss004(xl,xu,n,vx); end
%
% Array vx is used as temporary storage in the calculation
% of the term ((v - 1)*u )  which is finally stored in
% array uxx              x x
  for i=1:n
    vx(i)=(v(i)-1.0)*ux(i);
  end
  if(ncase==1)uxx=dss002(xl,xu,n,vx); end
  if(ncase==2)uxx=dss004(xl,xu,n,vx); end
%
```

```
% Two PDEs
  for i=2:n
    ex=exp(-4.0*x(i));
      ut(i)=uxx(i)+(16.0*x(i)*t-2.0*t-16.0*(v(i)-1.0))* ...
            (u(i)-1.0)+10.*x(i)*ex;vt(i)=vxx(i)+ux(i) ...
            +4.0*u(i)-4.0+x(i)^2-2.0*t-10.0*t*ex;
  end
%
% Two vectors to one vector
  for i=1:n
    yt(i)   =ut(i);
    yt(i+n) =vt(i);
  end
  yt=yt';
%
% Increment calls to pde_1
  ncall=ncall+1;
```

Listing 4.3. ODE routine pde_1

We can note the following points:

1. After the routine and the global parameters and variables are defined, the dependent-variable vector y is divided into the two problem-oriented arrays u and v (to facilitate programming in terms of these problem-oriented variables).

```
  function yt=pde_1(t,y)
%
% Function pde_1 defines the ODEs in the MOL solution of two
% nonlinear PDEs
%
  global ncall   ncase...
              u        v...
             ut       vt...
             ux      uxx       vx      vxx...
              n        x       e4     ndss
%
% One vector to two vectors
  for i=1:n
    u(i)=y(i);
    v(i)=y(i+n);
  end
```

2. BCs (4.3a) and (4.3b) at $x = 0$ are then set.

```
%
% Boundary conditions at x = 0
  u(1)=1.0;
  ut(1)=0.0;
  v(1)=1.0;
  vt(1)=0.0;
```

   Note that the derivatives in t, `ut(1)` and `vt(1)`, are set to zero to maintain the constant values of `u(1)` and `v(1)` at the boundary $x = 0$.
3. The first derivatives in x are computed by calls to `dss002` for second-order FDs or `dss004` for fourth-order FDs.

```
%
% First order spatial derivatives
  xl=x(1);
  xu=x(n);
%
% Three point centered differences
  if(ncase==1)ux=dss002(xl,xu,n,u); end
  if(ncase==1)vx=dss002(xl,xu,n,v); end
%
% Five point centered differences
  if(ncase==2)ux=dss004(xl,xu,n,u); end
  if(ncase==2)vx=dss004(xl,xu,n,v); end
```

4. BCs (4.3c) and (4.3d) are applied to give the first-order derivatives at $x = 1$.

```
%
% Boundary conditions at x = 1
  ux(n)=3.0-3.0*u(n);
  vx(n)=e4*(u(n)-1.0)/5.0;
```

5. The second-order derivatives in x are then computed by differentiating the first-order derivatives, that is, using *stagewise differentiation*.

```
%
% Second order spatial derivatives
%
% Three point centered  differences
```

```
  if(ncase==1)vxx=dss002(xl,xu,n,vx); end
%
% Five point centered differences
  if(ncase==2)vxx=dss004(xl,xu,n,vx); end
%
% Array vx is used as temporary storage in the calculation
% of the term ((v - 1)*u )  which is finally stored in
% array uxx              x x
  for i=1:n
    vx(i)=(v(i)-1.0)*ux(i);
  end
  if(ncase==1)uxx=dss002(xl,xu,n,vx); end
  if(ncase==2)uxx=dss004(xl,xu,n,vx); end
```

Note in particular how the nonlinear term $((v-1)u_x)_x$ in Eq. (4.1a) is computed.

6. Finally, the two PDEs, Eqs. (4.1a) and (4.1b), are programmed to give the derivative vectors `ut` and `vt`, which are in turn stored in the single-derivative array `yt`.

```
%
% Two PDEs
  for i=2:n
    ex=exp(-4.0*x(i));
      ut(i)=uxx(i)+(16.0*x(i)*t-2.0*t-16.0*(v(i)-1.0))* ...
            (u(i)-1.0)+10.*x(i)*ex;vt(i)=vxx(i)+ux(i)+4.0* ...
            u(i)-4.0+x(i)^2-2.0*t-10.0*t*ex;
  end
%
% Two vectors to one vector
  for i=1:n
    yt(i)   =ut(i);
    yt(i+n) =vt(i);
  end
  yt=yt';
%
% Increment calls to pde_1
  ncall=ncall+1;
```

Note in particular that `pde_1` has the input vector `y` and returns the derivative vector `yt` in accordance with the requirements of the ODE integrator `ode15s`.

This completes the MOL programming of Eqs. (4.1)–(4.3). Part of the numerical output from `pde_1_main` is given in Table 4.1.

Table 4.1. Abbreviated numerical output from `pde_1_main` and `pde_1`

```
 ncase =  1

 u =  numerical u(x,t)
ua = analytical u(x,t)
diff u = u - ua
 v =  numerical v(x,t)
va = analytical v(x,t)
diff v = v - uv

   t        x        u        ua       diff u
                     v        va       diff v
0.00     0.000    1.0000   1.0000   0.00e+000
                  1.0000   1.0000   0.00e+000

0.00     0.250    1.0000   1.0000   0.00e+000
                  1.0000   1.0000   0.00e+000

0.00     0.500    1.0000   1.0000   0.00e+000
                  1.0000   1.0000   0.00e+000

0.00     0.750    1.0000   1.0000   0.00e+000
                  1.0000   1.0000   0.00e+000

0.00     1.000    1.0000   1.0000   0.00e+000
                  1.0000   1.0000   0.00e+000

   t        x        u        ua       diff u
                     v        va       diff v
0.05     0.000    1.0000   1.0000   0.00e+000
                  1.0000   1.0000   0.00e+000

0.05     0.250    1.0460   1.0460  -7.26e-007
                  1.0031   1.0031   1.46e-005

0.05     0.500    1.0338   1.0338  -2.42e-007
                  1.0125   1.0125   6.58e-006

0.05     0.750    1.0187   1.0187  -3.14e-007
                  1.0281   1.0281   2.07e-006

0.05     1.000    1.0092   1.0092   2.81e-007
                  1.0500   1.0500   1.34e-006
   .                                    .
   .                                    .
   .                                    .
```

```
  t        x        u        ua       diff u
                    v        va       diff v
0.95    0.000   1.0000   1.0000   0.00e+000
                1.0000   1.0000   0.00e+000

0.95    0.250   1.8714   1.8737  -2.34e-003
                1.0598   1.0594   3.96e-004

0.95    0.500   1.6418   1.6428  -1.06e-003
                1.2377   1.2375   1.93e-004

0.95    0.750   1.3543   1.3547  -4.71e-004
                1.5342   1.5344  -1.51e-004

0.95    1.000   1.1738   1.1740  -1.68e-004
                1.9494   1.9500  -5.64e-004

  t        x        u        ua       diff u
                    v        va       diff v
1.00    0.000   1.0000   1.0000   0.00e+000
                1.0000   1.0000   0.00e+000

1.00    0.250   1.9171   1.9197  -2.58e-003
                1.0629   1.0625   4.05e-004

1.00    0.500   1.6755   1.6767  -1.15e-003
                1.2502   1.2500   1.69e-004

1.00    0.750   1.3729   1.3734  -5.02e-004
                1.5623   1.5625  -2.12e-004

1.00    1.000   1.1830   1.1832  -1.79e-004
                1.9993   2.0000  -6.57e-004
 ncall =   453


 ncase =  2

 u =  numerical u(x,t)
ua = analytical u(x,t)
diff u = u - ua
 v =  numerical v(x,t)
va = analytical v(x,t)
diff v = v - uv
```

(*continued*)

Table 4.1 (*continued*)

t	x	u v	ua va	diff u diff v
0.00	0.000	1.0000 1.0000	1.0000 1.0000	0.00e+000 0.00e+000
0.00	0.250	1.0000 1.0000	1.0000 1.0000	0.00e+000 0.00e+000
0.00	0.500	1.0000 1.0000	1.0000 1.0000	0.00e+000 0.00e+000
0.00	0.750	1.0000 1.0000	1.0000 1.0000	0.00e+000 0.00e+000
0.00	1.000	1.0000 1.0000	1.0000 1.0000	0.00e+000 0.00e+000

t	x	u v	ua va	diff u diff v
0.05	0.000	1.0000 1.0000	1.0000 1.0000	0.00e+000 0.00e+000
0.05	0.250	1.0460 1.0031	1.0460 1.0031	1.45e-008 -5.22e-008
0.05	0.500	1.0338 1.0125	1.0338 1.0125	1.38e-009 -2.82e-008
0.05	0.750	1.0187 1.0281	1.0187 1.0281	4.29e-010 -7.98e-009
0.05	1.000	1.0092 1.0500	1.0092 1.0500	3.89e-009 2.87e-009
. . .				. . .
0.95	0.250	1.8737 1.0594	1.8737 1.0594	2.76e-005 1.84e-006
0.95	0.500	1.6429 1.2375	1.6428 1.2375	1.39e-005 1.24e-006
0.95	0.750	1.3547 1.5344	1.3547 1.5344	9.96e-006 1.25e-007

```
 0.95     1.000    1.1740    1.1740   -1.03e-006
                   1.9500    1.9500    2.35e-006

    t         x         u        ua       diff u
                        v        va       diff v
 1.00     0.000    1.0000    1.0000    0.00e+000
                   1.0000    1.0000    0.00e+000

 1.00     0.250    1.9197    1.9197    2.97e-005
                   1.0625    1.0625    2.30e-006

 1.00     0.500    1.6767    1.6767    1.51e-005
                   1.2500    1.2500    1.98e-006

 1.00     0.750    1.3734    1.3734    1.13e-005
                   1.5625    1.5625    9.03e-007

 1.00     1.000    1.1832    1.1832   -1.30e-006
                   2.0000    2.0000    3.04e-006
  ncall =    363
```

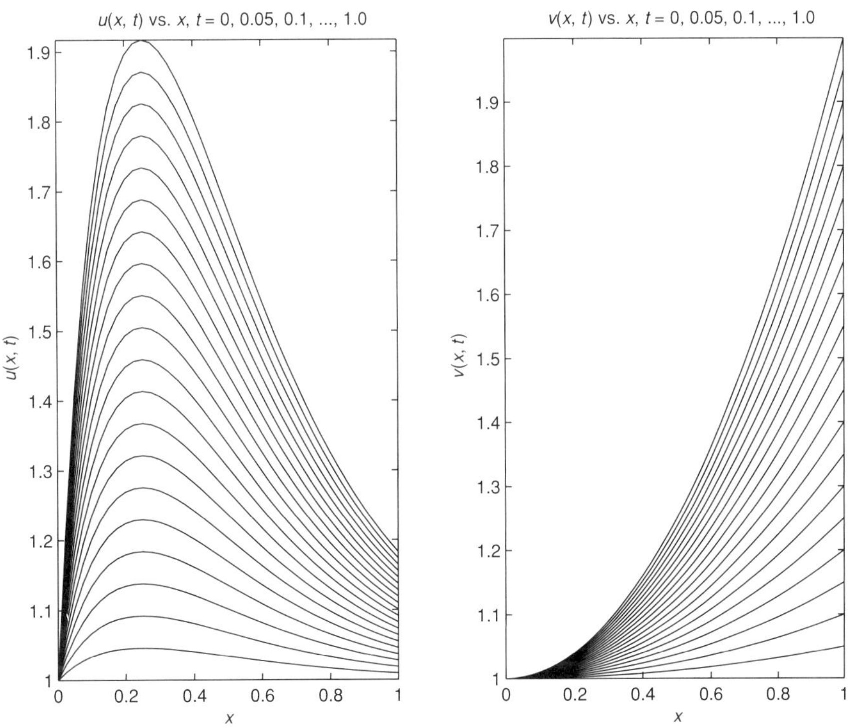

Figure 4.1. $u(x, t)$ and $v(x, t)$ profiles in x with t as a parameter

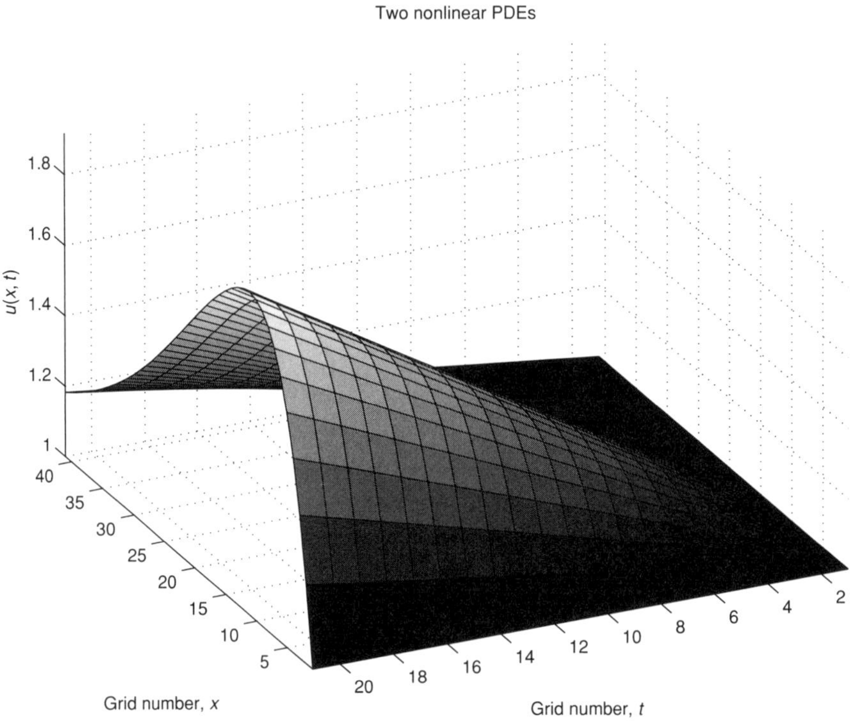

Figure 4.2. $u(x,t)$ in 3D perspective

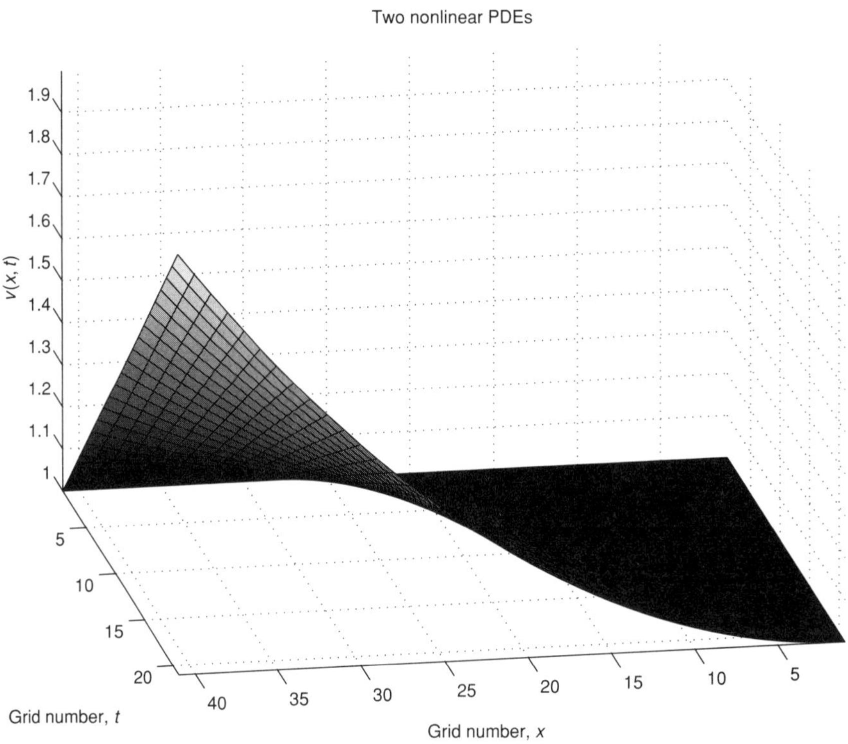

Figure 4.3. $v(x,t)$ in 3D perspective

We can note the following about the output provided in Table 4.1:

1. The agreement between the numerical and analytical solutions is better for the fourth-order FD approximations (`ncase = 2`).
2. The number of calls to `pde_1` is quite modest, and in fact, the fourth-order FDs required fewer (`ncall = 363`) than the second-order FDs (`ncall = 453`) so that for this problem at least, a more accurate solution was achieved with less computational effort.

The 2D plotted output from `pde_1_main` is shown in Figure 4.1. The solutions in Figure 4.1 are further elucidated by the 3D plots in Figures 4.2 and 4.3 (from `pde_1_main`).

In summary, this application based on Eqs. (4.1)–(4.3) demonstrates the applicability of the MOL to a system of PDEs with a variety of mathematical properties that are plainly programmed in `pde_1`.

REFERENCE

[1] Madsen, N. K. and R. F. Sincovec (1976), Software for Partial Differential Equations, In: L. Lapidus and W. E. Schiesser (Eds.), *Numerical Methods for Differential Systems*, Academic Press, New York, pp. 229–242

5

Euler, Navier Stokes, and Burgers Equations

This partial differential equation (PDE) problem introduces the following mathematical concepts and computational methods:

1. We start with a general PDE system in three dimensions (3D), with some simplifying assumptions, reduces to a 1D nonlinear PDE. Here is some terminology applied to this starting point and the final result:
   (a) We start with two classic (we might say without exaggeration "famous") PDE systems: the *Euler equations* and the *Navier Stokes equations* of fluid mechanics.
   (b) After reduction to one dimension, followed by some additional simplifications, we arrive at the *Burgers equation*, a PDE widely used as a test problem for numerical methods.
2. The analysis is in terms of coordinate-free PDEs that can then be specialized to a particular coordinate system; for the following analysis, this is Cartesian coordinates.
3. The 1D nonlinear PDE, Burgers' equation, has an analytical solution that can be used to test the numerical method of lines (MOL) algorithms developed and implemented in a series of computer routines.
4. The analytical solution is a traveling wave that has an increasingly steep moving front, usually termed *front sharpening*. The conditions under which front sharpening might occur are analyzed briefly.
5. Spatial convergence of the Burgers equation numerical solution is investigated by h- and p-refinement.
6. Stagewise differentiation is used to calculate higher-order spatial derivatives.
7. Dirichlet and Neumann boundary conditions are investigated as two cases of the numerical solution.

The starting point for the analysis is the *multidimensional, conservation equations written in conservative form*:

$$\mathbf{u}_t + \nabla \cdot [\mathbf{f}(\mathbf{u})] = \mathbf{0} \tag{5.1}$$

where $\mathbf{u}$ is the *vector of conserved quantities*, which typically includes mass, momentum, and/or energy. The terms in Eq. (5.1) represent

$\mathbf{u}_t$	time rate of accumulation (or depletion, depending on the sign of this term) of the conserved quantity
$\nabla \cdot [\mathbf{f}(\mathbf{u})]$	net flux of the conserved quantity into or out of a differential volume

Here we have used subscript notation for partial derivatives, so that, for example, $u_t = \partial u / \partial t$.

In the case of a simplified fluid mechanics model, $\mathbf{u}$ in Eq. (5.1) is usually mass and momentum, defined as

$$\mathbf{u} = \begin{bmatrix} \rho \\ \rho\mathbf{v} \end{bmatrix}, \qquad \mathbf{f}(\mathbf{u}) = \begin{bmatrix} \rho\mathbf{v} \\ \rho\mathbf{v}\mathbf{v} + p \end{bmatrix}$$

Variables in boldface represent a vector or matrix; for example, $\mathbf{v}$ is a velocity vector (introductory discussion of vector/tensor notation is given in Appendix 1). Substitution of $\mathbf{u}$ and $\mathbf{f}(\mathbf{u})$ into Eq. (5.1) gives the Euler equations

$$\rho_t + \nabla \cdot (\rho\mathbf{v}) = 0 \qquad \text{continuity} \tag{5.2a}$$

$$(\rho\mathbf{v})_t + \nabla \cdot (\rho\mathbf{v}\mathbf{v}) + \nabla p = 0 \qquad \text{momentum} \tag{5.2b}$$

where ∇ is a differential operator applied to the mass vector $\rho\mathbf{v}$ in the continuity equation, and to a scalar (pressure, p) and a tensor ($\rho\mathbf{v}\mathbf{v}$) in the momentum equation. Considering ∇ a bit further, with the dot $\cdot$ it forms the usual dot product with a vector (and the result is a scalar), while without the dot, it operates on a scalar to give a vector, or on a tensor as explained subsequently.

Also, ∇ is a *coordinate-free operator*; that is, it can be expressed in any *3D orthogonal coordinate system*. For example, in *Cartesian coordinates*, ∇ is

$$\nabla = \mathbf{i}\frac{\partial}{\partial x} + \mathbf{j}\frac{\partial}{\partial y} + \mathbf{k}\frac{\partial}{\partial z}$$

Since the fluid density ρ is a scalar, and with $\mathbf{v} = \mathbf{i}v_x + \mathbf{j}v_y + \mathbf{k}v_z$,

$$\begin{aligned} \nabla \cdot (\rho\mathbf{v}) &= \left(\mathbf{i}\frac{\partial}{\partial x} + \mathbf{j}\frac{\partial}{\partial y} + \mathbf{k}\frac{\partial}{\partial z}\right) \cdot (\mathbf{i}\rho v_x + \mathbf{j}\rho v_y + \mathbf{k}\rho v_z) \\ &= \frac{\partial(\rho v_x)}{\partial x} + \frac{\partial(\rho v_y)}{\partial y} + \frac{\partial(\rho v_z)}{\partial z} \end{aligned}$$

where we have made use of the dot product between orthogonal unit vectors

$$\mathbf{i} \cdot \mathbf{j} = \begin{cases} 0, & \mathbf{i} \neq \mathbf{j} \\ 1, & \mathbf{i} = \mathbf{j} \end{cases}$$

Then, the (scalar) continuity equation is from Eq. (5.2a):

$$\frac{\partial \rho}{\partial t}+\frac{\partial(\rho v_x)}{\partial x}+\frac{\partial(\rho v_y)}{\partial y}+\frac{\partial(\rho v_z)}{\partial z}=0 \tag{5.3}$$

Note that for a constant density (incompressible) fluid,

$$\frac{\partial \rho}{\partial t}=0$$

so that Eq. (5.2a) becomes

$$\frac{\partial(\rho v_x)}{\partial x}+\frac{\partial(\rho v_y)}{\partial y}+\frac{\partial(\rho v_z)}{\partial z}=\rho\left(\frac{\partial v_x}{\partial x}+\frac{\partial v_y}{\partial y}+\frac{\partial v_z}{\partial z}\right)+v_x\frac{\partial \rho}{\partial x}+v_y\frac{\partial \rho}{\partial x}+v_z\frac{\partial \rho}{\partial x}=0$$

or (with the derivatives of ρ zero),

$$\frac{\partial v_x}{\partial x}+\frac{\partial v_y}{\partial y}+\frac{\partial v_z}{\partial z}=0$$

$$\nabla\cdot\mathbf{v}=0 \tag{5.4}$$

The property of Eq. (5.4) for an incompressible fluid is termed *divergence free*. It can be used to check a numerical solution to a PDE system for an incompressible fluid (a numerical solution should be divergence free).

We now consider the momentum balance of Eq. (5.2), starting with the term

$$\nabla\cdot(\rho\mathbf{v}\mathbf{v})$$

The product **vv** is actually a second-order (nine-component) tensor. Thus,

$$\rho\mathbf{v}\mathbf{v}=\begin{bmatrix}\rho\mathbf{i}v_x\mathbf{i}v_x & \rho\mathbf{i}v_x\mathbf{j}v_y & \rho\mathbf{i}v_x\mathbf{k}v_z\\ \rho\mathbf{j}v_y\mathbf{i}v_x & \rho\mathbf{j}v_y\mathbf{j}v_y & \rho\mathbf{j}v_y\mathbf{k}v_z\\ \rho\mathbf{k}v_z\mathbf{i}v_x & \rho\mathbf{k}v_z\mathbf{j}v_y & \rho\mathbf{k}v_z\mathbf{k}v_z\end{bmatrix}$$

and applying $\nabla\cdot$ to this tensor,

$$\nabla\cdot(\rho\mathbf{v}\mathbf{v})=\begin{bmatrix}\mathbf{i}\left\{\frac{\partial}{\partial x}(\rho v_x v_x)+\frac{\partial}{\partial y}(\rho v_x v_y)+\frac{\partial}{\partial z}(\rho v_x v_z)\right\}\\ \mathbf{j}\left\{\frac{\partial}{\partial x}(\rho v_y v_x)+\frac{\partial}{\partial y}(\rho v_y v_y)+\frac{\partial}{\partial z}(\rho v_y v_z)\right\}\\ \mathbf{k}\left\{\frac{\partial}{\partial x}(\rho v_z v_x)+\frac{\partial}{\partial y}(\rho v_z v_y)+\frac{\partial}{\partial z}(\rho v_z v_z)\right\}\end{bmatrix}$$

Then, using this result (a vector) in the momentum equation,

$$(\rho\mathbf{v})_t + \nabla\cdot(\rho\mathbf{v}\mathbf{v}) + \nabla p = 0$$

and equating corresponding components,

$$\frac{\partial}{\partial t}(\rho v_x) + \frac{\partial}{\partial x}(\rho v_x v_x) + \frac{\partial}{\partial y}(\rho v_x v_y) + \frac{\partial}{\partial z}(\rho v_x v_z) + \frac{\partial p}{\partial x} = 0\ (\mathbf{i}\text{ component}) \tag{5.5a}$$

$$\frac{\partial}{\partial t}(\rho v_y) + \frac{\partial}{\partial x}(\rho v_y v_x) + \frac{\partial}{\partial y}(\rho v_y v_y) + \frac{\partial}{\partial z}(\rho v_y v_z) + \frac{\partial p}{\partial y} = 0\ (\mathbf{j}\text{ component}) \tag{5.5b}$$

$$\frac{\partial}{\partial t}(\rho v_z) + \frac{\partial}{\partial x}(\rho v_z v_x) + \frac{\partial}{\partial y}(\rho v_z v_y) + \frac{\partial}{\partial z}(\rho v_z v_z) + \frac{\partial p}{\partial z} = 0\ (\mathbf{k}\text{ component}) \tag{5.5c}$$

Equations (5.5a)–(5.5c) have several noteworthy properties. They are

1. Hyperbolic (first order in x, y, z, and t)
2. A source of sharp moving fronts and discontinuities (which are characteristic of hyperbolic PDEs)
3. Highly nonlinear (consider the products of the velocity components, e.g., $v_x v_x$, in Eqs. (5.5a)–(5.5c)
4. Difficult to solve numerically (and generally impossible analytically)
5. Limited in the sense that they apply only to isothermal, nonreacting systems

Equations (5.5a)–(5.5c) can be simplified through continuity equation (5.3); for example Eq. (5.5a) can be written as

$$\begin{aligned} &v_x\frac{\partial}{\partial t}(\rho) + v_x\frac{\partial}{\partial x}(\rho v_x) + v_x\frac{\partial}{\partial y}(\rho v_y) + v_x\frac{\partial}{\partial z}(\rho v_z) \\ &+ \rho\frac{\partial}{\partial t}(v_x) + \rho v_x\frac{\partial}{\partial x}(v_x) + \rho v_y\frac{\partial}{\partial y}(v_x) + \rho v_z\frac{\partial}{\partial z}(v_x) + \frac{\partial p}{\partial x} = 0 \end{aligned}$$

or

$$\begin{aligned} &v_x\left\{\frac{\partial\rho}{\partial t} + \frac{\partial(\rho v_x)}{\partial x} + \frac{\partial(\rho v_y)}{\partial y} + \frac{\partial(\rho v_z)}{\partial z}\right\} \\ &+ \rho\frac{\partial v_x}{\partial t} + \rho v_x\frac{\partial v_x}{\partial x} + \rho v_y\frac{\partial v_x}{\partial y} + \rho v_z\frac{\partial v_x}{\partial z} + \frac{\partial p}{\partial x} = 0 \end{aligned}$$

and the first row is zero through the continuity equation (5.3).

This result can be generalized to

$$\rho\mathbf{v}_t + \rho\mathbf{v}\cdot\nabla\mathbf{v} + \nabla p = 0$$

which is an alternate form of the Euler equations.

Finally, if a Newtonian viscosity term is added to the momentum equation

$$(\rho \mathbf{v})_t + \nabla \cdot (\rho \mathbf{v}\mathbf{v}) + \nabla p - \mu \nabla^2 \mathbf{v} = 0 \tag{5.6}$$

which along with Eq. (5.3) are the *Navier Stokes equations* for incompressible flow where gravitational effects are negligible. ∇^2 is the *Laplacian operator* that in Cartesian coordinates is

$$\nabla^2 = \frac{\partial^2}{\partial x^2} + \frac{\partial^2}{\partial y^2} + \frac{\partial^2}{\partial z^2} \tag{5.7}$$

Note that ∇^2 is a scalar, and it can be applied to a vector, as in Eq. (5.6), or it can be applied to a scalar, as in the parabolic heat equation $\partial u/\partial t = \nabla^2 u$.

We can use Eq. (5.6) as the starting point for a PDE analysis. For example, simplification of the x component of Eq. (5.6) (combined with Eq. (5.7)) gives

$$\rho \frac{\partial v_x}{\partial t} + \rho v_x \frac{\partial v_x}{\partial x} + \rho v_y \frac{\partial v_x}{\partial y} + \rho v_z \frac{\partial v_x}{\partial z} + \frac{\partial p}{\partial x} - \mu \left(\frac{\partial^2 v_x}{\partial x^2} + \frac{\partial^2 v_x}{\partial y^2} + \frac{\partial^2 v_x}{\partial z^2} \right) = 0 \tag{5.8}$$

If we consider a 1D problem with no pressure gradient, Eq. (5.8) reduces to

$$\rho \frac{\partial v_x}{\partial t} + \rho v_x \frac{\partial v_x}{\partial x} - \mu \frac{\partial^2 v_x}{\partial x^2} = 0 \tag{5.9}$$

which is *Burgers' equation*. This is an unusual (and widely studied) PDE because

1. It is nonlinear, yet has a known, exact (analytical) solution.
2. The solution exhibits moving fronts that can be made arbitrarily sharp by decreasing the *kinematic viscosity* $\nu = \mu/\rho$ (Eq. (5.9) can be divided by ρ to give the kinematic viscosity as a coefficient of the second-derivative term).

Burgers' equation for $\nu = \mu/\rho = 0$ reduces to the *inviscid Burgers equation* or *nonlinear advection equation*:

$$\frac{\partial v_x}{\partial t} + v_x \frac{\partial v_x}{\partial x} = 0 \tag{5.10}$$

We now consider the numerical solution of Burgers' equation (5.9) through a series of MOL routines. To evaluate the numerical solution, we will use an analytical solution to Eq. (5.9), v_{xa} [1].

$$v_{xa}(x, t) = \frac{0.1\, e^{-A} + 0.5\, e^{-B} + e^{-C}}{e^{-A} + e^{-B} + e^{-C}} \tag{5.11}$$

where

$$A = \frac{0.05}{\nu}(x - 0.5 + 4.95t) \tag{5.12a}$$

$$B = \frac{0.25}{\nu}(x - 0.5 + 0.75t) \tag{5.12b}$$

$$C = \frac{0.5}{\nu}(x - 0.375) \tag{5.12c}$$

For the initial condition (IC) and boundary conditions (BCs) for Eq. (5.9), we use Eqs. (5.11) and (5.12).

$$v_x(t = 0, x) = v_{xa}(t = 0, x) \tag{5.13a}$$

$$v_x(t, x = 0) = v_{xa}(t, x = 0) \tag{5.13b}$$

$$v_x(t, x = 1) = v_{xa}(t, x = 1) \tag{5.13c}$$

Note from Eqs. (5.13b) and (5.13c) $0 \le x \le 1$. For the numerical solution, we take $\nu = 0.003$, which, as we shall observe, produces a solution with a *steep moving front that sharpens with increasing t*.

A main program for the solution of Eq. (5.9) is given in Listing 5.1 (in the subsequent programming, we use the traditional variable u as the PDE dependent variable rather than v_x of Eqs. (5.10) and (5.13), and $\phi(x, t)$ (= `phi`) is the analytical solution of Eq. (5.11)).

```
%
% Clear previous files
  clear all
  clc
%
% Parameters shared with the ODE routine
  global n x xl xu ncall ncase ndss vis
%
% Select case
%
% ncase = 1 - analytical solution used in BCs
%
% ncase = 2 - homogeneous Neumann BCs
  ncase=1;
%
% Parameters
  n=201;
  xl=0.0;
  xu=1.0;
  vis=0.003;
%
% Initial condition
  t0=0.0;
  u0=inital_1(t0);
%
% Independent variable for ODE integration
```

```
  tf=1.0;
  tout=[t0:0.1:tf]';
  nout=11;
  ncall=0;
%
% ODE integration
  reltol=1.0e-04; abstol=1.0e-04;
  options=odeset('RelTol',reltol,'AbsTol',abstol);
  if(ncase==1)
    ndss=4; % ndss = 2, 4, 6, 8 or 10 required
    [t,u]=ode15s(@pde_1,tout,u0,options); end
  if(ncase==2)
    ndss=4; % ndss = 2, 4, 6, 8 or 10 required
    [t,u]=ode15s(@pde_2,tout,u0,options); end
%
% Store analytical solution, errors
  for it=1:nout
    for i=1:n
      u_anal(it,i)=phi(x(i),t(it));
      err(it,i)=u(it,i)-u_anal(it,i);
    end
  end
%
% Display selected output
  fprintf('\n vis = %8.4f   ndss = %2d\n\n',vis,ndss);
  for it=1:2:nout
    fprintf('   it   i   t(it)   x(i)       u(it,i)
            u_anal(it,i)    err(it,i)\n');
    for i=1:5:n
      fprintf('%5d%5d%8.2f%8.3f%15.6f%15.6f%15.6f\n', ...
              it,i,t(it),x(i),u(it,i),u_anal(it,i),err(it,i));
    end
    fprintf('\n');
  end
  fprintf('  ncall = %4d\n\n',ncall);
%
% Plot numerical and analytical solutions
  plot(x,u,'o',x,u_anal,'-')
  xlabel('x')
  ylabel('u(x,t)')
  title('Burgers equation; t = 0, 0.1, ..., 1;
         o - numerical; solid - analytical')
% print -deps -r300 pde.eps; print -dps -r300 pde.ps;
% print -dpng -r300 pde.png
```

Listing 5.1. Main program pde_1_main.m for the solution of Eqs. (5.10)–(5.13)

We can note the following points about this main program:

1. After defining a *global* area by which variables and parameters can be shared with other routines, a case for the solution is selected; for ncase=1, analytical solution (5.11) is used to define the BCs in the ordinary differential equation (ODE) routine pde_1 (discussed subsequently).

```
%
% Clear previous files
  clear all
  clc
%
% Parameters shared with the ODE routine
  global n x xl xu ncall ncase ndss vis
%
% Select case
%
% ncase = 1 - analytical solution used in BCs
%
% ncase = 2 - homogeneous Neumann BCs
  ncase=1;
```

2. A grid in x of 201 points, over the interval $0 \leq x \leq 1$, is defined and the kinematic viscosity in Eq. (5.9) set as $\nu = \mu/\rho = 0.003$ specifies a solution that has a steep front, as demonstrated subsequently.

```
%
% Parameters
  n=201;
  xl=0.0;
  xu=1.0;
  vis=0.003;
```

3. The IC for Eq. (5.9) is taken as the analytical solution, Eq. (5.11), with $t = 0$. This IC is in function inital_1, discussed subsequently.

```
%
% Initial condition
  t0=0.0;
  u0=inital_1(t0);
%
% Independent variable for ODE integration
```

```
  tf=1.0;
  tout=[t0:0.1:tf]';
  nout=11;
  ncall=0;
```

The time variation is then defined over the interval $0 \leq t \leq 1$ with an output interval of 0.1 so that there are a total of 11 outputs (counting the IC at $t = 0$).

4. The 201 ODEs are integrated by the stiff integrator `ode15s` for `ncase=1` or 2. These cases are explained when the corresponding ODE routines `pde_1` and `pde_2` are discussed subsequently.

```
%
% ODE integration
  reltol=1.0e-04; abstol=1.0e-04;
  options=odeset('RelTol',reltol,'AbsTol',abstol);
  if(ncase==1)
    ndss=4; % ndss = 2, 4, 6, 8 or 10 required
    [t,u]=ode15s(@pde_1,tout,u0,options); end
  if(ncase==2)
    ndss=4; % ndss = 2, 4, 6, 8 or 10 required
    [t,u]=ode15s(@pde_2,tout,u0,options); end
```

The spatial derivatives in Eq. (5.9) are computed by routine `dss004` as specified by `ndss=4` (passed as a global variable to the ODE routines `pde_1`, `pde_2`).

5. The analytical solution, Eq. (5.11), is then evaluated via function `phi` (discussed subsequently), and the difference between the numerical and analytical solutions is computed for subsequent output.

```
%
% Store analytical solution, errors
  for it=1:nout
    for i=1:n
      u_anal(it,i)=phi(x(i),t(it));
      err(it,i)=u(it,i)-u_anal(it,i);
    end
  end
```

6. Selected portions of the numerical and analytical solutions are displayed in tabular form, and these two solutions are plotted so that they can be compared graphically.

```
%
% Display selected output
  fprintf('\n vis = %8.4f    ndss = %2d\n\n',vis,ndss);
  for it=1:2:nout
    fprintf('  it   i   t(it)   x(i)      u(it,i)
          u_anal(it,i)    err(it,i)\n');
    for i=1:5:n
      fprintf('%5d%5d%8.2f%8.3f%15.6f%15.6f%15.6f\n',...
             it,i,t(it),x(i),u(it,i),u_anal(it,i),err(it,i));
    end
    fprintf('\n');
  end
  fprintf('  ncall = %4d\n\n',ncall);
%
% Plot numerical and analytical solutions
  plot(x,u,'o',x,u_anal,'-')
  xlabel('x')
  ylabel('u(x,t)')
  title('Burgers equation; t = 0, 0.1, ..., 1;
       o - numerical; solid - analytical')
% print -deps -r300 pde.eps; print -dps -r300 pde.ps;
% print -dpng -r300 pde.png
```

We briefly review the output before going on to discussions of the routines `inital_1`, `pde_1`, `pde_2`, `phi`. A portion of the output (for `ncase=1`) is given in Table 5.1.

We can note the following details about the output is given in Table 5.1:

1. The IC (at $t = 0$) is the same for the numerical and analytical solutions as expected (since the analytical solution, Eq. (5.11), is used to generate the numerical solution IC).
2. The IC has an important feature. In moving left to right (increasing x), the IC decreases (from 1 to 0.1). This decreasing value of $u(x, t = 0)$ with increasing x is the *condition that leads to front sharpening* for subsequent t. This can be noted in the output for $t = 1$ that has a considerably sharper front than at $t = 0$. In fact, if t were extended to higher values (than $t = 1$), a *shock or discontinuity would eventually develop*. This is one of the principal reasons why Burgers' equation (5.9) is a stringent test problem; that is, the solution steepens and therefore becomes increasingly difficult to resolve spatially (the accurate resolution of the solution $u(x, t)$ as a function of x as t increases becomes progressively more difficult).
3. For the spatial grid of 201 points, the agreement between the numerical and analytical solutions is quite satisfactory, and remains that way throughout the solution, for example, from $t = 0.2$ to $t = 1$. This is an important point since it indicates that *numerical errors did not grow or accumulate with increasing t.*

Table 5.1. Selected output from `pde_1_main` for `ncase=1`

vis = 0.0030 ndss = 4

it	i	t(it)	x(i)	u(it,i)	u_anal(it,i)	err(it,i)
1	1	0.00	0.000	1.000000	1.000000	0.000000
1	6	0.00	0.025	1.000000	1.000000	0.000000
1	11	0.00	0.050	1.000000	1.000000	0.000000
1	16	0.00	0.075	1.000000	1.000000	0.000000
1	21	0.00	0.100	0.999998	0.999998	0.000000
1	26	0.00	0.125	0.999985	0.999985	0.000000
1	31	0.00	0.150	0.999880	0.999880	0.000000
1	36	0.00	0.175	0.999037	0.999037	0.000000
1	41	0.00	0.200	0.992366	0.992366	0.000000
1	46	0.00	0.225	0.944636	0.944636	0.000000
1	51	0.00	0.250	0.750000	0.750000	0.000000
1	56	0.00	0.275	0.555364	0.555364	0.000000
1	61	0.00	0.300	0.507633	0.507633	0.000000
1	66	0.00	0.325	0.500960	0.500960	0.000000
1	71	0.00	0.350	0.500102	0.500102	0.000000
1	76	0.00	0.375	0.499919	0.499919	0.000000
1	81	0.00	0.400	0.499493	0.499493	0.000000
1	86	0.00	0.425	0.497323	0.497323	0.000000
1	91	0.00	0.450	0.486222	0.486222	0.000000
1	96	0.00	0.475	0.436452	0.436452	0.000000
1	101	0.00	0.500	0.300000	0.300000	0.000000
1	106	0.00	0.525	0.163548	0.163548	0.000000
1	111	0.00	0.550	0.113778	0.113778	0.000000
1	116	0.00	0.575	0.102677	0.102677	0.000000
1	121	0.00	0.600	0.100508	0.100508	0.000000
1	126	0.00	0.625	0.100096	0.100096	0.000000
1	131	0.00	0.650	0.100018	0.100018	0.000000
1	136	0.00	0.675	0.100003	0.100003	0.000000
1	141	0.00	0.700	0.100001	0.100001	0.000000
1	146	0.00	0.725	0.100000	0.100000	0.000000
1	151	0.00	0.750	0.100000	0.100000	0.000000
1	156	0.00	0.775	0.100000	0.100000	0.000000
1	161	0.00	0.800	0.100000	0.100000	0.000000
1	166	0.00	0.825	0.100000	0.100000	0.000000
1	171	0.00	0.850	0.100000	0.100000	0.000000
1	176	0.00	0.875	0.100000	0.100000	0.000000
1	181	0.00	0.900	0.100000	0.100000	0.000000
1	186	0.00	0.925	0.100000	0.100000	0.000000
1	191	0.00	0.950	0.100000	0.100000	0.000000
1	196	0.00	0.975	0.100000	0.100000	0.000000
1	201	0.00	1.000	0.100000	0.100000	0.000000

```
 it    i   t(it)    x(i)      u(it,i)   u_anal(it,i)    err(it,i)
  3    1    0.20   0.000     1.000000       1.000000    -0.000000
  3    6    0.20   0.025     1.000000       1.000000     0.000000
  3   11    0.20   0.050     1.000000       1.000000    -0.000000
  3   16    0.20   0.075     1.000000       1.000000    -0.000000
  3   21    0.20   0.100     1.000000       1.000000    -0.000000
  3   26    0.20   0.125     1.000000       1.000000    -0.000000
  3   31    0.20   0.150     1.000000       1.000000    -0.000000
  3   36    0.20   0.175     1.000000       1.000000    -0.000000
  3   41    0.20   0.200     1.000000       1.000000    -0.000000
  3   46    0.20   0.225     1.000000       1.000000    -0.000000
  3   51    0.20   0.250     0.999998       0.999998    -0.000000
  3   56    0.20   0.275     0.999985       0.999985    -0.000000
  3   61    0.20   0.300     0.999877       0.999880    -0.000003
  3   66    0.20   0.325     0.999015       0.999037    -0.000022
  3   71    0.20   0.350     0.992269       0.992366    -0.000097
  3   76    0.20   0.375     0.944801       0.944636     0.000165
  3   81    0.20   0.400     0.749763       0.749992    -0.000229
  3   86    0.20   0.425     0.555319       0.555314     0.000005
  3   91    0.20   0.450     0.507447       0.507371     0.000075
  3   96    0.20   0.475     0.499597       0.499584     0.000013
  3  101    0.20   0.500     0.492921       0.492925    -0.000004
  3  106    0.20   0.525     0.464658       0.464655     0.000003
  3  111    0.20   0.550     0.364335       0.364304     0.000031
  3  116    0.20   0.575     0.207536       0.207577    -0.000041
  3  121    0.20   0.600     0.126005       0.125988     0.000017
  3  126    0.20   0.625     0.105184       0.105181     0.000003
  3  131    0.20   0.650     0.100989       0.100989    -0.000000
  3  136    0.20   0.675     0.100187       0.100187    -0.000000
  3  141    0.20   0.700     0.100035       0.100035    -0.000000
  3  146    0.20   0.725     0.100007       0.100007    -0.000000
  3  151    0.20   0.750     0.100001       0.100001    -0.000000
  3  156    0.20   0.775     0.100000       0.100000    -0.000000
  3  161    0.20   0.800     0.100000       0.100000    -0.000000
  3  166    0.20   0.825     0.100000       0.100000    -0.000000
  3  171    0.20   0.850     0.100000       0.100000    -0.000000
  3  176    0.20   0.875     0.100000       0.100000    -0.000000
  3  181    0.20   0.900     0.100000       0.100000    -0.000000
  3  186    0.20   0.925     0.100000       0.100000    -0.000000
  3  191    0.20   0.950     0.100000       0.100000    -0.000000
  3  196    0.20   0.975     0.100000       0.100000    -0.000000
  3  201    0.20   1.000     0.100000       0.100000    -0.000000

                .                                          .
                .                                          .
                .                                          .
```

(continued)

Table 5.1 *(continued)*

```
                Output at t = 0.4, 0.6, 0.8 removed
        .                                                  .
        .                                                  .
        .                                                  .
```

it	i	t(it)	x(i)	u(it,i)	u_anal(it,i)	err(it,i)
11	1	1.00	0.000	1.000049	1.000000	0.000049
11	6	1.00	0.025	0.999999	1.000000	-0.000001
11	11	1.00	0.050	1.000000	1.000000	-0.000000
11	16	1.00	0.075	1.000000	1.000000	-0.000000
11	21	1.00	0.100	1.000000	1.000000	0.000000
11	26	1.00	0.125	1.000000	1.000000	-0.000000
11	31	1.00	0.150	1.000000	1.000000	0.000000
11	36	1.00	0.175	1.000000	1.000000	-0.000000
11	41	1.00	0.200	1.000000	1.000000	-0.000000
11	46	1.00	0.225	1.000000	1.000000	0.000000
11	51	1.00	0.250	1.000000	1.000000	-0.000000
11	56	1.00	0.275	1.000000	1.000000	0.000000
11	61	1.00	0.300	1.000000	1.000000	-0.000000
11	66	1.00	0.325	1.000000	1.000000	0.000000
11	71	1.00	0.350	1.000000	1.000000	-0.000000
11	76	1.00	0.375	1.000000	1.000000	0.000000
11	81	1.00	0.400	1.000000	1.000000	-0.000000
11	86	1.00	0.425	1.000000	1.000000	0.000000
11	91	1.00	0.450	1.000000	1.000000	-0.000000
11	96	1.00	0.475	1.000000	1.000000	0.000000
11	101	1.00	0.500	1.000000	1.000000	-0.000000
11	106	1.00	0.525	1.000000	1.000000	0.000000
11	111	1.00	0.550	1.000000	1.000000	-0.000000
11	116	1.00	0.575	1.000000	1.000000	0.000000
11	121	1.00	0.600	1.000000	1.000000	-0.000000
11	126	1.00	0.625	1.000000	1.000000	0.000000
11	131	1.00	0.650	1.000000	1.000000	-0.000000
11	136	1.00	0.675	1.000000	1.000000	0.000000
11	141	1.00	0.700	1.000000	1.000000	0.000000
11	146	1.00	0.725	1.000000	1.000000	0.000000
11	151	1.00	0.750	1.000000	1.000000	0.000000
11	156	1.00	0.775	1.000000	1.000000	0.000000
11	161	1.00	0.800	1.000001	1.000000	0.000001
11	166	1.00	0.825	1.000002	0.999998	0.000005
11	171	1.00	0.850	0.999920	0.999904	0.000015
11	176	1.00	0.875	0.995958	0.996005	-0.000048
11	181	1.00	0.900	0.858691	0.856946	0.001746
11	186	1.00	0.925	0.200038	0.199719	0.000319
11	191	1.00	0.950	0.102603	0.102646	-0.000043
11	196	1.00	0.975	0.100059	0.100065	-0.000006
11	201	1.00	1.000	0.100002	0.100002	-0.000000

ncall = 728

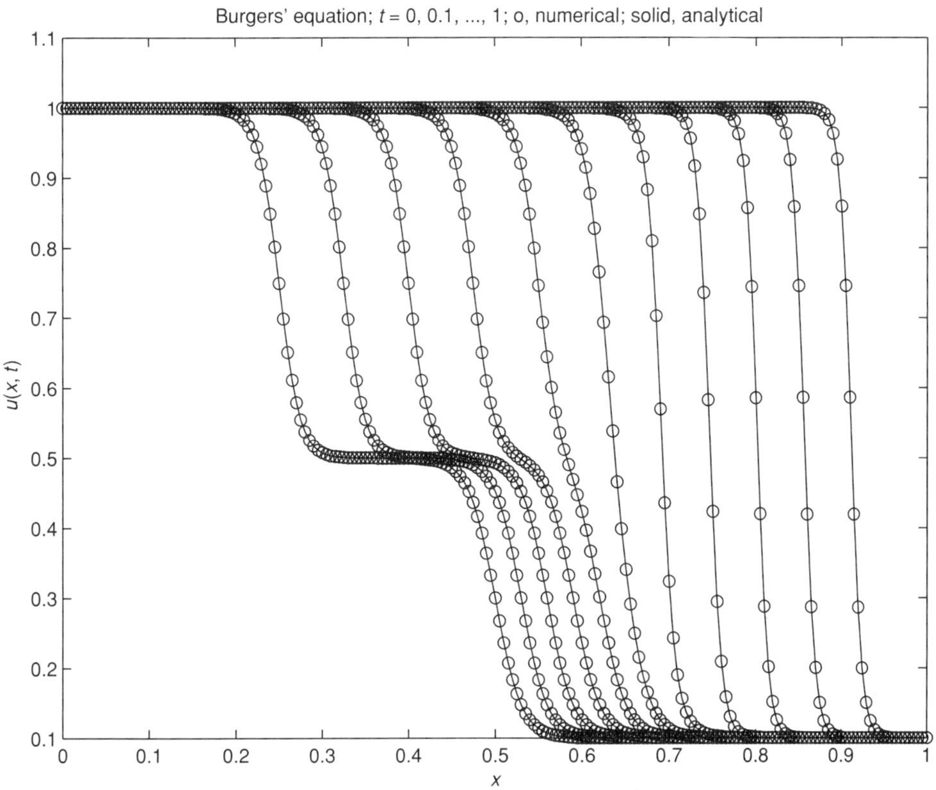

Figure 5.1. Plot of the numerical and analytical solutions of Burgers' equation (5.9) from `pde_1_main` and `pde_1` for `ncase=1`

> The agreement between the numerical and analytical solutions is illustrated by the plotted output indicated in Figure 5.1 (for `ncase=1`).

The front sharpening of the solution is evident in Figure 5.1 (the IC at $t = 0$ is the leftmost curve and the rightmost curve is the solution at $t = 1$).

We now discuss the remaining routines to produce the solution of Table 5.1 and Figure 5.1. The IC routine, `inital_1`, is given in Listing 5.2.

```
  function u0=inital_1(t0)
%
% Function inital_1 sets the initial condition for...
% Burgers' equation
%
  global n x xl xu ncall ncase ndss vis
%
% Analytical solution initially
  dx=(xu-xl)/(n-1);
  for i=1:n
```

```
    x(i)=xl+(i-1)*dx;
    u0(i)=phi(x(i),0.0);
  end
```

Listing 5.2. inital_1 for the IC of Eq. (5.9) based on the analytical solution of Eq. (5.11) with $t = 0$

This routine is straightforward. Note that it calls phi for analytical solution (5.11) with $t = 0$ over the 201-point grid in x.

Refer to Listing 5.3 for ODE routine pde_1.

```
  function ut=pde_1(t,u)
%
% Function pde_1 computes the t derivative vector for
% Burgers' equation
%
  global n x xl xu ncall ncase ndss vis
%
% BC at x = 0
  u(1)=phi(xl,t);
%
% BC at x = 1
  u(n)=phi(xu,t);
%
% ux
  if      (ndss== 2) ux=dss002(xl,xu,n,u); % second order
    elseif(ndss== 4) ux=dss004(xl,xu,n,u); % fourth order
    elseif(ndss== 6) ux=dss006(xl,xu,n,u); % sixth order
    elseif(ndss== 8) ux=dss008(xl,xu,n,u); % eighth order
    elseif(ndss==10) ux=dss010(xl,xu,n,u); % tenth order
  end
%
% uxx
  if      (ndss== 2) uxx=dss002(xl,xu,n,ux); % second order
    elseif(ndss== 4) uxx=dss004(xl,xu,n,ux); % fourth order
    elseif(ndss== 6) uxx=dss006(xl,xu,n,ux); % sixth order

    elseif(ndss== 8) uxx=dss008(xl,xu,n,ux); % eighth order
    elseif(ndss==10) uxx=dss010(xl,xu,n,ux); % tenth order
  end
%
% PDE
  for i=1:n
    ut(i)=vis*uxx(i)-u(i)*ux(i);
  end
```

```
  ut=ut';
%
% Increment calls to pde_1
  ncall=ncall+1;
```

Listing 5.3. pde_1 for the MOL solution of Eq. (5.9) based on BCs from analytical solution (5.11)

We can note the following points about pde_1:

1. After the call definition of the function and a global area, two Dirichlet BCs for Eq. (5.9) at $x = xl = 0$ and $x = xu = 1$ are defined by using the analytical solution of Eq. (5.11) at these values of x (through calls to phi for Eq. (5.11)).

```
  function ut=pde_1(t,u)
%
% Function pde_1 computes the t derivative vector for
% Burgers' equation
%
  global n x xl xu ncall ncase ndss vis
%
% BC at x = 0
  u(1)=phi(xl,t);
%
% BC at x = 1
  u(n)=phi(xu,t);
```

2. The first derivative ux in Eq. (5.9) is computed by a differentiation routine, in this case dss004 since ndss=4 from pde_1_main is passed as a global variable to pde_1.

```
%
% ux
  if       (ndss== 2) ux=dss002(xl,xu,n,u); % second order
    elseif(ndss== 4) ux=dss004(xl,xu,n,u); % fourth order
    elseif(ndss== 6) ux=dss006(xl,xu,n,u); % sixth order
    elseif(ndss== 8) ux=dss008(xl,xu,n,u); % eighth order
    elseif(ndss==10) ux=dss010(xl,xu,n,u); % tenth order
  end
```

3. The second derivative in Eq. (5.9), uxx, is then computed by differentiating ux; that is, *stagewise differentiation* is used.

```
%
% uxx
  if      (ndss== 2) uxx=dss002(xl,xu,n,ux); % second order
    elseif(ndss== 4) uxx=dss004(xl,xu,n,ux); % fourth order
    elseif(ndss== 6) uxx=dss006(xl,xu,n,ux); % sixth order
    elseif(ndss== 8) uxx=dss008(xl,xu,n,ux); % eighth order
    elseif(ndss==10) uxx=dss010(xl,xu,n,ux); % tenth order
  end
```

4. Equation (5.9) is then programmed, and the usual transpose is included at the end (as required by `ode15s`). The counter `ncall` has a modest final value of 728 from Table 5.1.

```
%
% PDE
  for i=1:n
    ut(i)=vis*uxx(i)-u(i)*ux(i);
  end
  ut=ut';
%
% Increment calls to pde_1
  ncall=ncall+1;
```

The close correspondence of the coding of Eq. (5.9) and its mathematical form indicates one of the salient features of the MOL. Note also how easily the nonlinear term `u(i)*ux(i)` can be included, which is also a salient feature of the numerical approach.

Routine `phi` for analytical solution (5.11) is given in Listing 5.4.

```
  function ua=phi(x,t)
%
% Function phi computes the exact solution of Burgers'
% equation for comparison with the numerical solution. It is
% also used to define the initial and boundary conditions for
% the numerical solution.
%
  global vis
%
% Analytical solution
  a=(0.05/vis)*(x-0.5+4.95*t);
  b=(0.25/vis)*(x-0.5+0.75*t);
  c=( 0.5/vis)*(x-0.375);
  ea=exp(-a);
```

```
  eb=exp(-b);
  ec=exp(-c);
  ua=(0.1*ea+0.5*eb+ec)/(ea+eb+ec);
```

Listing 5.4. Function phi for the analytical solution of Eqs. (5.11)–(5.13) (with $v_{xa}(x, t)$ = phi)

phi is a straightforward implementation of Eqs. (5.11)–(5.13).

An alternative approach to the solution of Eq. (5.9) is to note that the solution in Figure 5.1 has a zero slope at $x = 0$ and $x = 1$ (because the moving front does not reach either of these boundaries for $0 \leq t \leq 1$). Thus, we can make use of this property of the solution to program the BCs as *homogeneous Neumann BCs*. This is illustrated in pde_2 called by pde_1_main for ncase=2 (see Listing 5.5).

```
  function ut=pde_2(t,u)
%
% Function pde_2 computes the t derivative vector for Burgers'
% equation
%
  global n x xl xu ncall ncase ndss vis
%
% Calculate ux
  if       (ndss== 2) ux=dss002(xl,xu,n,u); % second order
    elseif(ndss== 4) ux=dss004(xl,xu,n,u); % fourth order
    elseif(ndss== 6) ux=dss006(xl,xu,n,u); % sixth order
    elseif(ndss== 8) ux=dss008(xl,xu,n,u); % eighth order
    elseif(ndss==10) ux=dss010(xl,xu,n,u); % tenth order
  end
%
% BC at x = 0
  ux(1)=0.0;
%
% BC at x = 1
  ux(n)=0.0;
%
% Calculate uxx
  if       (ndss== 2) uxx=dss002(xl,xu,n,ux); % second order
    elseif(ndss== 4) uxx=dss004(xl,xu,n,ux); % fourth order
    elseif(ndss== 6) uxx=dss006(xl,xu,n,ux); % sixth order
    elseif(ndss== 8) uxx=dss008(xl,xu,n,ux); % eighth order
    elseif(ndss==10) uxx=dss010(xl,xu,n,ux); % tenth order
  end
%
% PDE
  for i=1:n
```

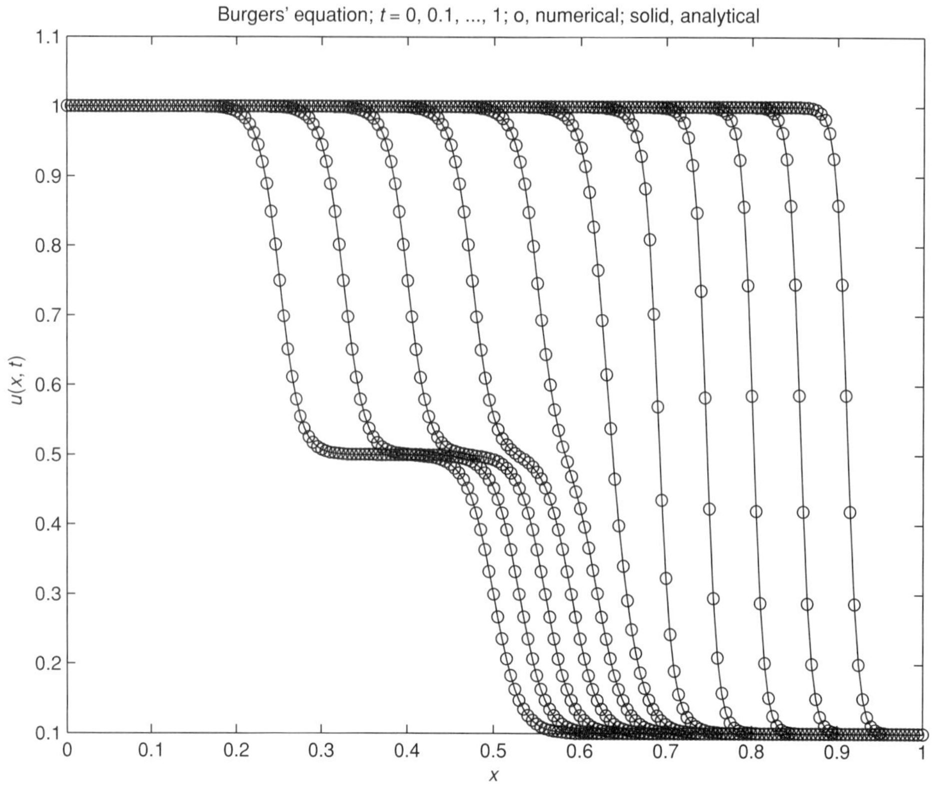

Figure 5.2. Plot of the numerical and analytical solutions of Burgers' equation (5.9) from pde_1_main and pde_2 for ncase=2

```
      ut(i)=vis*uxx(i)-u(i)*ux(i);
    end
    ut=ut';
%
% Increment calls to pde_2
    ncall=ncall+1;
```

Listing 5.5. pde_2 for the MOL solution of Eq. (5.9) based on homogeneous Neumann BCs

The only essential difference between pde_1 (of Listing 5.3) and pde_2 (of Listing 5.5) is the use of the Neumann BCs between the first and second calls to dss004 to set ux at the boundaries rather than u (from the analytical solution, Eq. (5.11)). Thus, in this second approach, the analytical solution is used only for the IC, Eq. (5.11), with $t = 0$.

The numerical solution from pde_2 is quite similar to the solution from pde_1 in Table 5.1. Therefore, we present only the plotted output in Figure 5.2. The close agreement between the numerical and analytical solutions is again evident.

Figures 5.1 and 5.2 indicate that the solution front sharpens (steepens) with increasing t. In other words, at $t = 0$, the IC consists of two connected fronts, and as t increases, these two fronts merge into a single front that then steepens with

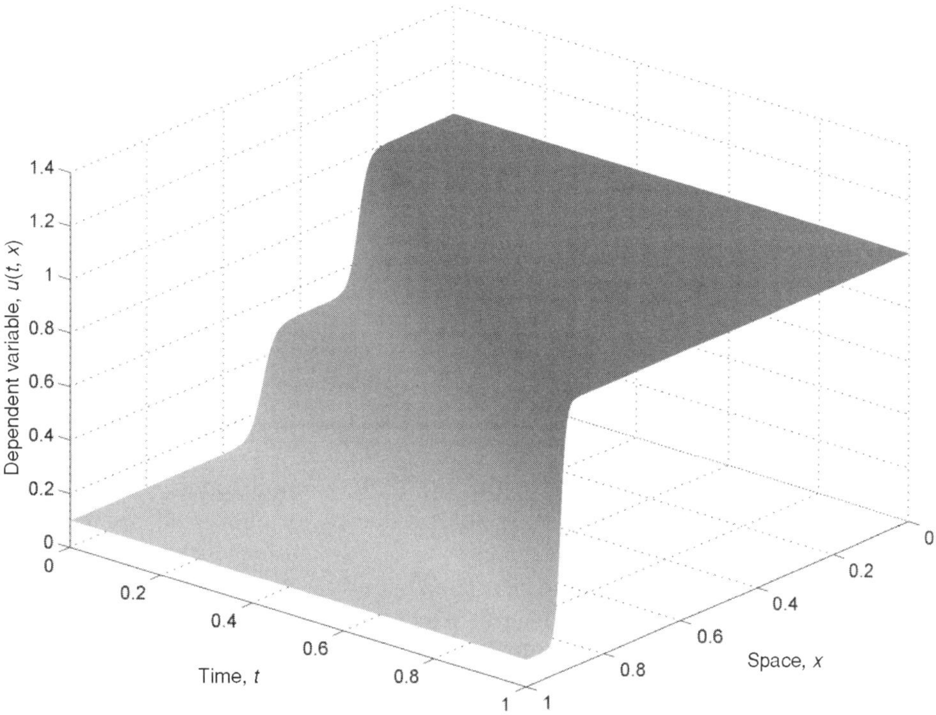

Figure 5.3. Three-dimensional plot of the numerical solution of Burgers' equation (5.9) from `pde_1_main` and `pde_1` for `ncase=1`

increasing t (note the relatively steep single front at $t = 1$). An explanation for this front sharpening can be inferred from the nonlinear first-derivative term of Eq. (5.9):

$$v_x \frac{\partial v_x}{\partial x}$$

This *convective term* has the important property that the velocity is v_x, the dependent variable of Eq. (5.9). If v_x decreases with increasing x as in the case of the solution of Figures 5.1 and 5.2, then the velocity is greater for small x. In other words, the solution "flows" more rapidly for small x, and this higher velocity leads to front sharpening. These properties of the solution of Eq. (5.9) are elucidated by the 3D plot in Figure 5.3.

Figure 5.3 was produced by using 101 points in t to produce better resolution for the 3D plotting (in `pde_1_main`). `t`

```
%
% Independent variable for ODE integration
  tf=1.0;
  tout=[t0:0.01:tf]';
  nout=101;
  ncall-0;
```

The 3D plot of Figure 5.3 was then produced by the following code (added after the plotting for Figure 5.1 in pde_1_main).

```
%
  figure(2);
  surf(x,t,u,'EdgeColor','none');
  set(get(gca,'XLabel'),'String','space, x')
  set(get(gca,'YLabel'),'String','time, t')
  set(get(gca,'ZLabel'),'String','dependent variable, u(t,x)')
  view(15,15);
  colormap('Cool')
```

Additional enhanced plotting of the Eq. (5.9) solution can be accomplished by animation (a movie). The details for producing an animation are given in Appendix 6.

Finally, an interesting study can be carried out by varying the kinematic viscosity $\nu = \mu/\rho$ in Eq. (5.9). As this parameter is reduced, the front sharpening of the solution increases (because $\nu = \mu/\rho$ multiplies the diffusive term $\partial^2 v_x/\partial x^2$ in Eq. (5.9)), and eventually the spatial grid will not be fine enough to resolve the solution. In this case, an alternative solution method may be required such as a special approximation of the derivatives in x in Eq. (5.9). The resolution of steep fronts is beyond the scope of the present discussion; it is the subject of an extensive published literature and available codes.

In summary, our intention in presenting this chapter is to indicate how:

1. A particular PDE, in this case Burgers' equation, can be obtained as a special case of a more general PDE system, in this case the Navier Stokes equations.
2. A numerical solution can be computed using either Dirichlet BCs (in this case set by the use of an analytical solution) or Neumann BCs. But the use of Neumann BCs required some knowledge of the solution (which might be through physical reasoning), for example, the solution does not change with x at the boundaries.
3. The use of the differentiation routines, for example, dss004 in pde_1 and pde_2, facilitates changing the order of the FD approximations; all that is required is to change the number of the routine since the arguments remain the same. Thus, the fourth-order FDs of dss004 can be changed to sixth-order FDs by changing dss004 to dss006 in Listings 5.3 and 5.5. Since the order of numerical approximations is often given the symbol p, this procedure of changing the order of the FDs is termed *p-refinement*.
4. Additionally, the number of grid points can be changed (in pde_1_main of Listing 5.1). Since the grid spacing in numerical approximations is often given the symbol h, changing the number of grid points is termed *h-refinement*.
5. The effect of h- and p-refinement on the numerical solution can be observed through repeated solutions. If the solution remains unchanged, for example, in the third figure, then three-figure accuracy can be inferred. Note that

this procedure does not require knowledge of an analytical solution (but, of course, it is not a mathematical proof of convergence).

6. To demonstrate an application of h- and p-refinement, we can consider a summary of some numerical output from `pde_1_main` and `pde_1` as given in Table 5.2.

   We can note the following points about this output

   (a) The kinematic viscosity, $\nu = \mu/\rho$ (`= nu`), is varied through the values 1.0, 0.1, 0.01, 0.003. As expected, this decrease in ν sharpens the front of the solution of Eq. (5.9) and therefore makes the calculation of the numerical solution more difficult, which is reflected in an increasing numerical error. For example, at `n = 201` grid points and a fourth-order FD approximation for the spatial derivatives in x, `ndss = 4`, the maximum errors in the solution are

nu	maxerror	ncall
1	$6.381e-006$	218
0.1	$1.703e-004$	228
0.01	$5.006e-004$	372
0.003	$3.135e-003$	728

   Although the maximum error increases by approximately 10^3 as ν is decreased from 1 to 0.003, this error is still quite acceptable for a solution that has a maximum value of the order of 1. Also, the computational effort increases with decreasing ν, but it is modest, even for $\nu = 0.003$ (`ncall = 728`).

   To complete the explanation of the output in Table 5.2, `xmax` is the value of x at which the maximum error occurs and `tmax` is the value of t at which this maximum error occurs.

   (b) As the number of grid points is increased, the maximum error decreases as expected. Thus, for $\nu = 0.003$ and `ndss = 4`,

n	maxerror	ncall
51	$2.356e-001$	451
101	$4.414e-002$	658
201	$3.135e-003$	728

   This reduction in error with increasing number of grid points is an example of h-refinement, which for the problem of Eq. (5.9) is quite effective (a reduction in the maximum error of approximately 10^{-2} with an increase in the number of grid points from `n = 51` to `n = 201`). In fact, this

Table 5.2. Maximum solution errors as a function of the kinematic viscosity, ν (= nu), number of grid points, n, and order of the FD approximation, ndss[1]

nu	n	ndss	maxerror	xmax	tmax	ncall
1.000	51	2	1.847e-006	0.0000	1.0	71
		4	6.379e-006	0.5400	0.3	68
		6	6.379e-006	0.5400	0.3	68
	101	2	7.825e-006	0.5800	0.3	122
		4	6.381e-006	0.5500	0.3	118
		6	6.381e-006	0.5500	0.3	118
	201	2	8.276e-006	0.5800	0.3	222
		4	6.381e-006	0.5500	0.3	218
		6	6.483e-006	0.5700	0.2	219
0.100	51	2	2.325e-004	0.0000	1.0	76
		4	1.702e-004	0.7200	0.2	78
		6	1.702e-004	0.7200	0.2	78
	101	2	9.413e-005	0.7700	0.7	130
		4	1.703e-004	0.7300	0.2	128
		6	1.703e-004	0.7300	0.2	128
	201	2	8.286e-005	0.7650	0.7	229
		4	1.703e-004	0.7300	0.2	228
		6	1.703e-004	0.7300	0.2	228
0.010	51	2	4.117e-002	0.9200	1.0	235
		4	5.725e-003	0.8800	1.0	218
		6	1.129e-002	1.0000	1.0	217
	101	2	9.256e-003	0.9300	1.0	273
		4	6.264e-004	0.8300	0.9	275
		6	4.967e-004	0.5700	0.5	272
	201	2	2.392e-003	0.9300	1.0	371
		4	5.006e-004	0.5750	0.5	372
		6	4.982e-004	0.5750	0.5	370
0.003	51	2	5.126e-001	0.7600	0.8	441
		4	2.356e-001	0.8800	1.0	451
		6	6.492e-001	0.0000	0.2	454
	101	2	1.738e-001	0.8900	1.0	669
		4	4.414e-002	0.8400	0.9	658
		6	2.409e-002	0.9000	1.0	635
	201	2	2.997e-002	0.9150	1.0	797
		4	3.135e-003	0.8500	0.9	728
		6	1.111e-003	0.8500	0.9	723

[1] Contributed by John Carroll, School of Mathematical Sciences, Dublin City University, Ireland.

reduction can be estimated approximately by considering the fourth-order FD approximation: $[(201-1)/(51-1)]^4 = 64$, while $2.356e - 001/3.135e - 003 = 75.2 \approx 64$.

Better agreement between the ratio of errors estimated from the FD order condition (`ndss = 4`) and the actual ratio of errors is not necessarily expected. Keep in mind that the FD order condition is for the derivatives in x in eq. (5.9) and the accuracy of these numerical derivatives is not necessarily the same as the accuracy of the numerical PDE solution. However, we would expect improvement in the numerical PDE solution with improved accuracy in the numerical derivatives in x, as observed in this case.

As a related point, the total computational effort does not increase substantially (`ncall = 451` to `ncall = 728`), which is due principally to the use of a stiff integrator (`ode15s`), since the stiffness of the ODEs most likely increased substantially with their number (`51` to `201`), but the stiff integrator handled the stiffness quite well (in general the ODE stiffness increases with decreasing grid spacing or increasing numbers of ODEs).

(c) As the order of the FD approximation is increased, the maximum error decreases as expected. Thus, for $\nu = 0.003$ and `n = 201`,

`ndss`	`maxerror`	`ncall`
2	$2.997e - 002$	797
4	$3.135e - 003$	728
6	$1.111e - 003$	723

This reduction in error with increasing order of the FD approximation is an example of p-refinement, which for the problem of Eq. (5.9) is quite effective (a reduction in the maximum error of approximately 10^{-1} with an increase in the order of the FD approximation from `ndss = 2` to `ndss = 6`). Equally important, this improvement was achieved without an increase in `ncall` (but the formulas in `dss006` require more calculations than those in `dss002`). The extension of this improvement could be expected by using the eighth- and tenth-order FD approximations of routines `dss008` and `dss010`.

REFERENCE

[1] Madsen, N. K. and R. F. Sincovec (1976), General Software for Partial Differential Equations, In: L. Lapidus and W. E. Schiesser (Eds.), *Numerical Methods for Differential Systems*, New York, San Diego, CA, pp. 229–242

6

The Cubic Schrödinger Equation

The following partial differential equation (PDE) problem introduces the following mathematical concepts and computational methods:

1. A nonlinear PDE with an exact solution that can be used to assess the accuracy of a numerical method of lines (MOL) solution.
2. A complex PDE. Specifically, the complex PDE is expressed as two real PDEs, so this problem also illustrates the numerical integration of simultaneous PDEs.
3. Some of the basic properties of *traveling waves* that have broad application in many areas of physics, engineering, and the biological sciences.
4. Illustrative programming of a traveling wave solution.
5. Sparse matrix integration of the ordinary differential equations (ODEs) for the MOL approximation of PDEs.
6. Computer analysis of simultaneous PDEs over an essentially infinite spatial domain.

The *one-dimensional* (1D) *cubic Schrödinger equation* (CSE) *equation* is [1]

$$i\frac{\partial u}{\partial t} + \frac{\partial^2 u}{\partial x^2} + q|u|^2 u = 0$$

or in subscript notation,

$$iu_t + u_{xx} + q|u|^2 u = 0 \tag{6.1}$$

where

u complex dependent variable
x boundary-value (spatial) independent variable
t initial-value independent variable
q arbitrary parameter (constant)
i $\sqrt{-1}$

The nonlinear term $q|u|^2 u = 0$ is the origin of the term "cubic" in the name *cubic Schrödinger equation* ($|u|$ denotes the absolute value of the complex variable u).

Some additional background information to the Schrödinger equation is given in Appendix A. As we will observe, including this nonlinearity in the MOL solution of Eq. (6.1) is quite straightforward.

Equation (6.1) is first order in t and second order in x. It therefore requires one *initial condition* (IC) and two *boundary conditions* (BCs). The IC is taken from the analytical solution (with $q = 1$) [1]

$$u(x, t) = \sqrt{2}\, e^{i(0.5x+0.75t)}\, \text{sech}(x - t) \tag{6.2a}$$

Since $u(x, t)$ is complex, we analyze it as $u(x, t) = v(x, t) + iw(x, t)$, where $v(x, t)$ and $w(x, t)$ are two real functions that are to be computed numerically.

It then follows from Eq. (6.2a) that

$$|u(x, t)| = \sqrt{2}\, \text{sech}(x - t) \tag{6.2b}$$

Equation (6.2b) will be used subsequently for comparison with the numerical solution.

Then with $t = 0$ in Eq. (6.2a), the IC is

$$u(x, t = 0) = \sqrt{2}\, e^{i(0.5x)}\, \text{sech}(x) \tag{6.3}$$

To complete the specification of the problem, we would naturally consider the two required BCs for Eq. (6.1). However, if the PDE is analyzed over an essentially infinite domain, $-\infty \le x \le \infty$, and if changes in the solution occur only over a finite interval in x, then *BCs at infinity have no effect*; in other words, we do not have to actually specify BCs (since they have no effect). This situation will be clarified through the PDE solution.

Consequently, Eqs. (6.1) and (6.3) constitute the complete PDE problem. The solution to this problem, Eqs. (6.2a) and (6.2b), is used in the subsequent programming and analysis to evaluate the numerical MOL solution. Note that Eqs. (6.2a) and (6.2b) are a *traveling wave* solution since the argument of the sech function is $x - t$.

We now consider some Matlab routines for a numerical MOL solution of Eqs. (6.1) and (6.3) with the analytical solution, Eqs. (6.2a) and (6.2b), included. A main program, pde_1_main, is given in Listing 6.1.

```
%
% Clear previous files
  clear all
  clc %
%
% Parameters shared with the ODE routine
  global  ncall     x     xl     xu      n
%
% Boundaries, number of grid points
  xl=-10.0;
  xu= 40.0;
  n=251;
%
% Initial condition
```

```
  t0=0.0;
  u0=inital_1(t0);
%
% Independent variable for ODE integration
  tf=30.0;
  tout=[t0:5.0:tf]';
  nout=7;
  ncall=0;
%
% ODE integration
  mf=1;
  reltol=1.0e-06; abstol=1.0e-06;
  options=odeset('RelTol',reltol,'AbsTol',abstol);
%
% Explicit (nonstiff) integration
  if(mf==1)[t,u]=ode45(@pde_1,tout,u0,options); end
%
% Implicit (sparse stiff) integration
  if(mf==2)
    S=jpattern_num;
    pause
    options=odeset(options,'JPattern',S)
    [t,u]=ode15s(@pde_1,tout,u0,options);
  end
%
% One vector to two vectors
  for it=1:nout
  for i=1:n
    v(it,i)=u(it,i);
    w(it,i)=u(it,i+n);
  end
  end
%
% Analytical solution and difference between the numerical and
% analytical solutions
  for it=1:nout
    fprintf('     t     x  x - t      v^2+w^2      v^2+w^2
                  err\n');
    fprintf('                                    num        anal
              \n')
    for i=1:n
      v2w2_anal(it,i)=ua(t(it),x(i));
      v2w2(it,i)=u(it,i)^2+u(it,i+n)^2;
      err(it,i)=v2w2(it,i)-v2w2_anal(it,i);
%
%     Output in the neighborhood of the soliton peak
      if(abs(x(i)-t(it))<=1.0)
```

```
        fprintf('%6.2f%8.1f%8.2f%15.6f%15.6f%15.6f\n', ...
                t(it),x(i),x(i)-t(it),v2w2(it,i), ...
                v2w2_anal(it,i),err(it,i));
      end
    end
%
%   Calculate and display three invariants
    ui=u(it,:);
    uint=simp(xl,xu,n,ui);
    fprintf('\n Invariants at t = %5.2f',t(it));
    fprintf('\n    I1 = %10.4f',    uint(1));
    fprintf('\n    I2 = %10.4f',    uint(2));
    fprintf('\n    I3 = %10.4f\n\n',uint(3));
    fprintf('\n');
  end
  fprintf('  ncall = %4d\n\n',ncall);
%
% Plot numerical and analytical solutions
  figure(1)
  plot(x,v2w2,'-',x,v2w2_anal,'o')
  xlabel('x')
  ylabel('u(x,t)')
  title('CSE; t = 0, 5,..., 30; o - numerical;
         solid - analytical')
  print -deps -r300 pde.eps; print -dps -r300 pde.ps; ...
  print -dpng -r300 pde.png
```

Listing 6.1. Main program pde_1_main

We can note the following points about this program:

1. After specifying some *global* variables, the spatial domain is defined as $-10 \leq x \leq 40$, and the MOL grid has 251 points.

```
%
% Clear previous files
  clear all
  clc
%
% Parameters shared with the ODE routine
  global ncall     x     xl     xu      n
%
% Boundaries, number of grid points
  xl=-10.0;
  xu= 40.0;
  n=251;
```

Since $u(x, t) = v(x, t) + iw(x, t)$, pde_1_main of Listing 6.1 will actually integrate $2 \times 251 = 502$ ODEs (251 for $v(x, t)$ and 251 for $w(x, t)$).

2. Routine inital_1 (discussed subsequently) is called to define IC (6.3) over $-10 \le x \le 40$. The t interval is then defined as $0 \le t \le 30$ with solution outputs at $t = 0, 5, \ldots, 30$.

```
%
% Initial condition
  t0=0.0;
  u0=inital_1(t0);
%
% Independent variable for ODE integration
  tf=30.0;
  tout=[t0:5.0:tf]';
  nout=7;
  ncall=0;
```

3. The $n = 2(251)$ ODEs are integrated by a call to either a nonstiff integrator ode45 (for mf=1) or a stiff integrator ode15s (for mf=2). For this problem, ode45 requires more calls to the ODE routine pde_1 than ode15s, which is not unusual. However, the additional computation required by ode15s, that is, the sparse matrix integrator, requires additional time, so the two integrators are not very different in their overall computational efficiency. In other words, a stiff integrator *does not always lead to greater computational efficiency* (depending on the characteristics of the problem). In the subsequent discussion, we consider the output from ode45.

```
%
% ODE integration
  mf=1;
  reltol=1.0e-06; abstol=1.0e-06;
  options=odeset('RelTol',reltol,'AbsTol',abstol);
%
% Explicit (nonstiff) integration
  if(mf==1)[t,u]=ode45(@pde_1,tout,u0,options); end
%
% Implicit (sparse stiff) integration
  if(mf==2)
    S=jpattern_num;
    pause
    options=odeset(options,'JPattern',S)
    [t,u]=ode15s(@pde_1,tout,u0,options);
  end
```

4. Since the Matlab integrators such as ode45 integrate a single vector of ODEs, the solution vector u returned from ode45 is then divided into two parts, $v(x,t)$ and $w(x,t)$, of $u(x,t) = v(x,t) + iw(x,t)$.

```
%
% One vector to two vectors
  for it=1:nout
  for i=1:n
    v(it,i)=u(it,i);
    w(it,i)=u(it,i+n);
  end
  end
```

5. The analytical solution of Eq. (6.2b) is then computed by function ua (discussed subsequently) and the difference between the analytical and numerical solutions is computed. Selected portions of the analytical and numerical solutions and their differences are then displayed.

```
%
% Analytical solution and difference between the numerical and
% analytical solutions
  for it=1:nout
    fprintf('    t      x     x - t        v^2+w^2        v^2+w^2
                 err\n');
    fprintf('                                  num            anal
                  \n')
    for i=1:n
      v2w2_anal(it,i)=ua(t(it),x(i));
      v2w2(it,i)=u(it,i)^2+u(it,i+n)^2;
      err(it,i)=v2w2(it,i)-v2w2_anal(it,i);
%
%     Output in the neighborhood of the soliton peak
      if(abs(x(i)-t(it))<=1.0)
        fprintf('%6.2f%8.1f%8.2f%15.6f%15.6f%15.6f\n', ...
                t(it),x(i),x(i)-t(it),v2w2(it,i), ...
                v2w2_anal(it,i),err(it,i));
      end
    end
```

Since v and w are computed over 251 points in x, the output when fully displayed would be quite lengthy. We therefore display the solution only in the neighborhood of the peak of the solution. This is accomplished by monitoring the argument $x - t$ in Eqs. (6.2a) and (6.2b). At $x = t$, the solution has a peak (maximum) value, and drops off in the neighborhood of this peak, which in the preceding code is the interval defined by $|x - t| \leq 1$. The traveling wave of Eqs. (6.2a) and (6.2b) is clear from the numerical output discussed subsequently.

6. Three invariants are evaluated by a call to `simp` that implements a numerical quadrature (integration) based on Simpson's rule (discussed subsequently).

```
%
%   Calculate and display three invariants
    ui=u(it,:);
    uint=simp(xl,xu,n,ui);
    fprintf('\n Invariants at t = %5.2f',t(it));
    fprintf('\n    I1 = %10.4f',    uint(1));
    fprintf('\n    I2 = %10.4f',    uint(2));
    fprintf('\n    I3 = %10.4f\n\n',uint(3));
    fprintf('\n');
  end
  fprintf('  ncall = %4d\n\n',ncall);
```

The integral invariants for the single soliton are [1]

$$u_1(t) = \int_{-\infty}^{\infty} |u(x,t)|^2 \, dx \tag{6.4a}$$

$$\begin{aligned} u_2(t) &= \int_{-\infty}^{\infty} i(u\bar{u}_x - \bar{u}u_x)dx = \int_{-\infty}^{\infty} i[(v+iw)(v_x - iw_x) - (v-iw)(v_x + iw_x)]dx \\ &= \int_{-\infty}^{\infty} i[(vv_x + iwv_x - ivw_x + ww_x) - (vv_x - iwv_x + ivw_x + ww_x)]dx \\ &= 2\int_{-\infty}^{\infty} (vw_x - wv_x)dx \end{aligned} \tag{6.4b}$$

$$u_3(t) = \int_{-\infty}^{\infty} [|u_x|^2 - \frac{1}{2}q|u|^4]dx \tag{6.4c}$$

The computed values of these invariants are discussed subsequently.

7. The numerical and analytical solutions are then plotted.

```
%
% Plot numerical and analytical solutions
  figure(1)
  plot(x,v2w2,'-',x,v2w2_anal,'o')
  xlabel('x')
  ylabel('u(x,t)')
  title('CSE; t = 0, 5,..., 30; o - numerical;
```

```
        solid - analytical')
print -deps -r300 pde.eps; print -dps -r300 pde.ps;
print -dpng -r300 pde.png
```

The selected numerical output and the plotted output are given in Table 6.1 and Figure 6.1.

Table 6.1. Partial output from pde_1_main and pde_1

```
   t        x    x - t         v^2+w^2        v^2+w^2         err
                                 num            anal
 0.00     -1.0   -1.00        0.839949       0.839949       0.000000
 0.00     -0.8   -0.80        1.118110       1.118110       0.000000
 0.00     -0.6   -0.60        1.423156       1.423156       0.000000
 0.00     -0.4   -0.40        1.711278       1.711278      -0.000000
 0.00     -0.2   -0.20        1.922086       1.922086       0.000000
 0.00      0.0    0.00        2.000000       2.000000       0.000000
 0.00      0.2    0.20        1.922086       1.922086       0.000000
 0.00      0.4    0.40        1.711278       1.711278       0.000000
 0.00      0.6    0.60        1.423156       1.423156       0.000000
 0.00      0.8    0.80        1.118110       1.118110       0.000000
 0.00      1.0    1.00        0.839949       0.839949       0.000000

Invariants at t =  0.00
   I1 =      6.7303
   I2 =      1.9997
   I3 =    -13.8880

   t        x    x - t         v^2+w^2        v^2+w^2         err
                                 num            anal
 5.00      4.0   -1.00        0.840692       0.839949       0.000743
 5.00      4.2   -0.80        1.119083       1.118110       0.000973
 5.00      4.4   -0.60        1.424426       1.423156       0.001271
 5.00      4.6   -0.40        1.712740       1.711278       0.001463
 5.00      4.8   -0.20        1.923553       1.922086       0.001467
 5.00      5.0    0.00        2.000991       2.000000       0.000991
 5.00      5.2    0.20        1.922272       1.922086       0.000186
 5.00      5.4    0.40        1.710552       1.711278      -0.000725
 5.00      5.6    0.60        1.421850       1.423156      -0.001305
 5.00      5.8    0.80        1.116591       1.118110      -0.001520
 5.00      6.0    1.00        0.838547       0.839949      -0.001402
Invariants at t =  5.00
   I1 =      6.7252
   I2 =      1.9997
   I3 =    -13.8635
```

(continued)

Table 6.1 (*continued*)

t	x	x - t	v^2+w^2 num	v^2+w^2 anal	err
10.00	9.0	-1.00	0.841776	0.839949	0.001827
10.00	9.2	-0.80	1.120379	1.118110	0.002269
10.00	9.4	-0.60	1.425802	1.423156	0.002647
10.00	9.6	-0.40	1.714002	1.711278	0.002724
10.00	9.8	-0.20	1.924345	1.922086	0.002260
10.00	10.0	0.00	2.001183	2.000000	0.001183
10.00	10.2	0.20	1.921711	1.922086	-0.000375
10.00	10.4	0.40	1.709649	1.711278	-0.001628
10.00	10.6	0.60	1.420661	1.423156	-0.002494
10.00	10.8	0.80	1.115437	1.118110	-0.002674
10.00	11.0	1.00	0.837441	0.839949	-0.002508

```
Invariants at t = 10.00
   I1 =      6.7089
   I2 =      1.9997
   I3 =    -13.7750
```

t	x	x - t	v^2+w^2 num	v^2+w^2 anal	err
15.00	14.0	-1.00	0.842866	0.839949	0.002917
15.00	14.2	-0.80	1.121625	1.118110	0.003515
15.00	14.4	-0.60	1.427007	1.423156	0.003851
15.00	14.6	-0.40	1.715080	1.711278	0.003803
15.00	14.8	-0.20	1.924948	1.922086	0.002862
15.00	15.0	0.00	2.001288	2.000000	0.001288
15.00	15.2	0.20	1.921148	1.922086	-0.000938
15.00	15.4	0.40	1.708711	1.711278	-0.002567
15.00	15.6	0.60	1.419372	1.423156	-0.003784
15.00	15.8	0.80	1.114223	1.118110	-0.003888
15.00	16.0	1.00	0.836325	0.839949	-0.003624

```
Invariants at t = 15.00
   I1 =      6.6814
   I2 =      1.9997
   I3 =    -13.6262
```

t	x	x - t	v^2+w^2 num	v^2+w^2 anal	err
20.00	19.0	-1.00	0.843951	0.839949	0.004002
20.00	19.2	-0.80	1.123048	1.118110	0.004938
20.00	19.4	-0.60	1.428368	1.423156	0.005213
20.00	19.6	-0.40	1.716169	1.711278	0.004892
20.00	19.8	-0.20	1.925484	1.922086	0.003398
20.00	20.0	0.00	2.000912	2.000000	0.000912
20.00	20.2	0.20	1.920351	1.922086	-0.001735
20.00	20.4	0.40	1.707293	1.711278	-0.003985
20.00	20.6	0.60	1.418163	1.423156	-0.004992

```
20.00   20.8    0.80      1.112913      1.118110     -0.005197
20.00   21.0    1.00      0.835423      0.839949     -0.004525

Invariants at t = 20.00
   I1 =     6.6427
   I2 =     1.9997
   I3 =   -13.4157

    t        x   x - t    v^2+w^2       v^2+w^2         err
                            num           anal
25.00   24.0   -1.00      0.844906      0.839949      0.004957
25.00   24.2   -0.80      1.124041      1.118110      0.005931
25.00   24.4   -0.60      1.429421      1.423156      0.006265
25.00   24.6   -0.40      1.717158      1.711278      0.005881
25.00   24.8   -0.20      1.926096      1.922086      0.004010
25.00   25.0    0.00      2.001210      2.000000      0.001210
25.00   25.2    0.20      1.919751      1.922086     -0.002335
25.00   25.4    0.40      1.706370      1.711278     -0.004907
25.00   25.6    0.60      1.416704      1.423156     -0.006451
25.00   25.8    0.80      1.111702      1.118110     -0.006408
25.00   26.0    1.00      0.834255      0.839949     -0.005694

Invariants at t = 25.00
   I1 =     6.5935
   I2 =     1.9997
   I3 =   -13.1521

    t        x   x - t    v^2+w^2       v^2+w^2         err
                            num           anal
30.00   29.0   -1.00      0.846284      0.839949      0.006335
30.00   29.2   -0.80      1.125482      1.118110      0.007371
30.00   29.4   -0.60      1.431189      1.423156      0.008033
30.00   29.6   -0.40      1.718323      1.711278      0.007046
30.00   29.8   -0.20      1.926872      1.922086      0.004786
30.00   30.0    0.00      2.000883      2.000000      0.000883
30.00   30.2    0.20      1.919042      1.922086     -0.003044
30.00   30.4    0.40      1.704914      1.711278     -0.006364
30.00   30.6    0.60      1.415336      1.423156     -0.007820
30.00   30.8    0.80      1.110307      1.118110     -0.007803
30.00   31.0    1.00      0.833140      0.839949     -0.006808

Invariants at t = 30.00
   I1 =     6.5337
   I2 =     1.9997
   I3 =   -12.8329

 ncall = 18523
```

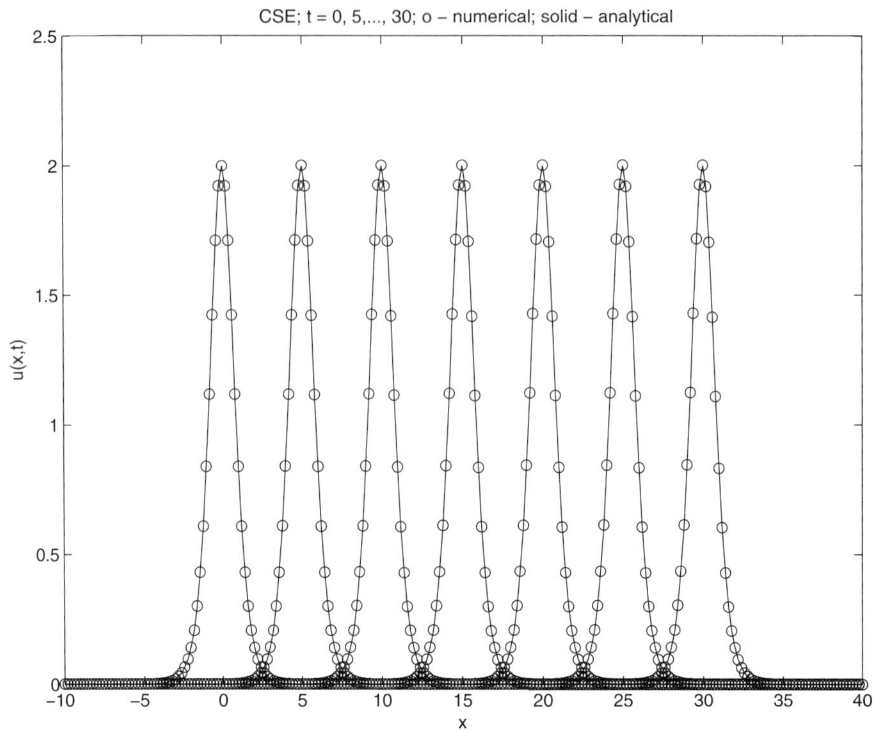

Figure 6.1. Output of main program `pde_1_main`

We can note the following points about this output:

1. The soliton moves left to right in x, as indicated by the movement of the peak. See Table 6.2 for a summary of the peak movement at $t = 0, 5, \ldots, 30$ corresponding to $x = 0, 5, \ldots, 30$.

Table 6.2. Partial Output from `pde_1_main` and `pde_1`, illustrating the movement of the solution with unit velocity

t	x	x - t	v^2+w^2 num	v^2+w^2 anal	err
0.00	0.0	0.00	2.000000	2.000000	0.000000
5.00	5.0	0.00	2.000991	2.000000	0.000991
10.00	10.0	0.00	2.001183	2.000000	0.001183
15.00	15.0	0.00	2.001288	2.000000	0.001288
20.00	20.0	0.00	2.000912	2.000000	0.000912
25.00	25.0	0.00	2.001210	2.000000	0.001210
30.00	30.0	0.00	2.000883	2.000000	0.000883

In this case, the maximum error in the peak amplitude is `0.001288`, which is 0.064% of the exact value of 2.

2. A similar summary of the three invariants is given in Table 6.3. `I3` has a maximum variation of $[-13.8880 - (-12.8329)]/(-13.8880) \times 100 = 7.6\%$. This variation probably reflects the accuracy of the MOL solution due to the limited number of grid points ($n = 251$) and possibly the associated limited accuracy of the Simpson's rule integration in `simp`. But increasing the number of grid points also increases the computer run time that is already rather substantial as reflected in the value `ncall = 18523`. The run time would increase with n because (a) more ODEs would be integrated, and (b) the ODE stiffness generally increases with more grid points; if the latter is the case, then a switch from `ode45` to `ode15s` would be logical.

Table 6.3. Partial output from `pde_1_main` and `pde_1`, illustrating the invariants `I1,I2,I3` of Eqs. (6.4a)–(6.4c)

```
( t = 0)   I1 =     6.7303
           I2 =     1.9997
           I3 =   -13.8880

( t = 5)   I1 =     6.7252
           I2 =     1.9997
           I3 =   -13.8635

(t = 10)   I1 =     6.7089
           I2 =     1.9997
           I3 =   -13.7750

(t = 15)   I1 =     6.6814
           I2 =     1.9997
           I3 =   -13.6262

(t = 20)   I1 =     6.6427
           I2 =     1.9997
           I3 =   -13.4157

(t = 25)   I1 =     6.5935
           I2 =     1.9997
           I3 =   -13.1521

(t = 30)   I1 =     6.5337
           I2 =     1.9997
           I3 =   -12.8329
```

3. However, the agreement between the numerical and analytical solutions is quite good as reflected in the differences err of Tables 6.1 and 6.2, and the plotted output shown in Figure 6.1.

The solution starts with the IC pulse (defined by Eq. (6.3)) centered at $x = 0$ (for $t = 0$). The movement of the traveling wave left to right according to Eq. (6.2b) for $t = 0, 5, \ldots, 30$ is evident in Figure 6.1.

We now consider the subordinate routines called by the main program, pde_1_main of Listing 6.1, inital_1, ua, simp and pde_1. inital_1 is listed first (see Listing 6.2).

```
  function u0=inital_1(t0)
%
% Function inital_1 sets the initial condition for the CSE
%
  global ncall    q    x    xl    xu   rt2    n
%
% q in cubic Schrodinger equation (CSE)
  q=1.0;
%
% Precompute 2^0.5
  rt2=sqrt(2.0);
%
% Grid spacing
  dx=(xu-xl)/(n-1);
%
% IC over the spatial grid
  for i=1:n
%
%   Uniform grid
    x(i)=xl+(i-1)*dx;
%
%   Initial real, imaginary parts (v(x,0), w(x,0)), absolute
%   value of u(x,0) = u0(x) = (2^0.5)exp(0.5*i*x)sech(x),
%   i = (-1)^0.5 over xl <= x <= xu
%
%   sech(x)
    sch=2.0/(exp(x(i))+exp(-x(i)));
%
%   Real part
    v(i)=rt2*cos(0.5*x(i))*sch;
%
%   Imaginary part
    w(i)=rt2*sin(0.5*x(i))*sch;
  end
```

```
%
% Two vectors to one vector
  for i=1:n
    u0(i)  =v(i);
    u0(i+n)=w(i);
  end
```

Listing 6.2. Initialization routine inital_1

We can note the following details about inital_1:

1. After the routine is defined and some parameters are declared as global, a spatial grid is defined for $-10 \le x \le 40$ (recall that the number of grid points, n, and the left and right boundary values of x, xl and xu, are set in main program pde_1_main as global parameters).

```
  function u0=inital_1(t0)
%
% Function inital_1 sets the initial condition for the CSE
%
  global ncall     q     x     xl     xu   rt2     n
%
% q in cubic Schrodinger equation (CSE)
  q=1.0;
%
% Precompute 2^0.5
  rt2=sqrt(2.0);
%
% Grid spacing
  dx=(xu-xl)/(n-1);
%
% IC over the spatial grid
  for i=1:n
%
%   Uniform grid
    x(i)=xl+(i-1)*dx;
```

2. Finally, the real and imaginary parts of the IC of Eq. (6.3) are computed.

```
%
%   Initial real, imaginary parts (v(x,0), w(x,0)), absolute
%   value of u(x,0) = u0(x) = (2^0.5)exp(0.5*i*x)sech(x),
```

```
%    i = (-1)^0.5 over xl <= x <= xu
%
%    sech(x)
     sch=2.0/(exp(x(i))+exp(-x(i)));
%
%    Real part
     v(i)=rt2*cos(0.5*x(i))*sch;
%
%    Imaginary part
     w(i)=rt2*sin(0.5*x(i))*sch;
   end
%
% Two vectors to one vector
  for i=1:n
    u0(i)  =v(i);
    u0(i+n)=w(i);
  end
```

Note that a single vector of ODE dependent variables u0 is defined as required by the calls to ode45 and ode15s in pde_1_main.

The analytical solution of Eq. (6.2b) is programmed in ua (see Listing 6.3).

```
  function uanal=ua(t,x)
%
% Function ua computes the analytical solution to the CSE
%
% The following exact solution is |u(x,t)|^2 which is compared
% with the numerical solution
%
  uanal=2.0*(2.0/(exp(x-t)+exp(-(x-t))))^2;
```

Listing 6.3. Analytical solution routine ua for Eq. (6.2b)

The code in ua is essentially self-explanatory when compared with Eq. (6.2b) (recall $\text{sech}(x) = 1/\cosh(x) = 2/(e^x + e^{-x})$ and note the argument for the traveling wave solution, $x - t$).

The routine for calculating the integrals of Eqs. (6.4a)–(6.4c) by Simpson's rule, simp, is given in Listing 6.4.

```
  function uint=simp(xl,xu,n,u)
%
% Function simp computes three integral invariants by
```

```
% Simpson's rule
%
  global q
%
% One vector to two vectors
  for i=1:n
    v(i)=u(i);
    w(i)=u(i+n);
  end
%
% Three invariants
  for int=1:3
    h=(xu-xl)/(n-1);
%
%   I1
    if(int==1)
      uabs(1)=u(1)^2+v(1)^2;
      uabs(n)=u(n)^2+v(n)^2;
      uint(1)=uabs(1)-uabs(n);
      for i=3:2:n
        uabs(i-1)=u(i-1)^2+v(i-1)^2;
        uabs(i  )=u(i  )^2+v(i  )^2;
        uint(1)=uint(1)+4.0*uabs(i-1)+2.0*uabs(i);
      end
      uint(1)=h/3.0*uint(1);
    end
%
%   I2
    if(int==2)
      vx=dss004(xl,xu,n,v);
      wx=dss004(xl,xu,n,w);
      int(1)=v(1)*wx(1)-w(1)*vx(1);
      int(n)=v(n)*wx(n)-w(n)*vx(n);
      uint(2)=int(1)-int(n);
      for i=3:2:n
        int(i-1)=v(i-1)*wx(i-1)-w(i-1)*vx(i-1);
        int(i  )=v(i  )*wx(i  )-w(i  )*vx(i  );
        uint(2)=uint(2)+4.0*int(i-1)+2.0*int(i);
      end
      uint(2)=h/3.0*uint(2);
%
%   I3
    if(int==3)
      uxabs(1)=vx(1)^2+wx(1)^2;
      uxabs(n)=vx(n)^2+wx(n)^2;
      uint(3)=uxabs(1)-0.5*q*uabs(1)^2 ...
                      -(uxabs(n)-0.5*q*uabs(n)^2);
```

```
        for i=3:2:n
          uxabs(i-1)=vx(i-1)^2+wx(i-1)^2;
          uxabs(i)  =vx(i  )^2+wx(i  )^2;
          uint(3)=uint(3)+4.0*(uxabs(i-1)-0.5*q*uabs(i-1)^2) ...
                         +2.0*(uxabs(i  )-0.5*q*uabs(i  )^2);
        end
        uint(3)=2.0*h/3.0*uint(3);
     end
   end
```

Listing 6.4. Numerical quadrature routine `simp` applied to Eqs. (6.4a)–(6.4c)

`simp` has three parts corresponding to the integrals `I1,I2,I3` of Eqs. (6.4a)–(6.4c).

1. The coding for `I1` is

```
  function uint=simp(xl,xu,n,u)
%
% Function simp computes three integral invariants by
% Simpson's rule
%
  global q
%
% One vector to two vectors
  for i=1:n
    v(i)=u(i);
    w(i)=u(i+n);
  end
%
% Three invariants
  for int=1:3
    h=(xu-xl)/(n-1);
%
%   I1
    if(int==1)
       uabs(1)=u(1)^2+v(1)^2;
       uabs(n)=u(n)^2+v(n)^2;
       uint(1)=uabs(1)-uabs(n);
       for i=3:2:n
         uabs(i-1)=u(i-1)^2+v(i-1)^2;
         uabs(i  )=u(i  )^2+v(i  )^2;
         uint(1)=uint(1)+4.0*uabs(i-1)+2.0*uabs(i);
       end
       uint(1)=h/3.0*uint(1);
    end
```

After defining the function, vector u is divided into two vectors, v and w, according to $u(x,t) = v(x,t) + iw(x,t)$. The integration interval h=(xl-xu)/(n-1) is then computed, where xl=-10 and xu=40 are the lower and upper limits of the integral (set in inital_1). The for loop for I1 is an implementation of the weighted sum for Simpson's rule applied to the function $|u(x,t)|^2$ (the integrand in Eq. (6.4a)).

$$\int_{-\infty}^{\infty} u(x,t)dx \approx \frac{h}{3}\left[|u_1|^2 + \sum_{i=2}^{n-2}\left(4|u_i|^2 + 2|u_{i+1}|^2\right) + |u_n|^2\right]$$

2. Similarly, the coding for I2 of Eq. (6.4b) is

```
%
%   I2
    if(int==2)
      vx=dss004(xl,xu,n,v);
      wx=dss004(xl,xu,n,w);
      int(1)=v(1)*wx(1)-w(1)*vx(1);
      int(n)=v(n)*wx(n)-w(n)*vx(n);
      uint(2)=int(1)-int(n);
      for i=3:2:n
        int(i-1)=v(i-1)*wx(i-1)-w(i-1)*vx(i-1);
        int(i  )=v(i  )*wx(i  )-w(i  )*vx(i  );
        uint(2)=uint(2)+4.0*int(i-1)+2.0*int(i);
      end
      uint(2)=h/3.0*uint(2);
    end
```

The only difference is the use of the integrand $v(x,t)w_x(x,t) - w(x,t)v_x(x,t)$ according to Eq. (6.4b). The derivatives $v_x(x,t)$ and $w_x(x,t)$ are computed by calls to the differentiation routine dss004.

3. Finally, the coding for I3 of Eq. (6.4c) is

```
%
%   I3
    if(int==3)
      uxabs(1)=vx(1)^2+wx(1)^2;
      uxabs(n)=vx(n)^2+wx(n)^2;
      uint(3)=uxabs(1)-0.5*q*uabs(1)^2 ...
                       -(uxabs(n)-0.5*q*uabs(n)^2);
      for i=3:2:n
        uxabs(i-1)=vx(i-1)^2+wx(i-1)^2;
        uxabs(i)   =vx(i   )^2+wx(i   )^2;
        uint(3)=uint(3)+4.0*(uxabs(i-1)-0.5*q*uabs(i-1)^2) ...
                       +2.0*(uxabs(i   )-0.5*q*uabs(i   )^2);
```

```
        end
        uint(3)=2.0*h/3.0*uint(3);
      end
    end
```

The integrand is $|u_x(x, t)|^2 - 0.5q|u(x, t)|^4$ according to Eq. (6.4c).

The MOL ODE routine, pde_1, called by ode45 or ode15s in pde_1_main of Listing 6.1, is given in Listing 6.5

```
  function ut=pde_1(t,u)
%
% Function yt computes the t derivative vector of the CSE
%
  global  ncall      q     xl     xu     n
%
% One vector to two vectors
  for i=1:n
    v(i)=u(i);
    w(i)=u(i+n);
  end
%
% PDEs
%
%   v
%    xx
%
  vx=zeros(n,1);
  v(1)=0.0;
  v(n)=0.0;
  nl=1;
  nu=1;
  vxx=dss044(xl,xu,n,v,vx,nl,nu);
%
%   w
%    xx
  wx=zeros(n,1);
  w(1)=0.0;
  w(n)=0.0;
  nl=1;
  nu=1;
  wxx=dss044(xl,xu,n,w,wx,nl,nu);
%
% ODEs at the boundaries
  vt(1)=0.0;
  wt(1)=0.0;
```

```
  vt(n)=0.0;
  wt(n)=0.0;
%
% ODEs at the interior points
  for i=2:n-1
%
%   v**2 + w**2
    v2w2(i)=v(i)^2+w(i)^2;
%
%   vt
    vt(i)=-wxx(i)-q*v2w2(i)*w(i);
%
%   wt
    wt(i)= vxx(i)+q*v2w2(i)*v(i);
  end
%
% Two vectors to one vector
  for i=1:n
    ut(i)  =vt(i);
    ut(i+n)=wt(i);
  end
  ut=ut';
%
% Increment calls to pde_1
  ncall=ncall+1;
```

Listing 6.5. MOL ODE routine pde_1 called by main program pde_1_main

We can note the following points about pde_1:

1. After the definition of the function and the global declaration of some parameters and variables, the dependent variable vector u is stored as two vectors, v and w, according to $u(x, t) = v(x, t) + iw(x, t)$.

```
  function ut=pde_1(t,u)
%
% Function yt computes the t derivative vector of the CSE
%
  global ncall     q     xl     xu      n
%
% One vector to two vectors
  for i=1:n
    v(i)=u(i);
    w(i)=u(i+n);
  end
```

2. The second derivatives v_{xx} and w_{xx} in Eq. (6.1) are then computed by a call to dss044 that uses a five-point finite-difference (FD) approximation for these second derivatives.

```
%
% PDEs
%
%   v
%    xx
%
  vx=zeros(n,1);
  v(1)=0.0;
  v(n)=0.0;
  nl=1;
  nu=1;
  vxx=dss044(xl,xu,n,v,vx,nl,nu);
%
%   w
%    xx
  wx=zeros(n,1);
  w(1)=0.0;
  w(n)=0.0;
  nl=1;
  nu=1;
  wxx=dss044(xl,xu,n,w,wx,nl,nu);
```

There are a few details about this calculation of second derivatives that require explanation:

(a) The first derivative arrays, vx and ux, which are input arguments to dss044, are not actually used in the calculation of uxx and wxx. However, Matlab requires that at least one element of an array that is an input to a function has a numerical value. This is accomplished by setting all of the elements of v and w to zero through the Matlab zeros utility.

(b) Again, BCs for Eq. (6.1) are not required since $-10 \leq x \leq 40$ is equivalent to $-\infty \leq x \leq \infty$. However, the calculation of v_{xx} and w_{xx} by dss044 requires BCs (since this is a general purpose routine that is typically applied to a finite spatial domain that requires BCs), so the Dirichlet BCs $v(x = -10, t) = v(x = 40, t) = 0$ and $w(x = -10, t) = w(x = 40, t) = 0$ are used (and programmed as v(1)=0.0, v(n)=0.0, w(1)=0.0, w(n)=0.0). Also, these Dirichlet BCs must be identified before calling dss044 by nl=1, nu=1 for the lower ($x = -10$) and upper ($x = 40$) boundary values of x.

3. Now that all of the spatial derivatives in Eq. (6.1) have been computed, the MOL programming of Eq. (6.1) can be added.

```
%
% ODEs at the boundaries
  vt(1)=0.0;
  wt(1)=0.0;
  vt(n)=0.0;
  wt(n)=0.0;
%
% ODEs at the interior points
  for i=2:n-1
%
%   v**2 + w**2
    v2w2(i)=v(i)^2+w(i)^2;
%
%   vt
    vt(i)=-wxx(i)-q*v2w2(i)*w(i);
%
%   wt
    wt(i)= vxx(i)+q*v2w2(i)*v(i);
  end
%
% Two vectors to one vector
  for i=1:n
    ut(i)  =vt(i);
    ut(i+n)=wt(i);
  end
  ut=ut';
%
% Increment calls to pde_1
  ncall=ncall+1;
```

Note that the derivatives in t at the boundaries are set to zero (to reflect the Dirichlet BCs). Then the real and imaginary parts of u are programmed for the interior points $2 \le i \le n-1$ according to Eq. (6.1). In particular, the programming of the nonlinear term $-q|u|^2u$ is easily accomplished. Then the two derivative vectors, `vt` and `wt`, are returned from `pde_1` as a single vector, `ut`, for integration by `ode45` or `ode15s`. The usual transpose of `ut` is included, as well as the updating of the counter for the calls to `pde_1`.

We conclude this chapter by mentioning the following concepts:

1. The complex CSE can be integrated numerically as a 2×2 system of real PDEs.
2. The traveling wave solution of Eqs. (6.2a) and (6.2b) is demonstrated by the numerical MOL solution.
3. The invariants of Eqs. (6.4a)–(6.4c) can be computed numerically from the MOL solution.

4. This calculation of the invariants demonstrates another important aspect of the numerical solution, namely the availability of all of the solution derivatives in x and t, for example, u_t, u_{xx}, as well as functions of the dependent variable, for example, $|u|^2u$. Thus, in addition to displaying the solution, u, the analysis can also be extended to a display of all of the terms in the problem PDE(s), Eq. (6.1) in this case. We have found that displaying these terms along with the solution often gives enhanced visualization and understanding of the solution (at essentially no cost since these terms are readily available from the MOL solution because they are computed as part of the MOL solution, generally in the ODE routine, e.g., pde_1).

Finally, to further elucidate the numerical solution we also include a 3D plot produced with the following code:

```
% 3D Plots
  figure(2)
  surfl(x,t,v2w2, 'light');
  shading interp
  title('Cubic Schrodinger Equation');
  set(get(gca,'XLabel'),'String','space, x')
  set(get(gca,'YLabel'),'String','time, t')
  set(get(gca,'ZLabel'),'String','dependent variable, u(t,x)')
  set(gca,'XLim',[-10 40],'YLim',[0 30],'ZLim', [0 2]);
  axis tight
  view(-10,45)
  colormap('cool');    % set color map
  print -dpng -r300 fig2.png;
```

The resulting 3D plot is shown in Figure 6.2.

Additional enhanced plotting of $|u|^2$ for the Eq. (6.1) solution can be accomplished by animation (a movie). The details for producing an animation are given in Appendix 6.

APPENDIX A: SOME BACKGROUND TO SCHRÖDINGER'S EQUATION

A.1. Introduction

De Broglie[1] was the first to develop a wave theory for the behavior of an electron, postulating in his Ph.D. thesis that the wavelength of any matter obeyed the relationship $\lambda = h/p$, where h is Planck's constant and p is momentum. His approach successfully solved some simple problems, but his theory did not adequately explain the behavior of an electron when subjected to different types of external potential fields. This led *Erwin Schrödinger* to investigate alternatives and, ultimately, to

[1] Full name *Louis-Victor, Pierre, Raymond 7th duc de Broglie*, usually shortened to de Broglie.

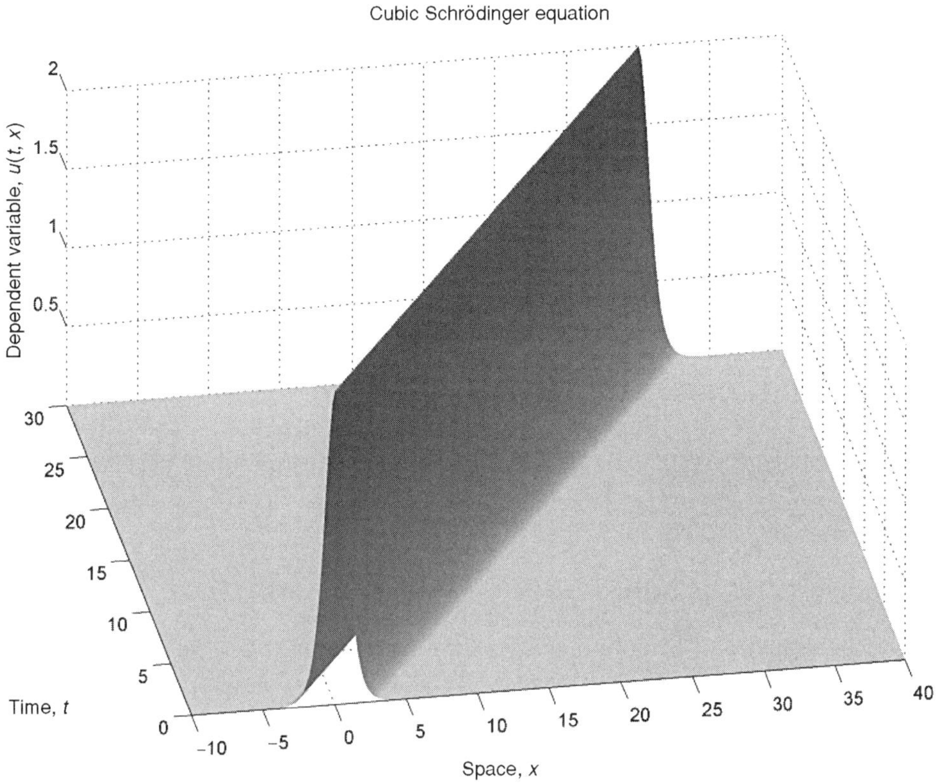

Figure 6.2. Three-dimensional output for main program `pde_1_main` with `nout=301, n=251`

derive a *wave function* solution to describe the electron that, crucially, conserved energy [2]. He then systematically applied his equation, the so-called *Schrödinger equation*, to problems involving different potentials applied to the electron, and the solutions were subsequently confirmed experimentally. It is the basis for our current understanding of the electron.

The *nonlinear Schrödinger equation* (NLS), or *cubic Schrödinger equation* (CSE), has the following *normalized* form in 3D space:

$$i\frac{\partial u(x,t)}{\partial t} + \frac{\partial^2 u(x,t)}{\partial x^2} + qu(x,t)\left|u(x,t)\right|^2 = 0, \quad x = (x_1, x_2, x_3) \tag{6.5}$$

where $q = \pm 1$. For $q = +1$, it is known as the focusing case and for $q = -1$, as the defocusing case (other nonzero values of q can be reduced to these two cases by rescaling the values of u [3]).

The cubic nonlinearity is one of the most common nonlinearities found in applications. It arises as a simplified model for studying Bose-Einstein condensates, Kerr media in nonlinear optics, and even freak waves in the ocean [3, 4].

An interesting property of the CSE is that it can sustain *solitary waves* or *solitons*, and it is this property that we have explored in the main text for the 1D case. A *soliton* is a *solitary wave* that has the additional property that other solitons can pass through it without changing its shape. The only difference after they emerge from the collision is that each exhibits a small phase shift.

A.2. A Nonrigorous derivation

A.2.1. The Hamiltonian

The Hamiltonian of an electron with mass m, momentum p, and coordinate q acted on by a conservative force is given by

$$H(q,p) = \frac{|p|^2}{2m} + V(q) \tag{6.6}$$

where physically, the quantity $|p|^2/2m$ represents *kinetic energy*, $V(q)$ represents *potential energy*, and $H(q,p)$ represents *total energy*. The motion of the electron then follows from *Hamiltonian's equations of motion*; that is,

$$\frac{\partial H}{\partial p} = \dot{q}(t); \qquad \frac{\partial H}{\partial q} = -\dot{p}(t) \tag{6.7}$$

where p and q are *vectors*, and $\dot{p}(t)$ and $\dot{q}(t)$ are the corresponding *gradients*. *Note:* This q should not be confused with the q in Eq. (6.5). Taking derivatives of Eq. (6.6), we obtain

$$\dot{q}(t) = \frac{1}{m}p(t); \qquad \dot{p}(t) = -\nabla V(q) \tag{6.8}$$

A consequence of Hamiltonian's equations of motion is that we obtain directly the principle of *conservation of energy*. Thus, we have $H(q(t), p(t)) = \text{constant}$, from which it follows that

$$\frac{d}{dt}H(q(t), p(t)) = 0, \quad \forall t \tag{6.9}$$

A.2.2. The Wave Function

Schrödinger's great insight was to make the assumption that the motion of an electron can be described by an appropriate *wave function*, which we choose to be of the following form:

$$u(x,t) \simeq \widehat{u}(x,t) = \sum_{\ell=1}^{\infty}\sum_{n=1}^{\infty} a_{\ell n}\, e^{i(k_\ell x - \omega_n t)} \tag{6.10}$$

where

$a_{\ell n}$ wave function coefficients
k_ℓ wave number ℓ
ω_n frequency n (r/s)
x position
t time

But we have the Planck-Einstein relation $E = h\nu = h\omega/2\pi = \hbar\omega$ and, also from de Broglie the relation, $k = 2\pi/\lambda = 2\pi(p/\hbar)$,
where

h Planck's constant
$\hbar$ reduced Planck's constant
ν frequency (1/s)
E energy
λ wave length

Thus, we obtain

$$\widehat{u}\,(x,t)=\sum_{\ell=1}^{\infty}\sum_{n=1}^{\infty}a_{\ell n}\,e^{i(p_\ell x/\hbar-E_n t/\hbar)} \tag{6.11}$$

A.2.3. Schrödinger's Equation

We now let $u_{\ell n}\,(x,t)=a_{\ell n}\,e^{i(p_\ell x/\hbar-E_n t/\hbar)}$ and for each term in the preceding summation we take the first derivative with respect to t and the second derivative with respect to x, to obtain

$$\frac{\partial u_{\ell n}}{\partial t}=u_{\ell n}\,(x,t)\left(-i\frac{E_n}{\hbar}\right) \tag{6.12}$$

$$\frac{\partial^2 u_{\ell n}}{\partial x^2}=u_{\ell n}\,(x,t)\left(-\frac{p_\ell^2}{\hbar^2}\right) \tag{6.13}$$

On rearranging and dropping the summation indices, we obtain

$$-\frac{\hbar}{i}\frac{\partial u}{\partial t}=u\,(x,t)\,E \tag{6.14}$$

$$-\hbar^2\frac{\partial^2 u}{\partial x^2}=u\,(x,t)\,p^2 \tag{6.15}$$

It should be noted that in quantum mechanics, unlike classical mechanics, a wave function $u\,(x,t)$ does not necessarily have specific position or momentum. Rather we consider it to possess an average speed and average position given by $\langle q(t)\rangle$ and $\langle p(t)\rangle$, respectively, where $\langle\cdot\rangle$ represents the *inner product* [5].

Now, on substituting Eqs. (6.14) and (6.15) into the Hamiltonian, Eq. (6.6), where we take $E=H$, $x=\langle q(t)\rangle$, and $p=\langle p(t)\rangle$, we obtain

$$H=E=-\frac{\hbar}{i}\frac{\partial u\,(x,t)}{\partial t}\left(\frac{1}{u\,(x,t)}\right)=-\frac{\hbar^2}{2m}\frac{\partial^2 u\,(x,t)}{\partial x^2}\left(\frac{1}{u\,(x,t)}\right)+V\,(x) \tag{6.16}$$

On rearranging we obtain the *Schrödinger equation* corresponding to a particular energy level,

$$i\hbar\frac{\partial u\,(x,t)}{\partial t}=-\frac{\hbar^2}{2m}\frac{\partial^2 u\,(x,t)}{\partial x^2}+u\,(x,t)\,V\,(x) \tag{6.17}$$

Finally, we now let $V(x)=-q\,|u|^2$ and normalize $\hbar$ and $2m$ to unity, when we obtain the preceding *normalized nonlinear* or *cubic Schrödinger equation*; that is,

$$i\frac{\partial u}{\partial t}+\frac{\partial^2 u}{\partial x^2}+qu\,|u|^2=0 \tag{6.18}$$

In Eq. (6.18) the term $|u|^2$ represents the *probability density* of the wave function. This means that the probability density for an electron being in a state defined by $u\,(x,t)$ is given by $p\,(x,t)=u^*\,(x,t)\,u\,(x,t)=|u|^2$, where u^* represents the complex conjugate of u (where, $p\,(x,t)$ should not be confused with p in the Hamiltonian).

The *normalization condition* implies that the spatial probability is given by

$$P\,(t)=\int_\Gamma p\,(x,t)\,dx=\int_\Gamma \left|u\,(x,t)\right|^2\,dx=1 \tag{6.19}$$

that is, the probability of the electron being located somewhere within the domain under consideration Γ is one.

Note: We can seek solutions to the Schrödinger equation (6.16) using the *separation of variables* (SOV) method, whereby we assume a solution of the form $u(x, t) = T(t)X(x)$. This leads to the *time-independent Schrödinger equation*

$$-\frac{\hbar^2}{2m}\nabla^2 u + Vu = Eu \tag{6.20}$$

and, under certain conditions, to the corresponding energy relationship

$$E = E_n = \frac{n^2\pi^2\hbar^2}{2ma^2} \tag{6.21}$$

where $0 \leq x \leq a$. This illustrates one of the extraordinary features of quantum mechanics, namely, that the energy of a system can take only certain discrete values [6].

For further details relating to the physics of Schrödinger's equation, readers are referred to references [6, 7].

REFERENCES

[1] Myint-U, T. and L. Debnath (1987), *Partial Differential Equations for Scientists and Engineers*, 3rd ed., North Holland, Amsterdam

[2] Schrodinger, E. (1926), Quantisierung als Eigenwertproblem, I–IV, *Annalen der Physik*, **79**:361–376, 489–527; **80**:437–490; **81**:109–139. Reprinted (English translation) in Erwin Schrodinger, *Collected Papers on Wave Mechanics*, Blackie & Son, London, 1928, pp. 325–432

[3] Killip, R., T. Tao, and M. Visan (2007), The Cubic Nonlinear Schrödinger Equation in Two Dimensions with Radial Data, *arXiv:0707.3188v1 [math.AP]*; available online at `http://arxiv.org/abs/0707.3188v`

[4] Antoine, X., C. Besse, and S. Descombes (2006), Artificial Boundary Conditions for One-Dimensional Cubic Nonlinear Schrodinger Equations, *SIAM J. Num. Anal.*, **43**(6): 2272–2293

[5] Tao, T. (2007), The Schrödinger Equation, In: *Princeton Companion to Mathematics*; available online at `http://www.math.ucla.edu/ tao/preprints/schrodinger.pdf`

[6] Hannabuss, K. (1997), *An Introduction to Quantum theory*, Oxford University Press, Oxford, UK

[7] French, A. P., and E. F. Taylor (1992), *An Introduction to Quantum Physics*, Chapman & Hall, Boca Raton, FL

7

The Korteweg–deVries Equation

This partial differential equation (PDE) problem introduces the following mathematical concepts and computational methods:

1. A nonlinear PDE with an exact solution that can be used to assess the accuracy of a numerical method of lines (MOL) solution.
2. Some of the basic properties of *solitons* that have broad application in many areas of physics, engineering, and the biological sciences.
3. Nonlinear interaction of two (or more) solitons.
4. Illustrative programming of a soliton solution.
5. Sparse matrix integration of the ordinary differential equations (ODEs) for the MOL approximation of a PDE.
6. Computer analysis of a PDE over an essentially infinite spatial domain.
7. Evaluation of *invariants of the solution* (integrals of functions of the solution that do not change with t).
8. Comparison of procedural and vectorized programming in Matlab.

The *one-dimensional (1D) Korteweg–deVries equation (KdV) equation* is [1]

$$\frac{\partial u}{\partial t} + 6u\frac{\partial u}{\partial x} + \frac{\partial^3 u}{\partial x^3} = 0$$

or in subscript notation,

$$u_t + 6uu_x + u_{xxx} = 0 \tag{7.1}$$

where

u dependent variable
x boundary-value (spatial) independent variable
t initial-value independent variable

Equation (7.1) is the conventional, nondimensional version of the equation originally derived by Korteweg and de Vries [2] (some background information is given in Appendix C). It is first order in t and third order in x, and therefore requires one *initial condition* (IC) and three *boundary conditions* (BCs). The IC is taken from the analytical solution (for a single soliton with velocity c)

$$u(x,t) = \frac{1}{2}c\,\text{sech}^2\left\{\frac{1}{2}\sqrt{c}(x-ct)\right\} \tag{7.2}$$

with $t = 0$; that is,

$$u(x,t=0) = \frac{1}{2}c\,\text{sech}^2\left\{\frac{1}{2}\sqrt{c}(x)\right\} \tag{7.3}$$

To complete the specification of the problem, we would naturally consider the three required BCs for Eq. (7.1). However, if the PDE is analyzed over an essentially infinite domain, $-\infty \le x \le \infty$, and if changes in the solution occur only over a finite interval in x, then *BCs at infinity have no effect*; in other words, we do not have to actually specify BCs (since they have no effect). This situation will be clarified through the PDE solution.

Consequently, Eqs. (7.1) and (7.3) constitute the complete PDE problem. The solution to this problem, Eq. (7.2), is used in the subsequent programming and analysis to evaluate the numerical MOL solution. Note that Eq. (7.2) is a *traveling wave* solution since the argument of the sech function is $x - ct$.

We now consider some Matlab routines for a numerical MOL solution of Eqs. (7.1) and (7.3) with the analytical solution, Eq. (7.2), included. A main program, `pde_1_main`, is given in Listing 7.1.

```
%
% Clear previous files
  clear all
  clc
%
% Parameters shared with other routines
  global  ncall ncase   c   c1    c2    xl    xu     x    n
%
% Select case; 1 - one soliton; 2 - two solitons;
  ncase=1;
  if(ncase==1)c=1.0; n=201; end
  if(ncase==2)c1=2.0; c2=0.5; n=301; end
%
% Initial condition
  t0=0.0;
  u0=inital_1(t0);
%
% Independent variable for ODE integration
  tf=30.0;
  tout=[t0:10.0:tf]';
  nout=4;
  ncall=0;
%
```

```
% ODE integration
  mf=2;
  reltol=1.0e-06; abstol=1.0e-06;
  options=odeset('RelTol',reltol,'AbsTol',abstol);
%
% Explicit (nonstiff) integration
  if(mf==1)[t,u]=ode45(@pde_1,tout,u0,options); end
%
% Implicit (sparse stiff) integration
  if(mf==2)
    S=jpattern_num;
    pause
    options=odeset(options,'JPattern',S)
    [t,u]=ode15s(@pde_1,tout,u0,options);
  end
%
% Store analytical solution, errors in numerical solution
  if(ncase==1)
    for it=1:nout
      for i=1:n
        u_anal(it,i)=ua(x(i),t(it));
        err(it,i)=u(it,i)-u_anal(it,i);
      end
    end
%
%   Display selected output
    fprintf('\n ncase = %2d   c = %5.2f\n',ncase,c);
    for it=1:nout
      fprintf('     t        x         u(it,i) u_anal(it,i)
              err(it,i)\n');
      for i=1:5:n
        fprintf('%6.2f%8.3f%15.6f%15.6f%15.6f\n',...
              t(it),x(i),u(it,i),u_anal(it,i),err(it,i));
      end
%
%     Calculate and display three invariants
      ui=u(it,:);
      uint=simp(xl,xu,n,ui);
      fprintf('\n Invariants at t = %5.2f',t(it));
      fprintf('\n     I1 = %10.4f   Mass conservation'   , ...
              uint(1));
      fprintf('\n     I2 = %10.4f   Energy conservation' , ...
              uint(2));
      fprintf('\n     I3 = %10.4f   Whitham invariant\n\n', ...
              uint(3));
    end
    fprintf('    ncall = %4d\n\n',ncall);
```

```
%
%   Plot numerical and analytical solutions
    plot(x,u,'o',x,u_anal,'-')
    xlabel('x')
    ylabel('u(x,t)')
    title('KdV equation; t = 0, 10, 20, 30; o - numerical;
          solid - analytical')
    print -deps -r300 pde2.eps; print -dps -r300 pde2.ps;
    print -dpng -r300 pde2.png
  end
%
% Store numerical solution
  if(ncase==2)
%
%   Display selected output
    fprintf('\n ncase = %2d  c1 = %5.2f  c2 = %5.2f\n', ...
            ncase,c1,c2);
    for it=1:nout
      fprintf('     t        x         u(it,i)\n');
      for i=1:5:n
        fprintf('%6.2f%8.3f%15.6f\n',t(it),x(i),u(it,i));
      end
      fprintf('\n');
%
%     Calculate and display three invariants
      ui=u(it,:);
      uint=simp(xl,xu,n,ui);
      fprintf('\n Invariants at t = %5.2f',t(it));
      fprintf('\n    I1 = %10.4f   Mass conservation'    , ...
              uint(1));
      fprintf('\n    I2 = %10.4f   Energy conservation' , ...
              uint(2));
      fprintf('\n    I3 = %10.4f   Whitham invariant\n\n', ...
              uint(3));
    end
    fprintf('  ncall = %4d\n\n',ncall);
%
%   Plot numerical solution
    for it=1:nout
      subplot(2,2,it)
      plot(x,u(it,:),'-')
      axis([-30 70 0 1])
      xlabel('x')
      if(it==1)
        ylabel('u(x,0)')
        title('KdV; t = 0')
      elseif(it==2)
```

```
      ylabel('u(x,10)')
      title('KdV; t = 10')
    elseif(it==3)
      ylabel('u(x,20)')
      title('KdV; t = 20')
    elseif(it==4)
      ylabel('u(x,30)')
      title('KdV; t = 30')
    end
  end
end
```

Listing 7.1. Main program pde_1_main

We can note the following points about this program:

1. After specifying some *global* variables, the program executes one of two cases with ncase=1 for one soliton and ncase=2 for two solitons. Note that for ncase=1, c=1 and $n = 201$ grid points in x are used, while for ncase=2, two soliton velocities are specified, c1=2, c2=0.5, and $n = 301$ grid points in x are used since the two solitons require greater spatial resolution (larger n) than for the single-soliton case (as will be demonstrated subsequently).

```
%
% Clear previous files
  clear all
  clc
%
% Parameters shared with other routines
  global   ncall ncase     c    c1     c2    xl    xu    x    n
%
% Select case; 1 - one soliton; 2 - two solitons;
  ncase=1;
  if(ncase==1)c=1.0; n=201; end
  if(ncase==2)c1=2.0; c2=0.5; n=301; end
```

2. The IC, Eq. (7.3), is defined by a call to routine inital_1 that defines the IC over the interval $-30 \le x \le 70$ for either one solton (ncase=1) or two solitons (ncase=2), as explained subsequently. The t interval is then defined as $0 \le t \le 30$ with solution outputs at $t = 0, 10, 20, 30$.

```
%
% Initial condition
  t0-0.0;
  u0=inital_1(t0);
```

```
%
% Independent variable for ODE integration
  tf=30.0;
  tout=[t0:10.0:tf]';
  nout=4;
  ncall=0;
```

3. The n ODEs are integrated by a call to either a nonstiff integrator, `ode45` (for `mf=1`), or a stiff integrator, `ode15s` (for `mf=2`). The stiff integrator is recommended since the nonstiff integrator requires rather long computer runs, typically an indication of stiffness. The MOL ODE routine is `pde_1` (discussed subsequently).

   Note also that the stiff integrator is run in a sparse matrix mode, as reflected in the second `options` command. This call to `ode15s` illustrates a general approach to sparse matrix ODE integration, which is a particularly effective approach in the MOL analysis of PDE systems.

```
%
% ODE integration
  mf=2;
  reltol=1.0e-06; abstol=1.0e-06;
  options=odeset('RelTol',reltol,'AbsTol',abstol);
%
% Explicit (nonstiff) integration
  if(mf==1)[t,u]=ode45(@pde_1,tout,u0,options); end
%
% Implicit (sparse stiff) integration
  if(mf==2)
    S=jpattern_num;
    pause
    options=odeset(options,'JPattern',S)
    [t,u]=ode15s(@pde_1,tout,u0,options);
  end
```

4. For the single-soliton case (`ncase=1`), the analytical solution of Eq. (7.2) is then evaluated by a call to `ua` (discussed subsequently) and selected numerical output for the numerical and analytical solutions is displayed by the two nested `for` loops (the first for t and the second for x).

```
%
% Store analytical solution, errors in numerical solution
  if(ncase==1)
    for it=1:nout
      for i=1:n
```

```
        u_anal(it,i)=ua(x(i),t(it));
        err(it,i)=u(it,i)-u_anal(it,i);
      end
    end
%
%   Display selected output
    fprintf('\n ncase = %2d   c = %5.2f\n',ncase,c);
    for it=1:nout
      fprintf('     t      x          u(it,i)   u_anal(it,i)
              err(it,i)\n');
      for i=1:5:n
        fprintf('%6.2f%8.3f%15.6f%15.6f%15.6f\n',...
              t(it),x(i),u(it,i),u_anal(it,i),err(it,i));
      end
```

5. Three invariants are evaluated by a call to `simp` that implements a numerical quadrature (integration) based on Simpson's rule (discussed subsequently).

```
%
%   Calculate and display three invariants
    ui=u(it,:);
    uint=simp(xl,xu,n,ui);
    fprintf('\n Invariants at t = %5.2f',t(it));
    fprintf('\n    I1 = %10.4f   Mass conservation'    , ...
            uint(1));
    fprintf('\n    I2 = %10.4f   Energy conservation'  , ...
            uint(2));
    fprintf('\n    I3 = %10.4f   Whitham invariant\n\n', ...
            uint(3));
  end
  fprintf('    ncall = %4d\n\n',ncall);
```

The integral invariants are

1. Conservation of mass:

$$u_1(t) = \int_{-\infty}^{\infty} u(x,t)dx \tag{7.4a}$$

2. Conservation of energy:

$$u_2(t) = \int_{-\infty}^{\infty} \frac{1}{2}u^2(x,t)dx \tag{7.4b}$$

3. Proposed by Whitham [3]:

$$u_3(t) = \int_{-\infty}^{\infty} 2u^3(x,t) - u_x^2(x,t)dx \tag{7.4c}$$

The exact values of these invariants for the case of Eq. (7.2) are $u_1(t) = 2$, $u_2(t) = 1/3$, and $u_3(t) = 0.4$, which are subsequently compared with the values computed by numerical integration from the routine simp.

6. The numerical and analytical solutions are then plotted.

```
%
%    Plot numerical and analytical solutions
     plot(x,u,'o',x,u_anal,'-')
     xlabel('x')
     ylabel('u(x,t)')
     title('KdV equation; t = 0, 10, 20, 30;
           o - numerical; solid - analytical')
    end
```

Selected numerical output and the plotted output for ncase=1 are given in Table 7.1.

Table 7.1. Partial output from pde_1_main and pde_1 (ncase=1)

ncase = 1   c = 1.00

t	x	u(it,i)	u_anal(it,i)	err(it,i)
0.00	-30.000	0.000000	0.000000	0.000000
0.00	-27.500	0.000000	0.000000	0.000000
0.00	-25.000	0.000000	0.000000	0.000000
0.00	-22.500	0.000000	0.000000	0.000000
0.00	-20.000	0.000000	0.000000	0.000000
0.00	-17.500	0.000000	0.000000	0.000000
0.00	-15.000	0.000001	0.000001	0.000000
0.00	-12.500	0.000007	0.000007	0.000000
0.00	-10.000	0.000091	0.000091	0.000000
0.00	-7.500	0.001105	0.001105	0.000000
0.00	-5.000	0.013296	0.013296	0.000000
0.00	-2.500	0.140207	0.140207	0.000000
0.00	0.000	0.500000	0.500000	0.000000
0.00	2.500	0.140207	0.140207	0.000000
0.00	5.000	0.013296	0.013296	0.000000
0.00	7.500	0.001105	0.001105	0.000000
0.00	10.000	0.000091	0.000091	0.000000
0.00	12.500	0.000007	0.000007	0.000000
0.00	15.000	0.000001	0.000001	0.000000
0.00	17.500	0.000000	0.000000	0.000000
0.00	20.000	0.000000	0.000000	0.000000
	.			.
	.			.
	.			.

```
  0.00  60.000        0.000000        0.000000        0.000000
  0.00  62.500        0.000000        0.000000        0.000000
  0.00  65.000        0.000000        0.000000        0.000000
  0.00  67.500        0.000000        0.000000        0.000000
  0.00  70.000        0.000000        0.000000        0.000000

 Invariants at t =  0.00
    I1 =      2.0000   Mass conservation
    I2 =      0.3333   Energy conservation
    I3 =      0.4005   Whitham invariant

    t        x          u(it,i)   u_anal(it,i)       err(it,i)
 10.00 -30.000        0.000000        0.000000        0.000000
 10.00 -27.500       -0.000417        0.000000       -0.000417
 10.00 -25.000        0.000103        0.000000        0.000103
 10.00 -22.500        0.000193        0.000000        0.000193
 10.00 -20.000       -0.000281        0.000000       -0.000281
           .                                             .
           .                                             .
           .                                             .
 10.00   0.000       -0.000275        0.000091       -0.000366
 10.00   2.500        0.001032        0.001105       -0.000073
 10.00   5.000        0.013595        0.013296        0.000299
 10.00   7.500        0.141836        0.140207        0.001629
 10.00  10.000        0.500778        0.500000        0.000778
 10.00  12.500        0.138160        0.140207       -0.002047
 10.00  15.000        0.013043        0.013296       -0.000253
 10.00  17.500        0.000580        0.001105       -0.000525
 10.00  20.000        0.000129        0.000091        0.000038
 10.00  22.500        0.000016        0.000007        0.000008
 10.00  25.000        0.000007        0.000001        0.000007
 10.00  27.500       -0.000331        0.000000       -0.000331
 10.00  30.000        0.000172        0.000000        0.000172
           .                                             .
           .                                             .
           .                                             .
 10.00  60.000        0.000009        0.000000        0.000009
 10.00  62.500       -0.000126        0.000000       -0.000126
 10.00  65.000        0.000305        0.000000        0.000305
 10.00  67.500        0.000004        0.000000        0.000004
 10.00  70.000        0.000000        0.000000       -0.000000

 Invariants at t = 10.00
    I1 =      2.0001   Mass conservation
    I2 =      0.3333   Energy conservation
    I3 =      0.4005   Whitham invariant
```

(continued)

Table 7.1 (*continued*)

```
    t        x          u(it,i)   u_anal(it,i)        err(it,i)
20.00 -30.000         0.000000       0.000000         0.000000
20.00 -27.500        -0.000151       0.000000        -0.000151
20.00 -25.000         0.000166       0.000000         0.000166
20.00 -22.500        -0.000617       0.000000        -0.000617
20.00 -20.000        -0.000194       0.000000        -0.000194
             .                                             .
             .                                             .
             .                                             .
20.00  10.000         0.000188       0.000091         0.000098
20.00  12.500         0.000271       0.001105        -0.000834
20.00  15.000         0.013899       0.013296         0.000603
20.00  17.500         0.143332       0.140207         0.003124
20.00  20.000         0.501271       0.500000         0.001271
20.00  22.500         0.137520       0.140207        -0.002688
20.00  25.000         0.013101       0.013296        -0.000195
20.00  27.500         0.002060       0.001105         0.000955
20.00  30.000         0.000285       0.000091         0.000194
             .                                             .
             .                                             .
             .                                             .
20.00  60.000        -0.000102       0.000000        -0.000102
20.00  62.500        -0.000177       0.000000        -0.000177
20.00  65.000        -0.000219       0.000000        -0.000219
20.00  67.500         0.000230       0.000000         0.000230
20.00  70.000         0.000000       0.000000        -0.000000

Invariants at t = 20.00
   I1 =      2.0000   Mass conservation
   I2 =      0.3333   Energy conservation
   I3 =      0.4002   Whitham invariant

    t        x          u(it,i)   u_anal(it,i)        err(it,i)
30.00 -30.000         0.000000       0.000000         0.000000
30.00 -27.500        -0.000736       0.000000        -0.000736
30.00 -25.000         0.000314       0.000000         0.000314
30.00 -22.500        -0.000321       0.000000        -0.000321
30.00 -20.000         0.000509       0.000000         0.000509
             .                                             .
             .                                             .
             .                                             .
30.00  20.000         0.000002       0.000091        -0.000089
30.00  22.500         0.000797       0.001105        -0.000308
30.00  25.000         0.013761       0.013296         0.000464
30.00  27.500         0.146476       0.140207         0.006269
30.00  30.000         0.499348       0.500000        -0.000652
30.00  32.500         0.134876       0.140207        -0.005331
```

```
30.00  35.000     0.011331     0.013296    -0.001965
30.00  37.500     0.000444     0.001105    -0.000661
30.00  40.000    -0.000318     0.000091    -0.000409
          .                                    .
          .                                    .
          .                                    .
30.00  60.000    -0.000789     0.000000    -0.000789
30.00  62.500     0.000678     0.000000     0.000678
30.00  65.000     0.000492     0.000000     0.000492
30.00  67.500    -0.001515     0.000000    -0.001515
30.00  70.000     0.000000     0.000000    -0.000000

Invariants at t = 30.00
   I1 =      1.9996   Mass conservation
   I2 =      0.3334   Energy conservation
   I3 =      0.4005   Whitham invariant

   ncall = 5076
```

We can note the following points about this output:

1. The soliton moves left to right in x, as indicated by the movement of the peak. Refer to Table 7.2 for a summary of the peak movement at `t=0,10,20,30` corresponding to `x=0,10,20,30`. In this case, the maximum error in the peak amplitude is `0.001271`, which is 0.25% of the exact value of 0.5.
2. A similar summary of the first invariant (with an exact value of `I1=2`) is given in Table 7.3.

Table 7.2. Partial output from `pde_1_main` and `pde_1` (`ncase=1`), illustrating the movement of the single-soliton peak with velocity $c = 1$

t	x	u(it,i)	u_anal(it,i)	err(it,i)
0.00	0.000	0.500000	0.500000	0.000000
10.00	10.000	0.500778	0.500000	0.000778
20.00	20.000	0.501271	0.500000	0.001271
30.00	30.000	0.499348	0.500000	-0.000652

Table 7.3. Partial output from `pde_1_main` and `pde_1` (`ncase=1`), illustrating the invariant `I1` of Eq. (7.4a)

```
( t = 0) I1 =      2.0000

(t = 10) I1 =      2.0001

(t = 20) I1 =      2.0000

(t = 30) I1 =      1.9996
```

The output of Table 7.3 illustrates two points: (a) the accuracy of the MOL numerical solution for `I1` of Eq. (7.4a), and (7.4b) the accuracy of the Simpson's rule numerical quadrature in routine `simp`. Similar conclusions follow for the invariants `I2` and `I3` of Eqs. (7.4b) and (7.4c).

The properties of the soliton can be visualized from the graphical output of `pde_1` (see Figure 7.1).

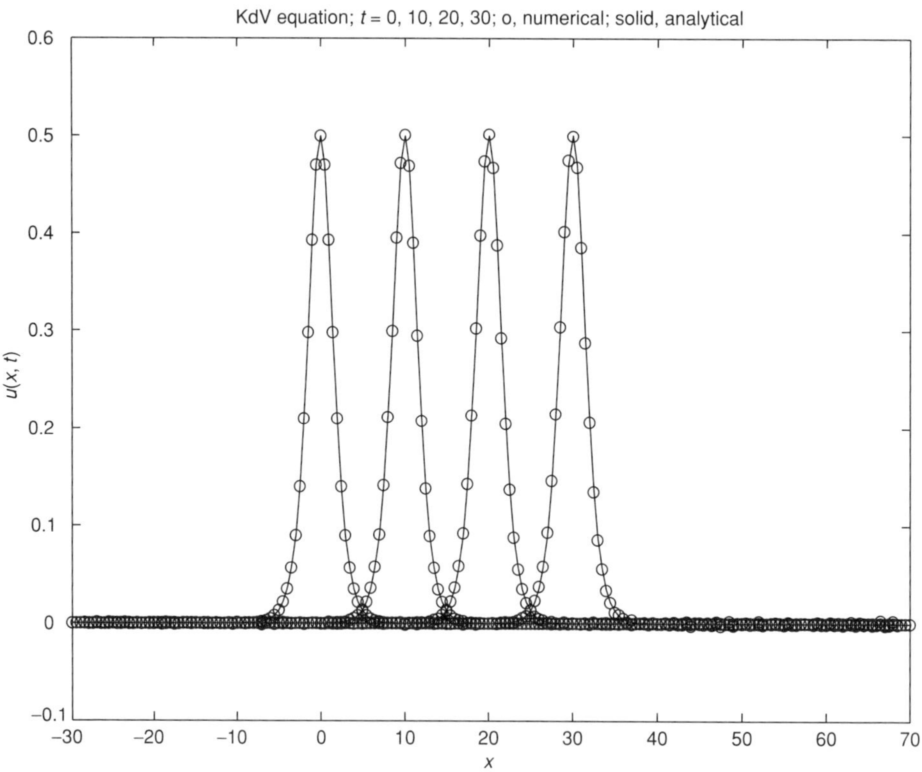

Figure 7.1. Output of main program `pde_1_main` for `ncase=1`

The solution starts with the IC pulse of Eq. (7.3) centered at $x = 0$. The movement of the soliton left to right is evident, as well as the close agreement between the numerical MOL solution and the analytical solution of Eq. (7.2). Note how the soliton retains its shape (a consequence of the traveling wave solution of Eq. (7.2)).

We now consider the subordinate routines called by the main program pde_1_main of Listing 7.1, inital_1, ua, simp and pde_1. inital_1 is listed first (see Listing 7.2).

```
  function u0=inital_1(t0)
%
% Function inital_1 sets the initial condition for the
% KdV equation
%
  global  ncase     c    c1    c2    xl    xu    x    n
%
% Spatial domain and initial condition
  xl=-30.0;
  xu= 70.0;
  dx=(xu-xl)/(n-1);
%
% Case 1 - single pulse
  if(ncase==1)
    for i=1:n
      x(i)=-30.0+(i-1)*dx;
      u0(i)=ua(x(i),0.0);
    end
  end
%
% Case 2 - two pulses
  if(ncase==2)
    for i=1:n
      x(i)=-30.0+(i-1)*dx;
      expm=exp(-1.0/2.0*sqrt(c1)*(x(i)+15.0));
      expp=exp( 1.0/2.0*sqrt(c1)*(x(i)+15.0));
      pulse1=(1.0/2.0)*c1*(2.0/(expp+expm))^2;
      expm=exp(-1.0/2.0*sqrt(c2)*(x(i)-15.0));
      expp=exp( 1.0/2.0*sqrt(c2)*(x(i)-15.0));
      pulse2=(1.0/2.0)*c2*(2.0/(expp+expm))^2;
      u0(i)=pulse1+pulse2;
    end
  end
```

Listing 7.2. Initialization routine inital_1

We can note the following details about inital_1:

1. After the routine is defined and some parameters are defined as global, a spatial grid is defined for $-30 \leq x \leq 70$ (recall that the number of grid points, n, is set in main program pde_1_main as a global parameter).

```
  function u0=inital_1(t0)
%
% Function inital_1 sets the initial condition for the
% KdV equation
%
  global ncase    c    c1    c2    xl    xu    x    n
%
% Spatial domain and initial condition
  xl=-30.0;
  xu= 70.0;
  dx=(xu-xl)/(n-1);
```

   This spatial interval is sufficiently long that the solution of Eq. (7.1) does not depart from the IC of Eq. (7.3) near $x = -30, 70$ so that this interval is effectively infinite. The consequence of this is that BCs for Eq. (7.1) are not required.
2. The IC for ncase=1 is then defined by a call to the analytical solution of Eq. (7.2) in ua with $t = 0$.

```
%
% Case 1 - single pulse
  if(ncase==1)
    for i=1:n
      x(i)=-30.0+(i-1)*dx;
      u0(i)=ua(x(i),0.0);
    end
  end
```

3. Finally, the two-soliton IC ncase=2 is defined.

```
%
% Case 2 - two pulses
  if(ncase==2)
    for i=1:n
      x(i)=-30.0+(i-1)*dx;
```

```
    expm=exp(-1.0/2.0*sqrt(c1)*(x(i)+15.0));
    expp=exp( 1.0/2.0*sqrt(c1)*(x(i)+15.0));
    pulse1=(1.0/2.0)*c1*(2.0/(expp+expm))^2;
    expm=exp(-1.0/2.0*sqrt(c2)*(x(i)-15.0));
    expp=exp( 1.0/2.0*sqrt(c2)*(x(i)-15.0));
    pulse2=(1.0/2.0)*c2*(2.0/(expp+expm))^2;
    u0(i)=pulse1+pulse2;
  end
end
```

Note the use of the velocity c1=2 for the first soliton (pulse1) centered at $x = -15$ and the velocity c2=0.5 for the second soliton (pulse2) centered at $x = 15$; these velocities are set as global variables in the main program pde_1_main.

The analytical solution of Eq. (7.2) is programmed in ua (see Listing 7.3).

```
  function uanal=ua(x,t)
%
% Function uanal computes the exact solution of the
% KdV equation for comparison with the numerical solution.
%
  global   c
%
% Analytical solution
  expm=exp(-1.0/2.0*sqrt(c)*(x-c*t));
  expp=exp( 1.0/2.0*sqrt(c)*(x-c*t));
  uanal=(1.0/2.0)*c*(2.0/(expp+expm))^2;
```

Listing 7.3. Analytical solution routine ua for Eq. (7.2)

The code in ua is essentially self-explanatory when compared with Eq. (7.2) (recall $\text{sech}(x) = 1/\cosh(x) = 2/(e^x + e^{-x})$ and note the argument for the traveling wave solution, $x - ct$, with $t = 0$ when ua is called from inital_1 for IC (7.3)).

The routine for calculating the integrals of Eqs. (7.4a)–(7.4c) by Simpson's rule, simp, is given in Listing 7.4.

```
  function uint=simp(xl,xu,n,u)
%
% Function simp computes three integral invariants by Simpson's
% rule
%
```

```
  for int=1:3
    h=(xu-xl)/(n-1);
%
%   Conservation of mass
    if(int==1)
       uint(1)=u(1)-u(n);
       for i=3:2:n
         uint(1)=uint(1)+4.0*u(i-1)+2.0*u(i);
       end
       uint(1)=h/3.0*uint(1);
    end
%
%   Conservation of energy
    if(int==2)
       uint(2)=u(1)^2-u(n)^2;
       for i=3:2:n
         uint(2)=uint(2)+4.0*u(i-1)^2+2.0*u(i)^2;
       end
       uint(2)=(1.0/2.0)*h/3.0*uint(2);
    end
%
%   Whitham conservation
    if(int==3)
       ux=dss004(xl,xu,n,u);
       uint(3)=2.0*u(1)^3-ux(1)^2-(2.0*u(n)^3-ux(n)^2);
       for i=3:2:n
         uint(3)=uint(3)+4.0*(2.0*u(i-1)^3-ux(i-1)^2)...
                        +2.0*(2.0*u(i)^3  -ux(i)^2);
       end
       uint(3)=h/3.0*uint(3);
    end
  end
```

Listing 7.4. Numerical quadrature routine simp applied to Eqs. (7.4a)–(7.4c)

simp has three parts corresponding to the integrals I1,I2,I3 of Eqs. (7.4a)–(7.4c).

1. The coding for I1 is

```
  function uint=simp(xl,xu,n,u)
%
% Function simp computes three integral invariants by Simpson's
% rule
%
```

```
    for int=1:3
      h=(xu-xl)/(n-1);
%
%     Conservation of mass
      if(int==1)
        uint(1)=u(1)-u(n);
        for i=3:2:n
          uint(1)=uint(1)+4.0*u(i-1)+2.0*u(i);
        end
        uint(1)=h/3.0*uint(1);
      end
```

After defining the function, the integration interval `h=(xu-xl)/(n-1)` is computed, where `xl=-30` and `xu=70` are the lower and upper limits of the integral (set in `inital_1`). The `for` loop for `I1` is an implementation of the weighted sum for Simpson's rule applied to the function $u(x, t)$ (the integrand in Eq. (7.4a)).

$$\int_{-\infty}^{\infty} u(x, t)dx \approx \frac{h}{3}\left[u_1 + \sum_{i=2}^{n-2}(4u_i + 2u_{i+1}) + u(n)\right]$$

2. Similarly, the coding for `I2` of Eq. (7.4b) is

```
%
%     Conservation of energy
      if(int==2)
        uint(2)=u(1)^2-u(n)^2;
        for i=3:2:n
          uint(2)=uint(2)+4.0*u(i-1)^2+2.0*u(i)^2;
        end
        uint(2)=(1.0/2.0)*h/3.0*uint(2);
      end
```

The only difference is the use of the integrand $u(x, t)^2$ according to Eq. (7.4b).

3. Finally, the coding for `I3` of Eq. (7.4c) is

```
%
%     Whitham conservation
      if(int==3)
        ux=dss004(xl,xu,n,u);
        uint(3)=2.0*u(1)^3-ux(1)^2-(2.0*u(n)^3-ux(n)^2);
```

```
      for i=3:2:n
        uint(3)=uint(3)+4.0*(2.0*u(i-1)^3-ux(i-1)^2)...
                       +2.0*(2.0*u(i)^3  -ux(i)^2);
      end
      uint(3)=h/3.0*uint(3);
    end
  end
```

The integrand is $2u(x,t)^3 - u_x(x,t)^2$ according to Eq. (7.4c). The derivative $u_x(x,t)$ is computed by a call to the differentiation routine dss004.

The MOL ODE routine, pde_1, is given in Listing 7.5.

```
  function ut=pde_1(t,u)
%
% Function pde_1 computes the t derivative vector for the
% KdV equation
%
  global  n xl xu ncall
%
% Calculate ux
  ux=dss004(xl,xu,n,u);
%
% Calculate uxxx
  uxxx=uxxx7c(xl,xu,n,u);
%
% PDE
  for i=1:n
    ut(i)=-uxxx(i)-6.0*u(i)*ux(i);
  end
  ut=ut';
%
% Increment calls to pde_1
  ncall=ncall+1;
```

Listing 7.5. MOL ODE routine pde_1 called by main program pde_1_main

We can note the following points about pde_1:

1. After the definition of the function and the global declaration of some parameters and variables, the first derivative u_x in Eq. (7.1) is computed by a call to dss004.

```
  function ut=pde_1(t,u)
%
% Function pde_1 computes the t derivative vector for the
% KdV equation
%
  global  n xl xu ncall
%
% Calculate ux
  ux=dss004(xl,xu,n,u);
```

2. The third derivative u_{xxx} in Eq. (7.1) is then computed by a call to uxxx7c, which has a seven-point finite-difference (FD) approximation for u_{xxx}.

```
%
% Calculate uxxx
  uxxx=uxxx7c(xl,xu,n,u);
```

An alternative to the use of uxxx7c would be to apply three successive differentiations to u using dss004; that is, $u \rightarrow u_x \rightarrow u_{xx} \rightarrow u_{xxx}$. This *stagewise differentiation* has been used to good effect in the MOL analysis of higher-order PDEs. However, we chose here to use uxxx7c to illustrate the rather specialized programming for a higher-order derivative. uxxx7c is listed in Appendix A at the end of this chapter.

3. Finally, the programming of Eq. (7.1) in a for loop over the spatial grid of n points is straightforward.

```
%
% PDE
  for i=1:n
    ut(i)=-uxxx(i)-6.0*u(i)*ux(i);
  end
  ut=ut';
%
% Increment calls to pde_1
  ncall=ncall+1;
```

Note in particular the close resemblance of the coding to Eq. (7.1), which is a major advantage of the MOL approach to the solution of PDE systems.

ncall is incremented and displayed at the end of the solution (in pde_1_main) to give an indication of the computational effort to produce the solution (the number of calls to pde_1).

The for loop for the PDE in pde_1 is an example of procedural programming that is common to most languages used for scientific computation, for example, Fortran, C, C++, Java. In Matlab, the subscripting in the for loop can be replaced by vector operations that are specific to Matlab. The advantages are basically twofold: The code is simplified and more efficient. To illustrate this approach, the preceding procedural code

```
%
% PDE
  for i=1:n
    ut(i)=-uxxx(i)-6.0*u(i)*ux(i);
  end
  ut=ut';
```

can be replaced with the vectorized code

```
%
% PDE
  ut=-uxxx'-6.0*u.*ux';
```

We note the following points:

1. The subscripting in the for loop (in terms of i) is eliminated and all of the variables are now vectors.
2. The nonlinear term uu_x in Eq. (7.1) is programmed using the vector multiplication .*, where "." specifies an *element-by-element* operation, in this case multiplication.
3. The RHS involves the formation of two *column vectors*, -6.0*u.*ux' and -uxxx', which are then subtracted (-uxxx'-6.0*u.*ux') to form the column vector ut (so that a transpose of ut is not required; i.e., u is already a column vector).
4. The differentiation routines dss004 and uxxx7c return row vectors, which then must be transposed to column vectors.

Of course, this code could be simplified further if dss004 and uxxx7c are programmed to return column vectors (so that the transposes are not required).

As a consequence of using the sparse version of ode15s for the integration of the n ODEs, a *map of the Jacobian matrix* of the ODE system is produced (by ode15s) (see Figure 7.2). We will not discuss the concept of the Jacobian matrix of an ODE system other than to point out that it displays the relationship between the derivative of the ith ODE, dy_i/dt, and the dependent variables on which this derivative depends; that is,

$$dy_i/dt = f_i(y_1, y_2, \ldots, y_{n-1}, y_n) \tag{7.5}$$

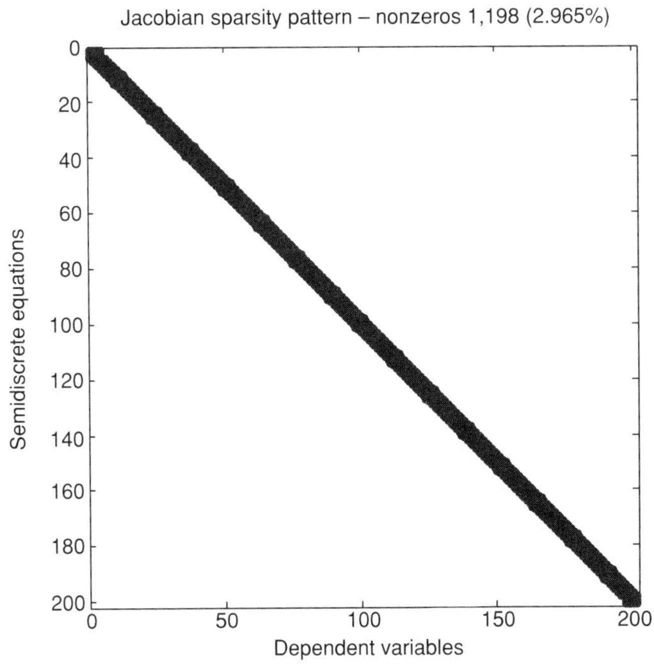

Figure 7.2. Jacobian matrix map for the system of n ODEs

For example, if dy_4/dt is a function of $y_1, y_2, y_3, y_4, y_5, y_6, y_7$ (e.g., seven values of y as would be the case for the seven-point FDs used in `uxxx7c`), then the fourth row of the map would have *seven* entries. This is the case of the map of Figure 7.2 because of the use of seven-point FDs in `uxxx7c`. Thus, the ODE system mapped in Figure 7.2 is said to be *banded* with a *bandwidth* of 7.

Note that only 2.965% of the $n \times n$ elements of the Jacobian map of Figure 7.2 are nonzero. This is not uncommon; that is, most of the elements are zero. This condition is the reason for the use of sparse matrix integrators since they in general will process only the *nonzero elements* and will not expend computer time processing the many zero elements. This saving in computer time can be very substantial, for example, *orders of magnitude*, so that the additional logic of the sparse matrix integrator (to detect the nonzero elements of the Jacobian matrix and then perform the numerical integration using only these nonzero elements) is well worth the additional complexity of the coding for the sparse matrix.

In order to produce a Jacobian map such as Figure 7.2, the sparse matrix option of `ode15s` requires a user-supplied routine `jpattern_num`. The name of this routine indicates that the $n \times n$ elements of the Jacobian matrix (n is the number of ODEs) are computed numerically (by FDs); the elements of the Jacobian matrix are actually the partial derivatives $((\partial dy_i/dt)/\partial y_j) = \partial f_i/\partial y_j$, where f_i is the derivative function from Eq. (7.5). An alternative approach to calculating the Jacobian matrix would be to analytically derive the partial derivatives $\partial f_i/\partial y_j$. However, this is generally impractical because of the number of such partial derivatives; for example, for $n = 1{,}000$, the Jacobian matrix has $1{,}000^2 = 1{,}000{,}000$ elements. In other words, numerical calculation of the Jacobian matrix is the only feasible approach.

The routine `jpattern_num` used in the present example, which produced the map of Figure 7.2, is listed in Appendix B at the end of this chapter. We chose to not go into the details of this because of space limitations. However, this routine is quite general and can therefore be applied to other ODE/PDE systems. `jpattern_num` has calls to several Matlab utilities. We will just mention the changes that might be required for other problems:

- The elements of the Jacobian matrix are computed by FDs. The FDs require a base point around which the numerical derivative is computed. The statement in `jpattern_num` that sets this base point is

```
ybase(i)=0.5;
```

  The value (0.5 in this case) need only be representative of the range of values of the ODE dependent variables. Since the solution in Figure 7.1 indicates a range of values of 0 to 1 in the solution, we chose 0.5. But again, a precise value is not required (but if `ode15s` fails, some experimentation with this value may be required).
- In order to calculate the partial derivatives in the Jacobian matrix by FD approximations, $((\partial dy_i/dt)/\partial y_j)$, the derivatives at the base point, dy_i/dt, are also required.

```
ytbase=pde_1(tbase,ybase);
```

  Note that the routine that calculates the derivatives dy_i/dt in Eq. (7.5) is called, in this case, `pde_1` of Listing 7.5.
- The elements of the Jacobian matrix are evaluated numerically by a call to the Matlab routine `numjac`. Since these elements are the partial derivatives of the derivative functions $dy_i/dt = f_i$ in Eq. (7.5), the name of the routine that evaluates these functions must be provided as the first argument of `numjac`, which is `pde_1` of Listing 7.5.

```
[Jac,fac]=numjac(@pde_1,tbase,ybase,ytbase,thresh,fac,
                 vectorized);
```

  Once the Jacobian matrix has been defined by this call to `numjac`, the map of the Jacobian can be plotted by the code that follows the call to `numjac` (see the listing in Appendix B).

We can add a few additional points about sparse matrix integration:

1. If PDE systems produced only banded Jacobians (with all of the nonzero elements concentrated around the main diagonal), then a *banded integrator* would be even more efficient than the *sparse integrator* since the banded integrator would know in advance where the nonzero elements occur (all are in the band) and it would therefore not have to search for the nonzero elements, and follow the resulting logic to use these nonzero elements, all of which add complexity to a sparse integrator.
2. However, the banded system of Figure 7.2 is not typical of a MOL PDE approximation. A more typical situation is to have some of the nonzero elements located along the main diagonal (as in a banded system), but to have others located outside the band (so that a banded integrator would miss using these *out-of-band elements*). Then the sparse integrator would be particularly effective since it would locate the out-of-band elements, the so-called *outliers*, and use them as well in the numerical integration, but it would not consume time processing the many zero elements. Thus, this combination of elements in a band along the main diagonal, plus outliers, leads to the efficiency of sparse matrix integration. Such a situation typically results from (a) PDE systems and (b) 2D and 3D PDEs (rather than the single, 1D problem considered here, Eq. (7.1)).
3. The processing of the ODE Jacobian matrix is a requirement for *stiff* (or *implicit*) integrators. Generally, this additional effort of processing the Jacobian matrix is well worth the effort if the ODEs are stiff, as they frequently are for PDE problems. However, if the ODEs are not stiff, then a *nonstiff* (or *explicit*) integrator should be used since they *do not require processing of the Jacobian matix*. This point is illustrated by the current PDE problem (Eq. (7.1)). If `ode45` is called in `pde_1_main` (`mf=1`), a nonstiff integrator, the number of calls to `pde_1` increases substantially relative to `ode15s`, a stiff integrator (as reflected in the larger final valuc of `ncall`). For borderline cases between stiff and nonstiff ODEs, some experimentation is generally required to determine which type of integrator is more efficient (or in other words, each type of integrator can be useful, depending on the stiffness characteristics of the problem; a common fallacy is to use a stiff integrator in all cases).
4. Returning to the idea of stagewise differentiation, each time a derivative is calculated, for example, u_{xx} from u_x, the bandwidth of the MOL ODEs increases. Thus, for example, if a five-point FD approximation is used to compute a derivative, the final bandwidth after three such numerical differentiations in calculating u_{xxx} would be substantially greater than the bandwidth of seven from the use of `uxx7c`. Since an increased bandwidth means more computational effort in processing the ODE Jacobian matrix, stagewise differentiation might not be as efficient as a direct calculation of the higher-order derivative (as in `uxxx7c`). There is also the matter of the accuracy of stagewise versus direct numerical differentiation, but we will not consider this further.

We now consider the execution of the previous routines for the case `ncase=2` (two solitons) to demonstrate a second quite remarkable property of solitons (in addition to maintaining their shape exactly with increasing t). All that is required to make the additional run described next is to change `ncase` to 2 in `pde_1_main`.

The numerical output for this case will not be considered in detail for two reasons:

1. The output is more complicated than for `ncase=1` (as listed in Table 7.1) and is therefore not easily examined in tabular form.
2. An analytical solution for the two-soliton case is not presented, so the errors in the numerical solution are not available.

We do, however, list the invariants at $t = 0, 10, 20, 30$:

```
Invariants at t =  0.00
   I1 =      4.2426   Mass conservation
   I2 =      1.0607   Energy conservation
   I3 =      2.3358   Whitham invariant

Invariants at t = 10.00
   I1 =      4.2424   Mass conservation
   I2 =      1.0607   Energy conservation
   I3 =      2.3365   Whitham invariant

Invariants at t = 20.00
   I1 =      4.2423   Mass conservation
   I2 =      1.0606   Energy conservation
   I3 =      2.3321   Whitham invariant

Invariants at t = 30.00
   I1 =      4.2433   Mass conservation
   I2 =      1.0630   Energy conservation
   I3 =      2.2826   Whitham invariant

 ncall = 14329
```

Note that the invariants have different values than for the one-soliton case. They do, however, remain essentially constant (since `I1,I2,I3` of Eqs. (7.4a)–(7.4c) are properties of Eq. (7.1), and are not dependent on the particular initial condition used, e.g., one or two solitons).

We now present the plotted output from `pde_1_main` (see Figure 7.3). We note the following points about Figure 7.3:

1. At $t = 0$, the higher soliton (with a peak value of 1 from `inital_1`) is to the left of the lower soliton (with a peak value of 0.25). This relative positioning is important since the higher soliton will travel left to right faster than the lower soliton.
2. At $t = 10$, the higher soliton has moved closer to the lower soliton.
3. At $t = 20$, the two solitons have merged.

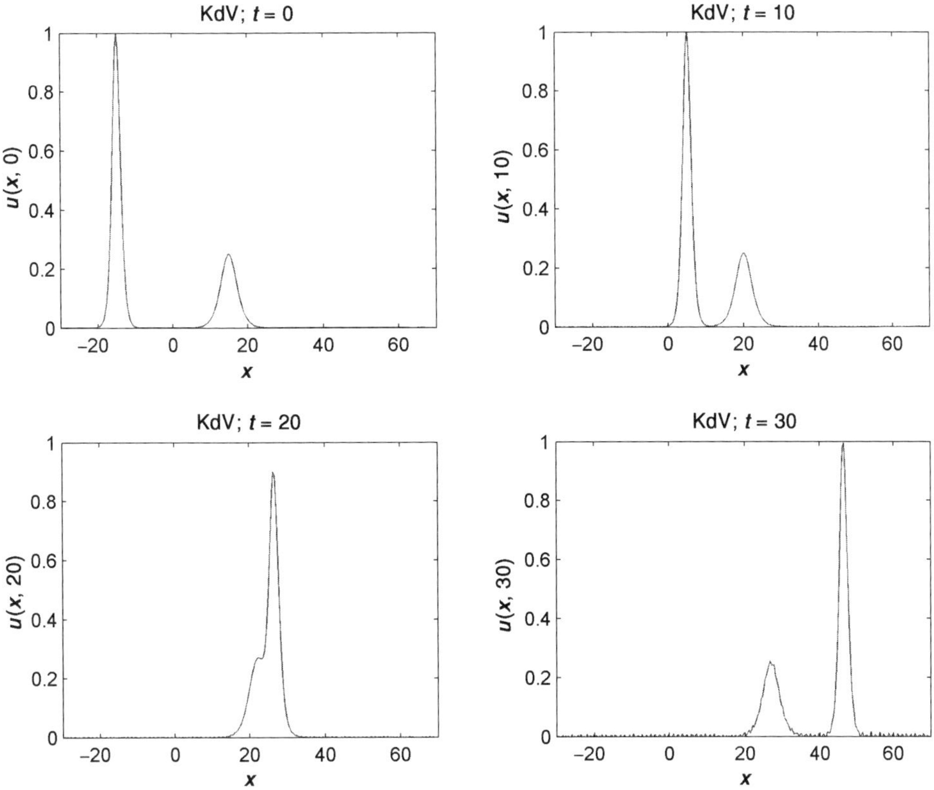

Figure 7.3. Output of main program `pde_1_main` for `ncase=2`

4. At $t = 30$, the solitons have again separated with the higher soliton now to the right of the lower soliton. Also, the two solitons have retained their shapes from $t = 0$.

Finally, to further elucidate the numerical solution for `ncase=2`, we also include a 3D plot produced with the following code:

```
% 3D Plots
 figure()
 surfl(x,t,u, 'light'); shading interp
 axis tight
 title('Korteweg-de Vries (KdV) equation');
 set(get(gca,'XLabel'),'String','space, x')
 set(get(gca,'YLabel'),'String','time, t')
 set(get(gca,'ZLabel'),'String','u(x,t)')
 set(gca,'XLim',[xl xu],'YLim',[t0 tf],'ZLim', [0 1]);
 view(-10,70);
 colormap('cool');
 print -dpng -r300 fig4.png;
```

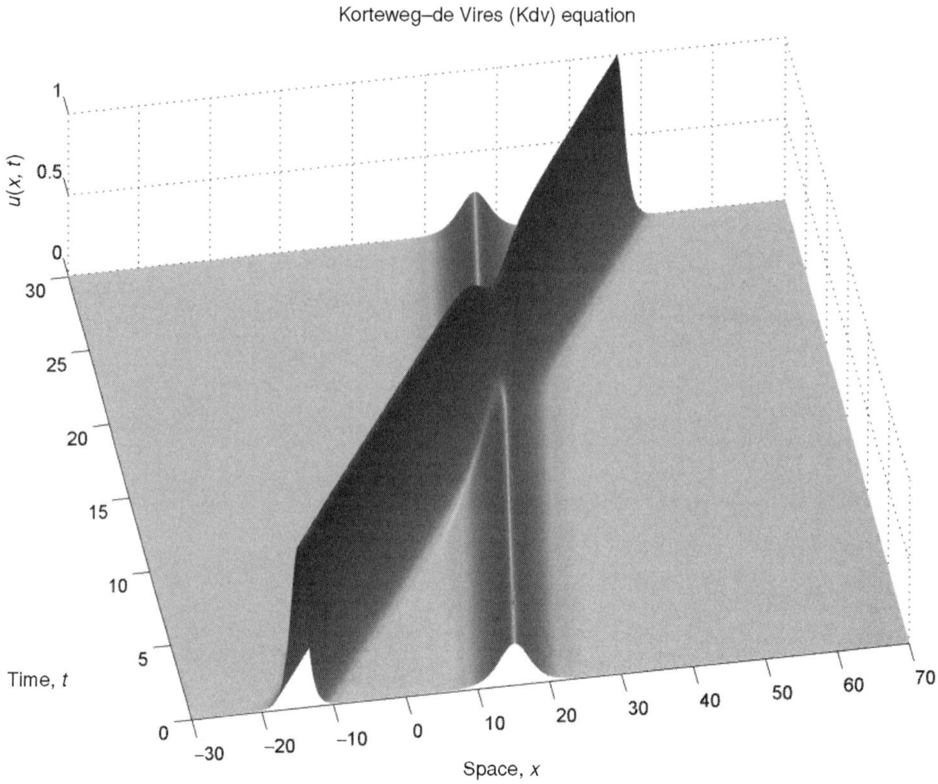

Figure 7.4. Three-dimensional output for main program `pde_1_main` with `ncase=2, nout=201, n=1001`

The resulting 3D plot is shown in Figure 7.4. Additional enhanced plotting of the solution of Eq. (7.1) can be accomplished by animation (a movie). The details for producing an animation are given in Appendix 6.

We conclude this chapter with this demonstration of the quite remarkable property of Eq. (7.1), the merging and subsequent emergence of the two solitons.

APPENDIX A

A.1. FD Routine uxxx7c

```
  function uxxx=uxxx7c(xl,xu,n,u)
%
% Function uxxx7c computes the derivative uxxx in the
% KdV equation
%
% Spatial increment
  dx=(xu-xl)/(n-1);
```

```
  r8dx3=1.0/(8.0*(dx^3));
%
% uxxx
  for i=1:n
%
%   At the left end, uxxx = 0
    if(i<4)uxxx(i)=0.0;
%
%   At the right end, uxxx = 0
    elseif(i>(n-3))uxxx(i)=0.0;
%
%   Interior points
    else
    uxxx(i)=r8dx3*...
       (   1.0*u(i-3)...
          -8.0*u(i-2)...
         +13.0*u(i-1)...
          +0.0*u(i  )...
         -13.0*u(i+1)...
          +8.0*u(i+2)...
          -1.0*u(i+3));
    end
  end
```

Note that the derivative u_{xxx} (= `uxxx(i)`) is calculated as a weighted sum of the seven values

`u(i-3),u(i-2),u(i-1),u(i),u(i+1),u(i+2),u(i+3)`

These values are centered around the point `i`, so this is a *seven-point, centered FD approximation*. The weighting coefficients are computed from standard formulas, but we will not consider the details here. At the ends of the interval in x (`i=1,2,3,n-2,n-1,n`), the derivative u_{xxx} is set to zero since we have assumed that these boundaries do not affect the solution (see Figures 7.1 and 7.3 to support this assumption).

APPENDIX B

B.1. Jacobian Matrix Routine `jpattern_num`

```
  function S=jpattern_num
%
  global n
%
% Sparsity pattern of the Jacobian matrix based on a
% numerical evaluation
```

```
%
% Set independent, dependent variables for the calculation
% of the sparsity pattern
  tbase=0;
  for i=1:n
    ybase(i)=0.5;
  end
  ybase=ybase';
%
% Compute the corresponding derivative vector
  ytbase=pde_1(tbase,ybase);
  fac=[];
  thresh=1e-16;
  vectorized='on';
  [Jac,fac]=numjac(@pde_1,tbase,ybase,ytbase,thresh,fac, ...
                   vectorized);
%
% Replace nonzero elements by "1" (so as to create a "0-1" map
% of the Jacobian matrix)
  S=spones(sparse(Jac));
%
% Plot the map
  figure
  spy(S);
  xlabel('dependent variables');
  ylabel('semi-discrete equations');
%
% Compute the percentage of non-zero elements
  [njac,mjac]=size(S);
  ntotjac=njac*mjac;
  non_zero=nnz(S);
  non_zero_percent=non_zero/ntotjac*100;
  stat=sprintf('Jacobian sparsity pattern - nonzeros %d
        (%.3f%%)', non_zero,non_zero_percent);
  title(stat);
```

APPENDIX C

C.1. Some Background to the KdV Equation

Equation (7.1) is the conventional, nondimensional version of the following equation originally derived by Korteweg and de Vries for a *moving (Lagrangian)* frame of reference [2, 4]:

$$\frac{\partial \eta}{\partial \tau} = \frac{3}{2}\sqrt{\frac{g}{h_o}}\frac{\partial}{\partial \chi}\left[\frac{1}{2}\eta^2 + \frac{2}{3}\alpha\eta + \frac{1}{3}\sigma\frac{\partial^2 \eta}{\partial \chi^2}\right]$$

It describes small-amplitude, shallow-water waves in a channel where $\sigma = h_o^3/3 - Th_o/(\rho g)$ and symbols have the following meaning:

g	gravitational acceleration (m/s^2)
h_o	nominal water depth (m)
T	capillary surface tension of fluid (N/m)
α	small arbitrary constant related to the uniform motion of the liquid (dimensionless)
η	wave height (m)
ρ	fluid density (kg/m^3)
τ	time (s)
χ	distance (m)

After rescaling and translating the dependent and independent variables to eliminate the physical constants using the transformations [5],

$$u = -\frac{1}{2}\eta - \frac{1}{3}\alpha; \qquad x = -\frac{\chi}{\sqrt{\sigma}}; \qquad t = \frac{1}{2}\sqrt{\frac{g}{h_o\sigma}}\tau$$

we arrive at the "canonical" form of the KdV equation, $u_t - 6uu_x + u_{\{xxx\}} = 0$, and this can be changed to the "alternative" form of Eq. (7.1) $u_t + 6uu_x + u_{\{xxx\}} = 0$ through the change of variable $u = -u$.

The KdV equation was found to have solitary wave solutions [6], which confirmed John Scott-Russell's account of the solitary wave phenomena [7] discovered during his experimental investigations into water flow in channels to determine the most efficient design for canal boats [4]. John Scott-Russell also described in poetic terms his first encounter with the solitary wave phenomena; thus,

> I was observing the motion of a boat which was rapidly drawn along a narrow channel by a pair of horses, when the boat suddenly stopped – not so the mass of water in the channel which it had put in motion; it accumulated round the prow of the vessel in a state of violent agitation, then suddenly leaving it behind, rolled forward with great velocity, assuming the form of a large solitary elevation, a rounded, smooth and well-defined heap of water, which continued its course along the channel apparently without change of form or diminution of speed. I followed it on horseback, and overtook it still rolling on at a rate of some eight or nine miles an hour, preserving its original figure some thirty feet long and a foot to a foot and a half in height. Its height gradually diminished, and after a chase of one or two miles I lost it in the windings of the channel. Such, in the month of August 1834, was my first chance interview with that singular and beautiful phenomenon which I have called the Wave of Translation [7].

The term *solitary wave* was also coined by Russell.

REFERENCES

[1] Debnath, L. (1997), Solitons and the Inverse Scattering Transform, *Nonlinear Partial Differential Equations for Scientists and Engineers*, Birkhauser, Boston, Chapter 9, pp. 331–404

[2] Korteweg, D. J. and F. de Vries (1895), On the Change of Form of Long Waves Advancing in a Rectangular Canal, and on a New Type of Long Stationary Waves, *Phil. Mag.* **39**:422–443

[3] Strang, G. (1986), *Introduction to Applied Mathematics*, Wellesley-Cambridge Press, Wellesley, MA, pp. 599–602

[4] de Jager, E. M. (2006), On the Origin of the Korteweg–deVries Equation, *arXiv e-print service*; available online at arXiv.org `http://arxiv.org/PS_cache/math/pdf/0602/0602661v1.pdf`

[5] Usman, M. (2007), *Forced Oscillations of the Korteweg–deVries Equation and Their Stability*, Ph.D. Thesis, McMicken College of Arts & Sciences, University of Cincinnati OH

[6] Lamb, H. (1993), *Hydrodynamics,* 6th ed., Cambridge University Press, Cambridge, UK, pp. 422–424

[7] Scott-Russell, J. (1844), Report on Waves, In: *14th Meeting of the British Association for the Advancement of Science*, John Murray, London, pp. 311–391, plates XLVII–LVII

8

The Linear Wave Equation

This partial differential equation (PDE) application introduces the following mathematical concepts and computational methods:

1. A PDE with an exact solution that can be used to assess the accuracy of a numerical method of lines (MOL) solution.
2. Some of the basic properties of *traveling waves*, a fundamental concept in the theory of PDEs and an important idea in many areas of physics, engineering, and the biological sciences.
3. The *characteristics of a hyperbolic PDE.*
4. Illustrative programming of a traveling wave solution.
5. The effect of smoothness (continuity in derivatives) in determining the effectiveness of numerical methods for computing PDE solutions.
6. Computer analysis of a PDE over an essentially infinite spatial domain.
7. Programming of a *PDE* second order in *t*.

The *one-dimensional (1D), second-order, linear wave equation* is [1]

$$\frac{\partial^2 u}{\partial t^2} = c^2 \frac{\partial^2 u}{\partial x^2}$$

or in subscript notation,

$$u_{tt} = c^2 u_{xx} \tag{8.1}$$

where

u dependent variable
x boundary-value (spatial) independent variable
t initial-value independent variable
c velocity

c is the *velocity of wave propagation* with the representative units `m/s` (meters/second). Note how these units are consistent with those of the derivatives in Eq. (8.1). u is typically a displacement, for example, the height of a water wave or amplitude of an electromagnetic wave.

Equation (8.1) is second order in t and x. It therefore requires two *initial conditions* (ICs) and two *boundary conditions* (BCs). For the ICs, we use

$$u(x, t = 0) = f(x), \tag{8.2}$$

$$u_t(x, t = 0) = 0 \tag{8.3}$$

Thus, the PDE starts with an initial displacement $f(x)$ and zero initial velocity.

To complete the specification of the problem, we would naturally consider the two required BCs for Eq. (8.1). However, if the PDE is analyzed over an essentially infinite domain, $-\infty \le x \le \infty$, and if changes in the solution occur only over a finite interval in x, then *BCs at infinity have no effect*; in other words, we do not have to actually specify BCs (since they have no effect). This situation will be clarified through the PDE solution.

Consequently, Eqs. (8.1)–(8.3) constitute the complete PDE problem. The solution to Eqs. (8.1)–(8.3) was first reported by the French mathematician *Jean-le-Rond d'Alembert* (1717–1783) in 1747 in a treatise on *vibrating strings* [1, 2]. d'Alembert's remarkable solution that used a method specific to the wave equation (based on the chain rule for differentiation) is

$$u(x, t) = \frac{1}{2}[f(x - ct) + f(x + ct)] + \frac{1}{2c}\int_{x-ct}^{x+ct} g(\xi)d\xi \tag{8.4}$$

It can also be obtained by the *Fourier transform* method or by the *separation of variables* method, which are more general [3].

Equation (8.4) is actually for a somewhat more general case than Eqs. (8.1)–(8.3). In particular, IC (8.3) is generalized to $u_t(x, t = 0) = g(x)$. For most of the subsequent discussion, we consider the homogeneous IC $u_t(x, t = 0) = 0$; that is, $g(t) = 0$.

We now consider some Matlab routines for a numerical MOL solution of Eqs. (8.1)–(8.3). A main program, pde_1_main, is given in Listing 8.1.

```
%
% Clear previous files
  clear all
  clc
%
% Parameters shared with the ODE routine
  global  ncall ncase  ndss     c     x    xl    xu     n
  c=1.0;
%
% Select case
%
%   Case 1 - rectangular pulse
%
%   Case 2 - triangular pulse
%
%   Case 3 - sine pulse
%
%   Case 4 - Gaussian pulse
```

```
%
  for ncase=1:4
%
% Boundaries, number of grid points
  xl=0.0;
  xu=100.0;
  n=401;
%
% Initial condition
  t0=0.0;
  u0=inital_1(t0);
%
% Independent variable for ODE integration
  tf=30.0;
  tout=[t0:10.0:tf]';
  nout=4;
  ncall=0;
%
% ODE integration
  mf=3;
  reltol=1.0e-04; abstol=1.0e-04;
  options=odeset('RelTol',reltol,'AbsTol',abstol);
  if(mf==1) ndss=0; % explicit FDs
    [t,u]=ode45(@pde_1,tout,u0,options); end
  if(mf==2) ndss=4; % ndss = 2, 4, 6, 8 or 10 required
    [t,u]=ode45(@pde_2,tout,u0,options); end
  if(mf==3) ndss=44; % ndss = 42, 44, 46, 48 or 50 required
    [t,u]=ode45(@pde_3,tout,u0,options); end
%
% One vector to two vectors
  for it=1:nout
  for i=1:n
    u1(it,i)=u(it,i);
    u2(it,i)=u(it,i+n);
  end
  end
%
% Display selected output
  fprintf('\n  ncase = %2d,   c = %2d,   ndss = %2d\n\n', ...
          ncase,c,ndss);
  for it=1:nout
    fprintf('    t        x        u(i,j)  u_anal(i,j)
            err(i,j)\n');
    u_anal(it,:)=ua(t(it),x);
    for i=1:10:n
      err(it,i)=u1(it,i)-u_anal(it,i);
      fprintf('%6.2f%8.3f%15.6f%15.6f%15.6f\n',...
```

```
            t(it),x(i),u1(it,i),u_anal(it,i),err(it,i));
    end
    fprintf('\n');
  end
  fprintf('  ncall = %4d\n\n',ncall);
%
% Plot numerical and analytical solutions
  figure(ncase)
  plot(x,u1,'-',x,u_anal,'o')
  xlabel('x')
  ylabel('u(x,t)')
  title('Wave equation; t = 0, 10, 20, 30; o - numerical;
        solid - analytical')
  print -deps -r300 pde.eps; print -dps -r300 pde.ps;
  print -dpng -r300 pde.png
%
% Next case
  end
```

Listing 8.1. Main program pde_1_main

We can note the following points about this program:

1. After specifying some *global* variables, the program cycles through four cases for different ICs ($f(x)$ in Eq. (8.2)). These ICs are discussed subsequently. Note also that the velocity c is set to 1.

```
%
% Clear previous files
  clear all
  clc
%
% Parameters shared with the ODE routine
  global  ncall ncase  ndss     c     x    xl    xu    n
  c=1.0;
%
% Select case
%
%   Case 1 - rectangular pulse
%
%   Case 2 - triangular pulse
%
%   Case 3 - sine pulse
%
%   Case 4 - Gaussian pulse
%
  for ncase=1:4
```

2. The spatial interval is defined as $0 \le x \le 100$ over 401 points; this is effectively an infinite interval ($-\infty \le x \le \infty$), as explained subsequently. The IC ($f(x)$ in Eq. (8.2)) is then specified through a call to `inital_1`. Finally, the t interval is defined as $0 \le t \le 30$ with solution outputs at $t = 0, 10, 20, 30$.

```
%
% Boundaries, number of grid points
  xl=0.0;
  xu=100.0;
  n=401;
%
% Initial condition
  t0=0.0;
  u0=inital_1(t0);
%
% Independent variable for ODE integration
  tf=30.0;
  tout=[t0:10.0:tf]';
  nout=4;
  ncall=0;
```

3. The 401 ordinary differential equations (ODEs) are integrated by a call to `ode45`; the three variations (`mf = 1,2,3`) correspond to differences in the programming of Eq. (8.1) in the ODE routines `pde_1`, `pde_2`, `pde_3`, as explained subsequently.

```
%
% ODE integration
  mf=3;
  reltol=1.0e-04; abstol=1.0e-04;
  options=odeset('RelTol',reltol,'AbsTol',abstol);
  if(mf==1) ndss=0; % explicit FDs
    [t,u]=ode45(@pde_1,tout,u0,options); end
  if(mf==2) ndss=4; % ndss = 2, 4, 6, 8 or 10 required
    [t,u]=ode45(@pde_2,tout,u0,options); end
  if(mf==3) ndss=44; % ndss = 42, 44, 46, 48 or 50 required
    [t,u]=ode45(@pde_3,tout,u0,options); end
%
% One vector to two vectors
  for it=1:nout
  for i=1:n
    u1(it,i)=u(it,i);
    u2(it,i)=u(it,i+n);
  end
  end
```

The solution vector in array u is then transferred to two arrays u1, u2 that are used because Eq. (8.1) is second order in t. This approach is explained when inital_1 is discussed.

4. Selected numerical output is then displayed followed by plotted output, and the program begins the next case (through the change in ncase).

```
%
% Display selected output
  fprintf('\n    ncase = %2d,    c = %2d,    ndss = %2d\n\n', ...
           ncase,c,ndss);
  for it=1:nout
    fprintf('    t        x          u(i,j)    u_anal(i,j)
             err(i,j)\n');
    u_anal(it,:)=ua(t(it),x);
    for
      err(it,i)=u1(it,i)-u_anal(it,i);
      fprintf('%6.2f%8.3f%15.6f%15.6f%15.6f\n',...
             t(it),x(i),u1(it,i),u_anal(it,i),err(it,i));
    end
    fprintf('\n');
  end
  fprintf('  ncall = %4d\n\n',ncall);
%
% Plot numerical and analytical solutions
  figure(ncase)
  plot(x,u1,'-',x,u_anal,'o')
  xlabel('x')
  ylabel('u(x,t)')
  title('Wave equation; t = 0, 10, 20, 30; o - numerical;
       solid - analytical')
  print -deps -r300 pde.eps; print -dps -r300 pde.ps;
  print -dpng -r300 pde.png
%
% Next case
  end
```

Before going on to the other routines called by main program pde_1_main, we will consider the output that helps to explain these routines. For ncase = 1 (the first pass of for ncase=1:4), a rectangular pulse is programmed in inital_1 as demonstrated in Table 8.1.

Table 8.1. Partial output from pde_1_main and pde_1 (ncase = 1)

```
 ncase =  1,    c =  1

    t        x          u(i,j)    u_anal(i,j)      err(i,j)
  0.00    0.000       0.000000       0.000000      0.000000
  0.00    2.500       0.000000       0.000000      0.000000
  0.00    5.000       0.000000       0.000000      0.000000
  0.00    7.500       0.000000       0.000000      0.000000
  0.00   10.000       0.000000       0.000000      0.000000
             .                          .
             .                          .
             .                          .
  0.00   40.000       0.000000       0.000000      0.000000
  0.00   42.500       0.000000       0.000000      0.000000
  0.00   45.000       1.000000       1.000000      0.000000
  0.00   47.500       1.000000       1.000000      0.000000
  0.00   50.000       1.000000       1.000000      0.000000
  0.00   52.500       1.000000       1.000000      0.000000
  0.00   55.000       1.000000       1.000000      0.000000
  0.00   57.500       0.000000       0.000000      0.000000
  0.00   60.000       0.000000       0.000000      0.000000
             .                          .
             .                          .
             .                          .
  0.00   90.000       0.000000       0.000000      0.000000
  0.00   92.500       0.000000       0.000000      0.000000
  0.00   95.000       0.000000       0.000000      0.000000
  0.00   97.500       0.000000       0.000000      0.000000
  0.00  100.000       0.000000       0.000000      0.000000

             .                          .
             .                          .
             .                          .

        Output at t=10,20 removed from here.

             .                          .
             .                          .
             .                          .

    t        x          u(i,j)    u_anal(i,j)      err(i,j)
 30.00    0.000       0.000000       0.000000      0.000000
 30.00    2.500       0.000000       0.000000      0.000000
 30.00    5.000       0.000000       0.000000      0.000000
 30.00    7.500      -0.000000       0.000000     -0.000000
 30.00   10.000      -0.000000       0.000000     -0.000000
 30.00   12.500       0.000025       0.000000      0.000025
```

(continued)

Table 8.1 *(continued)*

```
30.00  15.000      0.278997     0.500000    -0.221003
30.00  17.500      0.441472     0.500000    -0.058528
30.00  20.000      0.541156     0.500000     0.041156
30.00  22.500      0.525924     0.500000     0.025924
30.00  25.000      0.396482     0.500000    -0.103518
30.00  27.500     -0.021875     0.000000    -0.021875
30.00  30.000     -0.001194     0.000000    -0.001194
30.00  32.500      0.025994     0.000000     0.025994
30.00  35.000     -0.025431     0.000000    -0.025431
30.00  37.500      0.003114     0.000000     0.003114
30.00  40.000     -0.034373     0.000000    -0.034373
         .                          .
         .                          .
         .                          .
30.00  60.000     -0.034373     0.000000    -0.034373
30.00  62.500      0.003114     0.000000     0.003114
30.00  65.000     -0.025431     0.000000    -0.025431
30.00  67.500      0.025994     0.000000     0.025994
30.00  70.000     -0.001194     0.000000    -0.001194
30.00  72.500     -0.021875     0.000000    -0.021875
30.00  75.000      0.396482     0.500000    -0.103518
30.00  77.500      0.525924     0.500000     0.025924
30.00  80.000      0.541156     0.500000     0.041156
30.00  82.500      0.441472     0.500000    -0.058528
30.00  85.000      0.278997     0.500000    -0.221003
30.00  87.500      0.000025     0.000000     0.000025
30.00  90.000     -0.000000     0.000000    -0.000000
30.00  92.500     -0.000000     0.000000    -0.000000
30.00  95.000      0.000000     0.000000     0.000000
30.00  97.500      0.000000     0.000000     0.000000
30.00 100.000      0.000000     0.000000     0.000000

 ncall = 2677
```

We can note the following points about this output:

1. Considering first the IC (at $t = 0$), a rectangular pulse is programmed with the properties given in Table 8.2. The pulse is defined numerically in `pde_1_main` over 401 points in x so that the grid spacing in x is $100/(401 - 1) = 0.25$ rather than the spacing of 2.5 in the preceding output (since only every 10th point is displayed in the output from `for i=1:10:n` in `pde_1_main`); still, the IC is only an approximation to a rectangular pulse since the discontinuous jump from 0 to 1 and back to 0 is approximated within an interval of 0.25.

Table 8.2. Properties of the rectangular pulse IC $f(x)$ of Eq. (8.2) (from Table 8.1) for t=0

$0 \leq x \leq 42.5$	$f(x) = 0$
$45 \leq x \leq 55$	$f(x) = 1$
$57.5 \leq x \leq 100$	$f(x) = 0$

2. The numerical MOL and analytical solutions agree at $t = 0$ (as they should); any subsequent differences between these two solutions (for $t > 0$) will be due to errors in the numerical solution.
3. These differences (errors) are evident in the output at $t = 30$. These errors are elucidated in the plotted output in Figure 8.1.
4. The analytical solution has the important properties at $t = 30$, as given in Table 8.3.

   Some additional elaboration is required to explain the significance of these properties:

   (a) The analytical solution is now in two parts: one part with $u(x, t) = 0.5$ is centered around $x = 20$ and a second part with $u(x, t) = 0.5$ is centered around $x = 80$; note, in particular, that the amplitude for both pulses is 0.5 (while the original pulse at $t = 0$ had an amplitude of 1). Outside these two pulses, the solution is zero ($u(x, t) = 0$). These features are a

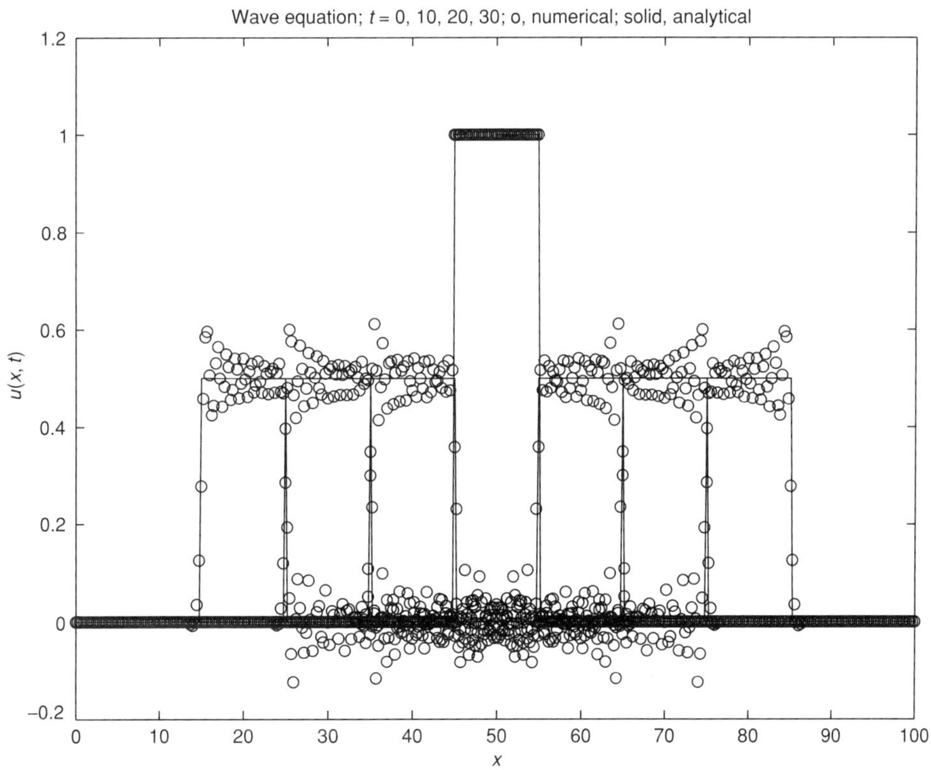

Figure 8.1. Output of main program `pde_1_main` for `ncase = 1`

Table 8.3. Properties of the solution of Eqs. (8.1)–(8.3) at $t = 30$ (from Eq. (8.4) and Table 8.1) for the IC $f(x)$ of Table 8.2

$0 \le x \le 12.5$	$u(x, t) = 0$
$15 \le x \le 25$	$u(x, t) = 0.5$
$27.5 \le x \le 72.5$	$u(x, t) = 0$
$75 \le x \le 85$	$u(x, t) = 0.5$
$87.5 \le x \le 100$	$u(x, t) = 0$

consequence of the analytical solution of Eq. (8.4) (with $g(x) = 0$ so the integral is zero).

(b) The pulse centered at $x = 20$ comes from the term $(1/2)f(x + ct)$ in Eq. (8.4) for which $45 \le x + ct \le 55$. Since $ct = (1)(30)$, we have that the position of the pulse is $45 - (1)(30) \le x \le 55 - (1)(30)$ or $15 \le x \le 25$, which is the interval in x we observe, for which the exact solution is 0.5; that is, the left pulse is traveling right to left at velocity c, and at $t = 30$, it is centered at $x = 20$. Thus, $f(x + ct)$ can be thought of as a *traveling wave* and PDE solutions with arguments such as $x - ct$ and $x + ct$ are generally termed *traveling wave solutions*.

(c) For the maximum value of t of $t = 30$, the pulse moving right to left does not reach the boundary at $x = 0$. Thus, this boundary has no effect on the solution and is effectively at $x = -\infty$; a BC is therefore not required at the left end of the interval $0 \le x \le 100$.

(d) The term $(1/2)f(x - ct)$ in Eq. (8.4) is zero for $15 \le x \le 25$ since it has the values $(1/2)f(15 - (1)(30))$ to $(1/2)f(25 - (1)(30))$ or $(1/2)f(-15)$ to $(1/2)f(-5)$. In other words, $(1/2)f(x - ct)$ has a negative argument over the interval $15 \le x \le 25$ that is outside the defined interval $0 \le x \le 100$, so we take $f(x - ct) = 0$ (as implied by the properties of Table 8.2). Thus, for the interval $15 \le x \le 25$, $f(x + ct) = 1$ and $f(x - ct) = 0$, which when substituted in Eq. (8.4) gives $u(x, t) = 0.5$ over $15 \le x \le 25$ (as observed in Table 8.1).

(e) The preceding discussion can be cast in a slightly different way. If we consider the relation $x - ct = c_1$, where c_1 is a prescribed constant, the property $f(x - ct) =$ constant follows. In other words, for x and t related by $x - ct = c_1$, the term $f(x - ct)$ in Eq. (8.4) is constant. Also, for $x + ct = c_2$, the property $f(x + ct) =$ constant follows. The conditions $x - ct = c_1$ and $x + ct = c_2$ are termed the *characteristics* of Eq. (8.1), and if we follow the evolution of the solution to Eq. (8.1), $u(x, t)$, along its characteristics, this approach to producing the solution is called the *method of characteristics*; the important feature of this approach to a PDE solution is that along the characteristics, the *solution is usually simplified*, for example, often a constant. Typically, in an analysis based on the method of characteristics, the evolution of the solution is depicted in an x–t plane with t as the ordinate (vertical coordinate) and x as the abscissa (horizontal coordinate); the numerical value of the solution appears in parametric form along with the characteristics in the x–t plane.

(f) We can view this approach to a PDE solution as *moving through the x–t plane along the characteristics* that is generally termed a *Lagrangian coordinate system*; if we consider the solution at fixed (stationary) values of x, this is generally termed an *Eulerian coordinate system*. The MOL solution of Eq. (8.1) discussed here is Eulerian since it is implemented on a fixed grid in x; that is, $0 \leq x \leq 100$. A method of characteristics version of MOL has also been reported, sometimes termed the *pseudo-characteristic method of lines*. We will not consider further the method of characteristics, but it is discussed extensively in references pertaining to the solution of hyperbolic PDEs (such as Eq. (8.1)).

(g) The preceding reasoning applies in the same way to the solution over $75 \leq x \leq 85$ for the pulse traveling left to right with amplitude 0.5 (as demonstrated by the properties of Table 8.3 and the numerical solution at $t = 30$ in Table 8.1). $f(x - ct) = 1$ is now the term in Eq. (8.4) that defines the pulse moving left to right and $f(x + ct) = 0$.

5. Finally, we can note that the characteristics $x - ct = c_1$ and $x + ct = c_2$ are *straight lines* in the x–t plane; this is a property *resulting from the linearity of the PDE*, for example, Eq. (8.1). If the PDE is nonlinear, the characteristics in the x–t plane generally *will not be linear*, but will be a function of the solution. Thus, as the solution evolves in x and t, it is necessary to compute the characteristics in using the method of characteristics.

The graphical output from `pde_1_main` in Listing 8.1 is shown in Figure 8.1. Figure 8.1 is somewhat difficult to interpret because the MOL solution is not very accurate. In fact, because the *rectangular pulse is discontinuous*, the MOL solution oscillates markedly at the points of discontinuity (which can be considered a form of the *Gibbs phenomenon*); this oscillation is also evident in the numerical output of Table 8.1. About the best we do for the interpretation of Figure 8.1 is to conclude that the *MOL does not perform satisfactorily in the case of solutions with discontinuities if finite differences (FDs) are used to approximate the spatial derivatives.* In other words, for solutions with discontinuities, the PDE spatial derivatives require special treatment, such as approximations based on *flux limiters* or *weighted essentially nonoscillatory methods*, which will not be considered in the present discussion. Numerical methods for the resolution of discontinuities and moving fronts are discussed extensively in the literature, for example, references [4–7], and are implemented in available computer codes.

The coding that produced the rectangular pulse with the properties of Table 8.2 is in the IC function `inital_1` (as might be expected since it is $f(x)$ in IC (8.2)) (see Listing 8.2).

```
  function u0=inital_1(t0)
%
% Function inital_1 sets the initial conditions for the wave
% equation
%
  global ncall ncase  ndss    c    x    xl    xu    n
%
```

```
% Spatial domain and initial condition
  dx=(xu-xl)/(n-1);
  for i=1:n
    x(i)=(i-1)*dx;
  end
  u1=ua(0.0,x);
  for i=1:n
    u2(i)=0.0;
    u0(i)  =u1(i);
    u0(i+n)=u2(i);
  end
```

Listing 8.2. Routine inital_1 for ICs (8.2) and (8.3)

We can note the following details about inital_1:

1. After definition of the function and the specification of some global variables, the spatial grid in x is defined over $n = 401$ points (with `xl = 0, xu = 100` set in the main program of Listing 8.1).

```
  function u0=inital_1(t0)
%
% Function inital_1 sets the initial conditions for the wave
% equation
%
  global ncall ncase  ndss    c     x    xl    xu     n
%
% Spatial domain and initial condition
  dx=(xu-xl)/(n-1);
  for i=1:n
    x(i)=(i-1)*dx;
  end
```

2. The analytical solution at $t = 0$ is defined by a call to function ua (discussed subsequently).

```
  u1=ua(0.0,x);
  for i=1:n
    u2(i)=0.0;
    u0(i)  =u1(i);
    u0(i+n)=u2(i);
  end
```

The numerical solution consists of two parts: $u(x, t = 0)$ stored as u1, and $u_t(x, t = 0)$ stored as u2. Note that the homogeneous IC of Eq. (8.3) is used.

Thus, since Eq. (8.1) is second order in t, we need two ICs (Eqs. (8.2) and (8.3)), and we will actually be integrating $2n = 802$ ODEs, with their ICs stored in u0.

The subordinate routine ua for the analytical solution of Eq. (8.4) (with $g(x) = 0$) is given in Listing 8.3.

```
  function uanal=ua(t,x)
%
% Function ua computes the analytical solution to the wave
% equation
%
  global  ncall ncase  ndss     c    xl    xu     n
%
% The following coding is specific to the interval
% 0 le x le 100
%
% Rectangular pulse
  if(ncase==1)
    for i=1:n
      if     x(i)<(45.0-c*t)  pulse1(i)=0.0;
      elseif x(i)>(55.0-c*t)  pulse1(i)=0.0;
      elseif x(i)>=(45.0-c*t) & x(i)<=(50.0-c*t)
             pulse1(i)=0.5;
      elseif x(i)>=(50.0-c*t) & x(i)<=(55.0-c*t)
             pulse1(i)=0.5;
      end
      if     x(i)<(45.0+c*t)  pulse2(i)=0.0;
      elseif x(i)>(55.0+c*t)  pulse2(i)=0.0;
      elseif x(i)>=(45.0+c*t) & x(i)<=(50.0+c*t)
             pulse2(i)=0.5;
      elseif x(i)>=(50.0+c*t) & x(i)<=(55.0+c*t)
             pulse2(i)=0.5;
      end
      uanal(i)=pulse1(i)+pulse2(i);
    end
  end
%
% Triangular pulse
  if(ncase==2)
    for i=1:n
      if     x(i)<(45.0-c*t)  pulse1(i)=0.0;
      elseif x(i)>(55.0-c*t)  pulse1(i)=0.0;
      elseif x(i)>=(45.0-c*t) & x(i)<=(50.0-c*t)
             pulse1(i)=(x(i)-(45.0-c*t))/10.0;
      elseif x(i)>=(50.0-c*t) & x(i)<=(55.0-c*t)
             pulse1(i)=((55.0-c*t)-x(i))/10.0;
```

```
      end
      if     x(i)<(45.0+c*t)  pulse2(i)=0.0;
      elseif x(i)>(55.0+c*t)  pulse2(i)=0.0;
      elseif x(i)>=(45.0+c*t) & x(i)<=(50.0+c*t)
             pulse2(i)=(x(i)-(45.0+c*t))/10.0;
      elseif x(i)>=(50.0+c*t) & x(i)<=(55.0+c*t)
             pulse2(i)=((55.0+c*t)-x(i))/10.0;
      end
      uanal(i)=pulse1(i)+pulse2(i);
    end
  end
%
% Sine pulse
  if(ncase==3)
    for i=1:n
      if     x(i)<(45.0-c*t)  pulse1(i)=0.0;
      elseif x(i)>(55.0-c*t)  pulse1(i)=0.0;
      elseif x(i)>=(45.0-c*t) & x(i)<=(50.0-c*t)
             pulse1(i)=(1.0/2.0)*sin(pi*(x(i)-(45.0-c*t))/10.0);
      elseif x(i)>=(50.0-c*t) & x(i)<=(55.0-c*t)
             pulse1(i)=(1.0/2.0)*sin(pi*((55.0-c*t)-x(i))/10.0);
      end
      if     x(i)<(45.0+c*t)  pulse2(i)=0.0;
      elseif x(i)>(55.0+c*t)  pulse2(i)=0.0;
      elseif x(i)>=(45.0+c*t) & x(i)<=(50.0+c*t)
             pulse2(i)=(1.0/2.0)*sin(pi*(x(i)-(45.0+c*t))/10.0);
      elseif x(i)>=(50.0+c*t) & x(i)<=(55.0+c*t)
             pulse2(i)=(1.0/2.0)*sin(pi*((55.0+c*t)-x(i))/10.0);
      end
      uanal(i)=pulse1(i)+pulse2(i);
    end
  end
%
% Gaussian pulse
  if(ncase==4)
    a=2.0;
    for i=1:n
      if     x(i)<(45.0-c*t)  pulse1(i)=0.0;
      elseif x(i)>(55.0-c*t)  pulse1(i)=0.0;
      elseif x(i)>=(45.0-c*t) & x(i)<=(55.0-c*t)
             pulse1(i)=(1.0/2.0)* ...
                        exp(-a*(x(i)-(50.0-c*t))^2/10.0);
      end
      if     x(i)<(45.0+c*t)  pulse2(i)=0.0;
      elseif x(i)>(55.0+c*t)  pulse2(i)=0.0;
      elseif x(i)>=(45.0+c*t) & x(i)<=(55.0+c*t)
             pulse2(i)=(1.0/2.0)* ...
```

```
                        exp(-a*(x(i)-(50.0+c*t))^2/10.0);
      end
      uanal(i)=pulse1(i)+pulse2(i);
    end
  end
```

Listing 8.3. Function ua for the analytical solution, Eq. (8.4), (with $g(x) = 0$)

We can note the following points about ua:

1. After the function and some global variables are defined, the coding for the rectangular pulse is executed if `ncase = 1` (from `for ncase=1:4` in the main program of Listing 8.1).

```
  function uanal=ua(t,x)
%
% Function ua computes the analytical solution to the wave
% equation
%
  global  ncall ncase  ndss     c    xl    xu     n
%
% The following coding is specific to the interval
% 0 le x le 100
%
% Rectangular pulse
  if(ncase==1)
    for i=1:n
      if     x(i)<(45.0-c*t)  pulse1(i)=0.0;
      elseif x(i)>(55.0-c*t)  pulse1(i)=0.0;
      elseif x(i)>=(45.0-c*t) & x(i)<=(50.0-c*t)...
             pulse1(i)=0.5;
      elseif x(i)>=(50.0-c*t) & x(i)<=(55.0-c*t)...
             pulse1(i)=0.5;
      end
      if     x(i)<(45.0+c*t)  pulse2(i)=0.0;
      elseif x(i)>(55.0+c*t)  pulse2(i)=0.0;
      elseif x(i)>=(45.0+c*t) & x(i)<=(50.0+c*t)...
             pulse2(i)=0.5;
      elseif x(i)>=(50.0+c*t) & x(i)<=(55.0+c*t)...
             pulse2(i)=0.5;
      end
      uanal(i)=pulse1(i)+pulse2(i);
    end
  end
```

Table 8.4. Properties of the triangular pulse IC $f(x)$ of Eq. (8.2) for $t = 0$

$0 \le x \le 42.5$	$f(x) = 0$
$45 \le x \le 50$	$f(x) = (x - 45)/(50 - 45)$
$50 \le x \le 55$	$f(x) = 1 - (x - 50)/(55 - 50)$
$57.5 \le x \le 100$	$f(x) = 0$

The `for` loop computes $f(x)$ of Eq. (8.2) over the 401-point grid in `x` (the second argument of `ua`) when `t=0` is the input value (through the first argument of `ua`).

2. The calculation of $f(x)$ is in two parts: `pulse1` is $f(x - ct)$ in Eq. (8.4), while `pulse2` is $f(x + ct)$ (with `t = 0`). Note that these two variables have the nonzero value 0.5 for x corresponding to the properties of Table 8.2 and elsewhere they are zero, thereby defining the rectangular pulse.
3. In other words, the statement `uanal(i)=pulse1(i)+pulse2(i);` is the coding of Eq. (8.4), and for `t = 0`, it is `uanal(i)= 0.5 + 0.5 = 1` for `i` corresponding to $45 \le x \le 55$ as shown in Table 8.1; elsewhere, `uanal(i) = 0`.
4. For $t > 0$, the coding of `pulse1` and `pulse2` corresponds to the properties of $f(x - ct)$ and $f(x + ct)$ of Table 8.3.

We can consider other functions $f(x)$ in Eq. (8.4) (than the rectangular pulse); for example, we can consider the piecewise linear function given in Table 8.4. The coding for this triangular pulse in function `ua` (Listing 8.3, `ncase = 2`) is as follows:

```
%
% Triangular pulse
  if(ncase==2)
    for i=1:n
      if     x(i)<(45.0-c*t)  pulse1(i)=0.0;
      elseif x(i)>(55.0-c*t)  pulse1(i)=0.0;
      elseif x(i)>=(45.0-c*t) & x(i)<=(50.0-c*t)
             pulse1(i)=(x(i)-(45.0-c*t))/10.0;
      elseif x(i)>=(50.0-c*t) & x(i)<=(55.0-c*t)
             pulse1(i)=((55.0-c*t)-x(i))/10.0;
      end
      if     x(i)<(45.0+c*t)  pulse2(i)=0.0;
      elseif x(i)>(55.0+c*t)  pulse2(i)=0.0;
      elseif x(i)>=(45.0+c*t) & x(i)<=(50.0+c*t)
             pulse2(i)=(x(i)-(45.0+c*t))/10.0;
      elseif x(i)>=(50.0+c*t) & x(i)<=(55.0+c*t)
             pulse2(i)=((55.0+c*t)-x(i))/10.0;
      end
```

```
            uanal(i)=pulse1(i)+pulse2(i);
        end
    end
```

When the main program of Listing 8.1 is executed in the second pass through `for ncase=1:4`, the numerical and graphical output given in Table 8.5 results.

We can note the following points about the output given in Table 8.5:

1. For `t = 0`, the triangular pulse is centered at `x = 50` (as was the rectangular pulse) and the amplitude is unity.
2. For `t = 30`, the analytical solution is again in two parts as expected from Eq. (8.4), with the triangular pulses centered at `x = 20` and `x = 80` and each with a maximum value of 0.5.
3. For `t = 30`, the numerical solution agrees much better with the analytical solution than it did for the rectangular pulse. However, there is some appreciable difference between the numerical and analytical solutions; for example, the difference is -0.015836 at $x = 20, 80$; as might be expected, this difference is a maximum at the peaks of the two pulses at $x = 20, 80$ because the *first derivative has a discontinuity*, and switches from $+1$ to -1 (this is clear in the plotted output shown in Figure 8.2).

However, these differences are relatively small, and do not show up substantially in the plotted output of Figure 8.2. In other words, the plotted output shows relatively close agreement between the numerical and analytical solutions.

Note again that the solution starts as a single triangular pulse at $t = 0$ and centered at $x = 50$ in accordance with IC (8.2). This pulse then separates into two pulses that travel left and right with velocity `c=1` according to Eq. (8.4).

Figure 8.2 suggests that the smoothness of the IC function $f(x)$ of Eq. (8.2) plays an important role in the agreement between the numerical and analytical solutions. Before we go on to the case of a still smoother function to test this idea, we can note one other detail about the solutions for the rectangular and triangular pulses.

If the ICs for Eq. (8.1) are taken as

$$u(x, t = 0) = 0, \tag{8.5}$$

$$u_t(x, t = 0) = g(x) \tag{8.6}$$

then the analytical solution from Eq. (8.4) is

$$u(x, t) = \frac{1}{2c} \int_{x-ct}^{x+ct} g(\xi) d\xi \tag{8.7}$$

If $g(x)$ is the rectangular pulse considered previously, Eq. (8.7) indicates that it will be integrated to produce the analytical solution. Integration of the rectangular pulse

Table 8.5. Partial output from pde_1_main and pde_1 (ncase = 2)

```
 ncase =  2,   c =  1

    t        x          u(i,j)    u_anal(i,j)      err(i,j)
  0.00    0.000       0.000000       0.000000      0.000000
  0.00    2.500       0.000000       0.000000      0.000000
  0.00    5.000       0.000000       0.000000      0.000000
  0.00    7.500       0.000000       0.000000      0.000000
  0.00   10.000       0.000000       0.000000      0.000000
              .                         .
              .                         .
              .                         .
  0.00   40.000       0.000000       0.000000      0.000000
  0.00   42.500       0.000000       0.000000      0.000000
  0.00   45.000       0.000000       0.000000      0.000000
  0.00   47.500       0.500000       0.500000      0.000000
  0.00   50.000       1.000000       1.000000      0.000000
  0.00   52.500       0.500000       0.500000      0.000000
  0.00   55.000       0.000000       0.000000      0.000000
  0.00   57.500       0.000000       0.000000      0.000000
  0.00   60.000       0.000000       0.000000      0.000000
              .                         .
              .                         .
              .                         .
  0.00   90.000       0.000000       0.000000      0.000000
  0.00   92.500       0.000000       0.000000      0.000000
  0.00   95.000       0.000000       0.000000      0.000000
  0.00   97.500       0.000000       0.000000      0.000000
  0.00  100.000       0.000000       0.000000      0.000000

              .                         .
              .                         .
              .                         .

         Output at t=10,20 removed from here.

              .                         .
              .                         .
              .                         .

    t        x          u(i,j)    u_anal(i,j)      err(i,j)
 30.00    0.000      -0.000000       0.000000     -0.000000
 30.00    2.500      -0.000000       0.000000     -0.000000
 30.00    5.000      -0.000000       0.000000     -0.000000
 30.00    7.500      -0.000000       0.000000     -0.000000
 30.00   10.000       0.000000       0.000000      0.000000
 30.00   12.500      -0.000000       0.000000     -0.000000
 30.00   15.000       0.007346       0.000000      0.007346
 30.00   17.500       0.251863       0.250000      0.001863
```

```
30.00  20.000      0.484164     0.500000    -0.015836
30.00  22.500      0.245241     0.250000    -0.004759
30.00  25.000      0.009375     0.000000     0.009375
30.00  27.500      0.004243     0.000000     0.004243
30.00  30.000     -0.000070     0.000000    -0.000070
         .                         .
         .                         .
         .                         .
30.00  70.000     -0.000070     0.000000    -0.000070
30.00  72.500      0.004243     0.000000     0.004243
30.00  75.000      0.009375     0.000000     0.009375
30.00  77.500      0.245241     0.250000    -0.004759
30.00  80.000      0.484164     0.500000    -0.015836
30.00  82.500      0.251863     0.250000     0.001863
30.00  85.000      0.007346     0.000000     0.007346
30.00  87.500     -0.000000     0.000000    -0.000000
30.00  90.000      0.000000     0.000000     0.000000
30.00  92.500     -0.000000     0.000000    -0.000000
30.00  95.000     -0.000000     0.000000    -0.000000
30.00  97.500     -0.000000     0.000000    -0.000000
30.00 100.000     -0.000000     0.000000    -0.000000

 ncall = 1489
```

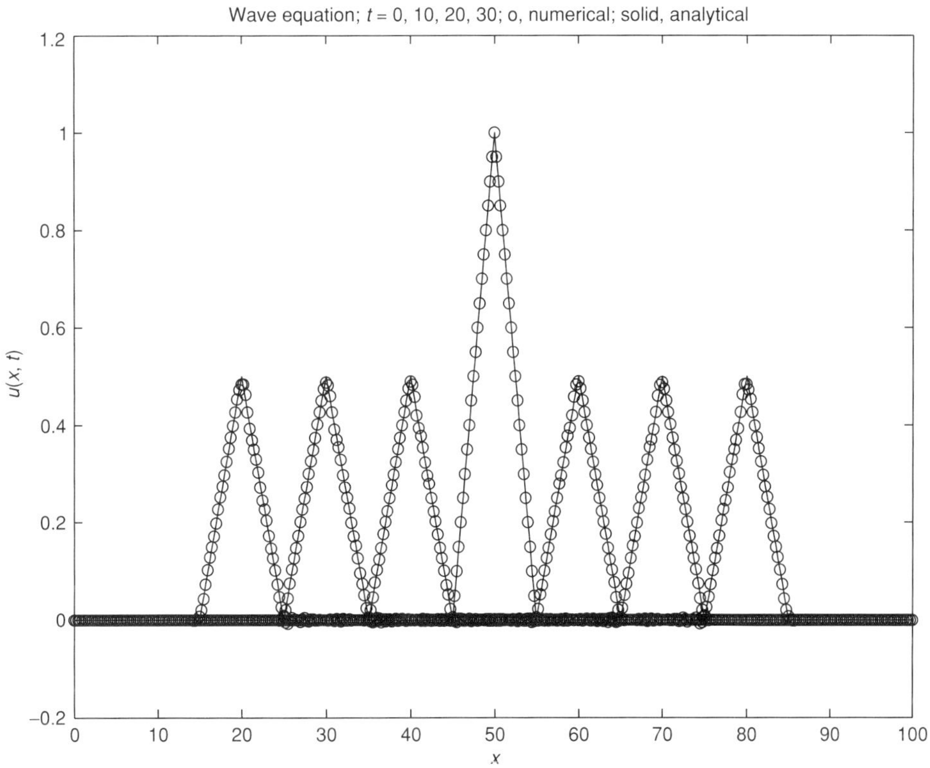

Figure 8.2. Output of main program `pde_1_main` for `ncase = 2`

Table 8.6. Properties of the sine pulse IC $f(x)$ of Eq. (8.2) for $t = 0$

$0 \le x \le 42.5$	$f(x) = 0$
$45 \le x \le 50$	$f(x) = \sin(\pi(x - 45)/(55 - 45)$
$57.5 \le x \le 100$	$f(x) = 0$

produces a linear function that is smoother than the rectangular pulse so that we would expect the analytical and numerical solutions to better agree for ICs (8.5) and (8.6) than for (8.2) and (8.3).

An IC function $f(x)$ that is smoother than either the rectangular or triangular pulses can be constructed based on the use of a sine function. Refer to Table 8.6 for an example. Now the derivatives of $f(x)$ of all orders will be smooth (a property of the sine function) except at $x = 45, 55$, where the first derivative is discontinuous due to the switch from the sine function to the constant function $f(x) = 0$.

The programming of the sine pulse in *ua* of Listing 8.3 is the traveling wave solution of Eq. (8.1) for the IC function of Table 8.6, produced by the main program of Listing 8.1 with `ncase = 3`.

```
%
% Sine pulse
  if(ncase==3)
    for i=1:n
      if      x(i)<(45.0-c*t)  pulse1(i)=0.0;
      elseif x(i)>(55.0-c*t)  pulse1(i)=0.0;
      elseif x(i)>=(45.0-c*t) & x(i)<=(50.0-c*t)
              pulse1(i)=(1.0/2.0)*sin(pi*(x(i)-(45.0-c*t))/10.0);
      elseif x(i)>=(50.0-c*t) & x(i)<=(55.0-c*t)
              pulse1(i)=(1.0/2.0)*sin(pi*((55.0-c*t)-x(i))/10.0);
      end
      if      x(i)<(45.0+c*t)  pulse2(i)=0.0;
      elseif x(i)>(55.0+c*t)  pulse2(i)=0.0;
      elseif x(i)>=(45.0+c*t) & x(i)<=(50.0+c*t)
              pulse2(i)=(1.0/2.0)*sin(pi*(x(i)-(45.0+c*t))/10.0);
      elseif x(i)>=(50.0+c*t) & x(i)<=(55.0+c*t)
              pulse2(i)=(1.0/2.0)*sin(pi*((55.0+c*t)-x(i))/10.0);
      end
      uanal(i)=pulse1(i)+pulse2(i);
    end
  end
```

Again, the traveling wave characteristic of this solution is evident from the use of the arguments $x - ct$ and $x + ct$.

Abbreviated numerical output from the main program of Listing 8.1 is given in Table 8.7.

Table 8.7. Partial output from pde_1_main and pde_1 (ncase = 3)

```
 ncase =  3,    c =  1

    t       x          u(i,j)     u_anal(i,j)        err(i,j)
  0.00    0.000      0.000000        0.000000        0.000000
  0.00    2.500      0.000000        0.000000        0.000000
  0.00    5.000      0.000000        0.000000        0.000000
  0.00    7.500      0.000000        0.000000        0.000000
  0.00   10.000      0.000000        0.000000        0.000000
            .                           .
            .                           .
            .                           .
  0.00   40.000      0.000000        0.000000        0.000000
  0.00   42.500      0.000000        0.000000        0.000000
  0.00   45.000      0.000000        0.000000        0.000000
  0.00   47.500      0.707107        0.707107        0.000000
  0.00   50.000      1.000000        1.000000        0.000000
  0.00   52.500      0.707107        0.707107        0.000000
  0.00   55.000      0.000000        0.000000        0.000000
  0.00   57.500      0.000000        0.000000        0.000000
  0.00   60.000      0.000000        0.000000        0.000000
            .                           .
            .                           .
            .                           .
  0.00   90.000      0.000000        0.000000        0.000000
  0.00   92.500      0.000000        0.000000        0.000000
  0.00   95.000      0.000000        0.000000        0.000000
  0.00   97.500      0.000000        0.000000        0.000000
  0.00  100.000      0.000000        0.000000        0.000000


            .                           .
            .                           .
            .                           .

      Output at t=10,20 removed from here.

            .                           .
            .                           .
            .                           .

    t       x          u(i,j)     u_anal(i,j)        err(i,j)
 30.00    0.000     -0.000000        0.000000       -0.000000
 30.00    2.500      0.000000        0.000000        0.000000
 30.00    5.000     -0.000000        0.000000       -0.000000
 30.00    7.500     -0.000000        0.000000       -0.000000
```

(continued)

Table 8.7 (*continued*)

```
30.00  10.000      0.000000     0.000000     0.000000
30.00  12.500     -0.000001     0.000000    -0.000001
30.00  15.000      0.011563     0.000000     0.011563
30.00  17.500      0.356483     0.353553     0.002929
30.00  20.000      0.498205     0.500000    -0.001795
30.00  22.500      0.351926     0.353553    -0.001628
30.00  25.000      0.011172     0.000000     0.011172
30.00  27.500      0.003404     0.000000     0.003404
30.00  30.000     -0.000901     0.000000    -0.000901
          .                        .
          .                        .
          .                        .
30.00  70.000     -0.000901     0.000000    -0.000901
30.00  72.500      0.003404     0.000000     0.003404
30.00  75.000      0.011172     0.000000     0.011172
30.00  77.500      0.351926     0.353553    -0.001628
30.00  80.000      0.498205     0.500000    -0.001795
30.00  82.500      0.356483     0.353553     0.002929
30.00  85.000      0.011563     0.000000     0.011563
30.00  87.500     -0.000001     0.000000    -0.000001
30.00  90.000      0.000000     0.000000     0.000000
30.00  92.500     -0.000000     0.000000    -0.000000
30.00  95.000     -0.000000     0.000000    -0.000000
30.00  97.500      0.000000     0.000000     0.000000
30.00 100.000     -0.000000     0.000000    -0.000000

  ncall = 1423
```

We again observe that the sine pulse centered at $x = 50$ at $t = 0$ divides into two pulses centered at $x = 20, 80$. Also, the agreement between the numerical and analytical solutions is better than that for the triangular pulse; note, for example, that the numerical peak height is `0.498205`, while the analytical value is `0.500000`. These properties are evident in Figure 8.3 produced by the main program in Listing 8.1.

Finally, we can consider a still smoother solution for a Gaussian pulse that has derivatives of all orders throughout the spatial domain $0 \le x \le 100$.

$$0 \le x \le 100 \quad f(x) = \exp(-a(x-50)^2) \tag{8.8}$$

where a is a prescribed constant.

The programming of the Gaussian pulse in *ua* of Listing 8.3 is the traveling wave solution of Eq. (8.1) for the IC function of Eq. (8.8), produced by the main program of Listing 8.1 with `ncase = 4`.

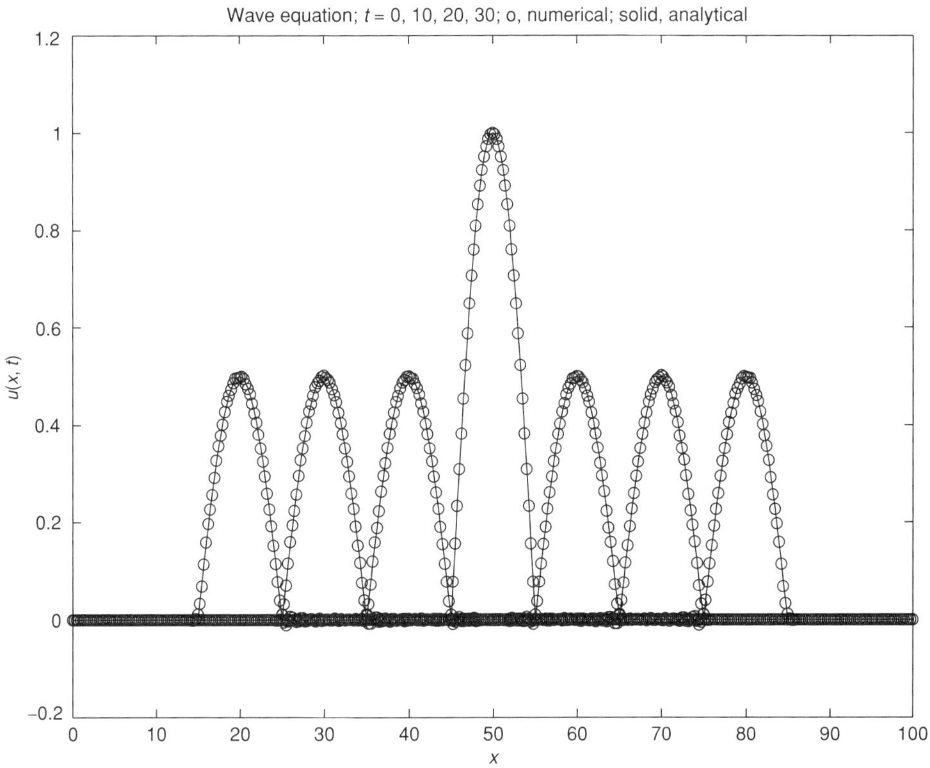

Figure 8.3. Output of main program `pde_1_main` for `ncase = 3`

```
%
% Gaussian pulse
  if(ncase==4)
    a=2.0;
    for i=1:n
      if     x(i)<(45.0-c*t)  pulse1(i)=0.0;
      elseif x(i)>(55.0-c*t)  pulse1(i)=0.0;
      elseif x(i)>=(45.0-c*t) & x(i)<=(55.0-c*t)
             pulse1(i)=(1.0/2.0)*exp(-a*(x(i)-(50.0-c*t))^2/10.0);
      end
      if     x(i)<(45.0+c*t)  pulse2(i)=0.0;
      elseif x(i)>(55.0+c*t)  pulse2(i)=0.0;
      elseif x(i)>=(45.0+c*t) & x(i)<=(55.0+c*t)
             pulse2(i)=(1.0/2.0)*exp(-a*(x(i)-(50.0+c*t))^2/10.0);
      end
      uanal(i)=pulse1(i)+pulse2(i);
    end
  end
```

Table 8.8. Partial Output from pde_1_main and pde_1 (ncase = 4)

```
 ncase =  4,    c =  1

     t        x           u(i,j)    u_anal(i,j)        err(i,j)
  0.00    0.000         0.000000       0.000000        0.000000
  0.00    2.500         0.000000       0.000000        0.000000
  0.00    5.000         0.000000       0.000000        0.000000
  0.00    7.500         0.000000       0.000000        0.000000
  0.00   10.000         0.000000       0.000000        0.000000
                  .                          .
                  .                          .
                  .                          .
  0.00   40.000         0.000000       0.000000        0.000000
  0.00   42.500         0.000000       0.000000        0.000000
  0.00   45.000         0.006738       0.006738        0.000000
  0.00   47.500         0.286505       0.286505        0.000000
  0.00   50.000         1.000000       1.000000        0.000000
  0.00   52.500         0.286505       0.286505        0.000000
  0.00   55.000         0.006738       0.006738        0.000000
  0.00   57.500         0.000000       0.000000        0.000000
  0.00   60.000         0.000000       0.000000        0.000000
                  .                          .
                  .                          .
                  .                          .
  0.00   90.000         0.000000       0.000000        0.000000
  0.00   92.500         0.000000       0.000000        0.000000
  0.00   95.000         0.000000       0.000000        0.000000
  0.00   97.500         0.000000       0.000000        0.000000
  0.00  100.000         0.000000       0.000000        0.000000

                  .                          .
                  .                          .
                  .                          .

            Output at t=10,20 removed from here.

                  .                          .
                  .                          .
                  .                          .

     t        x           u(i,j)    u_anal(i,j)        err(i,j)
 30.00    0.000         0.000000       0.000000        0.000000
 30.00    2.500         0.000000       0.000000        0.000000
 30.00    5.000         0.000000       0.000000        0.000000
 30.00    7.500        -0.000000       0.000000       -0.000000
```

```
30.00  10.000     -0.000000      0.000000     -0.000000
30.00  12.500      0.000000      0.000000      0.000000
30.00  15.000      0.002408      0.003369     -0.000961
30.00  17.500      0.143043      0.143252     -0.000209
30.00  20.000      0.500208      0.500000      0.000208
30.00  22.500      0.143357      0.143252      0.000104
30.00  25.000      0.003029      0.003369     -0.000340
30.00  27.500      0.000008      0.000000      0.000008
30.00  30.000     -0.000209      0.000000     -0.000209
          .                          .
          .                          .
          .                          .
30.00  70.000     -0.000209      0.000000     -0.000209
30.00  72.500      0.000008      0.000000      0.000008
30.00  75.000      0.003029      0.003369     -0.000340
30.00  77.500      0.143357      0.143252      0.000104
30.00  80.000      0.500208      0.500000      0.000208
30.00  82.500      0.143043      0.143252     -0.000209
30.00  85.000      0.002408      0.003369     -0.000961
30.00  87.500      0.000000      0.000000      0.000000
30.00  90.000     -0.000000      0.000000     -0.000000
30.00  92.500     -0.000000      0.000000     -0.000000
30.00  95.000      0.000000      0.000000      0.000000
30.00  97.500      0.000000      0.000000      0.000000
30.00 100.000      0.000000      0.000000      0.000000

 ncall = 1039
```

The constant a is 2, which determines the rate at which the Gaussian pulse Eq. (8.8) decays and for this value, ensures that it is effectively zero near the boundaries $x = 0, 100$ so that for Eq. (8.1) BCs are not required; also, to maintain continuity with the programming of the rectangular, triangular, and sine pulses in Listing 8.3, three sections for `pulse1(i)`, `pulse2(i)` are programmed, but a single section over all `i` could have been used just as well. Again, the traveling wave characteristic of this solution is evident from the use of the arguments $x - ct$ and $x + ct$. Abbreviated numerical output from the main program of Listing 8.1 is given in Table 8.8.

We again observe that the Gaussian pulse centered at $x = 50$ at $t = 0$ divides into two pulses centered at $x = 20, 80$. Also, the agreement between the numerical and analytical solutions is better than that for the triangular and sine pulses; note, for example, that the numerical peak height is `0.500208`, while the analytical value is `0.500000`. These properties are evident in Figure 8.4, produced by the main program in Listing 8.1.

Finally, the ODE routines for the MOL approximation of Eqs. (8.1)–(8.3) remain for some discussion. These routines, `pde_1`, `pde_2`, `pde_3`, called by `ode15s`

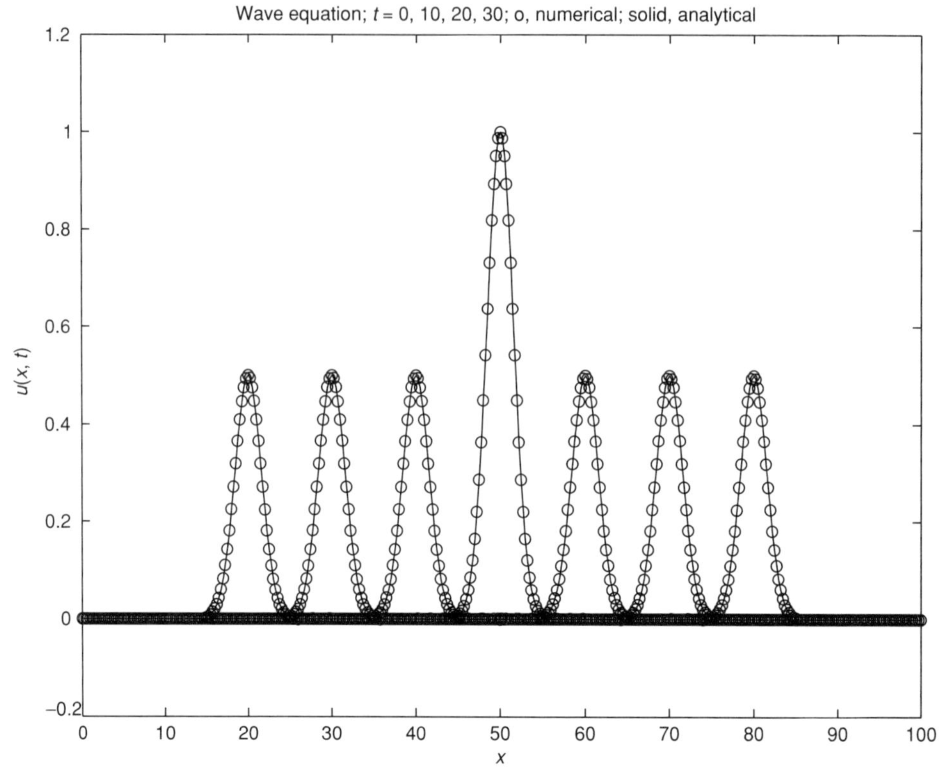

Figure 8.4. Output of main program `pde_1_main` for `ncase = 4`

from the main program of Listing 8.1 (for `mf=1,2,3`, respectively), are, of course, a central part of the MOL solutions discussed previously. However, these routines have been discussed in some detail in earlier PDE chapters, for example, the chapter on the diffusion equation, and they are therefore listed in an appendix at the end of this chapter, with a few explanatory comments.

In summary, we have observed the following:

1. Traveling wave solutions of the linear wave equation (8.1), as illustrated by the d'Alembert solution of Eq. (8.4).
2. The numerical MOL solutions with accuracy that is directly tied to the smoothness of the IC function, $f(x)$, of Eq. (8.2).
3. In particular, for an IC function that is discontinuous, for example, the rectangular pulse, the MOL solution based on FDs has severe numerical distortions (oscillations). This type of problem (with a discontinuous IC) is generally termed a *Riemann problem*, which, as the preceding discussion illustrates, can present significant numerical difficulties that require special numerical techninqes (in the form of a *Riemann solver*). In recent years major advances have been made in the development of so-called *high-resolution methods* that are able to resolve discontinuous solutions, often involving shock waves, to an increasingly high degree of accuracy and without oscillations [4, 5, 7]. A very

interesting and readable review of the state of the art in *computational fluid dynamics* is given in reference [8].

You may then question why a numerical MOL solution is considered for a Riemann problem even if it performs so poorly. The answer is that frequently all we have is a numerical solution since analytical solutions are generally unavailable for realistic physical problems. For example, the Euler equations of fluid mechanics (a system of hyperbolic PDEs) are often solved with a discontinuous IC, or under conditions for which a discontinuity develops (a *shock*). The only tractable approach to a solution to the Euler equations is usually numerical (because of their complexity, especially their nonlinearity). Thus, the development of numerical methods for this important class of (Riemann) problems is a very active area of research.

APPENDIX A

A.1. ODE Routines pde_1, pde_2, pde_3

```
  function ut=pde_1(t,u)
%
% Function pde_1 computes the t derivative vector for the wave
% equation
%
  global ncall     c    xl    xu     n
%
% One vector to two vectors
  for i=1:n
    u1(i)=u(i);
    u2(i)=u(i+n);
  end
%
% Calculate u1xx with ux(1) = ux(n) = 0 BCs
  dxs=((xu-xl)/(n-1))^2;
  for i=1:n
     if(i==1)      u1xx(i)=2.0*(u1(i+1)-u1(i))/dxs;
     elseif(i==n) u1xx(i)=2.0*(u1(i-1)-u1(i))/dxs;
     else          u1xx(i)=(u1(i+1)-2.0*u1(i)+u1(i-1))/dxs;
     end
  end
%
% PDE
  cs=c^2;
  for i=1:n
    u1t(i)=u2(i);
    u2t(i)=cs*u1xx(i);
  end
```

```
%
% Two vectors to one vector
  for i=1:n
    ut(i)  =u1t(i);
    ut(i+n)=u2t(i);
  end
  ut=ut';
%
% Increment calls to pde_1
  ncall=ncall+1;
```

Listing 8.4. Function pde_1 for the MOL Solution of Eq. (8.1) based on explicit FD approximations of u_{xx}

```
  function ut=pde_2(t,u)
%
% Function pde_2 computes the t derivative vector for the wave
% equation
%
  global ncall  ndss    c    xl    xu    n
%
% One vector to two vectors
  for i=1:n
    u1(i)=u(i);
    u2(i)=u(i+n);
  end
%
% Calculate u1x
  if    (ndss== 2) u1x=dss002(xl,xu,n,u1); % second order
  elseif(ndss== 4) u1x=dss004(xl,xu,n,u1); % fourth order
  elseif(ndss== 6) u1x=dss006(xl,xu,n,u1); % sixth order
  elseif(ndss== 8) u1x=dss008(xl,xu,n,u1); % eighth order
  elseif(ndss==10) u1x=dss010(xl,xu,n,u1); % tenth order
  end
%
% BC at x = 0
  u1x(1)=0.0;
%
% BC at x = 1
  u1x(n)=0.0;
%
% Calculate u1xx
  if    (ndss== 2) u1xx=dss002(xl,xu,n,u1x); % second order
  elseif(ndss== 4) u1xx=dss004(xl,xu,n,u1x); % fourth order
```

```
  elseif(ndss== 6) u1xx=dss006(xl,xu,n,u1x); % sixth order
  elseif(ndss== 8) u1xx=dss008(xl,xu,n,u1x); % eighth order
  elseif(ndss==10) u1xx=dss010(xl,xu,n,u1x); % tenth order
  end
%
% PDE
  cs=c^2;
  for i=1:n
    u1t(i)=u2(i);
    u2t(i)=cs*u1xx(i);
  end
%
% Two vectors to one vector
  for i=1:n
    ut(i)  =u1t(i);
    ut(i+n)=u2t(i);
  end
  ut=ut';
%
% Increment calls to pde_2
  ncall=ncall+1;
```

Listing 8.5. Function pde_2 for the MOL solution of Eq. (8.1) based on FD approximations of u_{xx} in dss002 to dss010

```
  function ut=pde_3(t,u)
%
% Function pde_3 computes the t derivative vector for the wave
% equation
%
  global ncall  ndss     c    xl    xu     n
%
% One vector to two vectors
  for i=1:n
    u1(i)=u(i);
    u2(i)=u(i+n);
  end
%
% Calculate u1xx with ux(1) = ux(n) = 0 as BCs
  u1x=zeros(n,1);
  u1x(1)=0.0;
  u1x(n)=0.0;
  nl=2;
  nu-2;
```

```
  if    (ndss==42) u1xx=dss042(xl,xu,n,u1,u1x,nl,nu);
  % second order
  elseif(ndss==44) u1xx=dss044(xl,xu,n,u1,u1x,nl,nu);
  % fourth order
  elseif(ndss==46) u1xx=dss046(xl,xu,n,u1,u1x,nl,nu);
  % sixth order
  elseif(ndss==48) u1xx=dss048(xl,xu,n,u1,u1x,nl,nu);
  % eighth order
  elseif(ndss==50) u1xx=dss050(xl,xu,n,u1,u1x,nl,nu);
  % tenth order
  end
%
% PDE
  cs=c^2;
  for i=1:n
    u1t(i)=u2(i);
    u2t(i)=cs*u1xx(i);
  end
%
% Two vectors to one vector
  for i=1:n
    ut(i)  =u1t(i);
    ut(i+n)=u2t(i);
  end
  ut=ut';
%
% Increment calls to pde_3
  ncall=ncall+1;
```

Listing 8.6. Function pde_3 for the MOL solution of Eq. (8.1) based on FD approximations of u_{xx} in dss042 to dss050

In each of these three routines, there are three sections of code that pertain specifically to Eq. (8.1).

1. The transfer of the incoming dependent-variable vector u into two arrays u1 and u2 (again, because Eq. (8.1) is second order in t).

```
%
% One vector to two vectors
  for i=1:n
    u1(i)=u(i);
    u2(i)=u(i+n);
  end
```

2. The coding of Eq. (8.1) (recall $u \leftrightarrow u1$, $u_t \leftrightarrow u2$).

```
%
% PDE
  cs=c^2;
  for i=1:n
    u1t(i)=u2(i);
    u2t(i)=cs*u1xx(i);
  end
```

3. The transfer of the two derivative arrays `u1t` and `u2t` to the dependent-variable derivative array `ut`.

```
%
% Two vectors to one vector
  for i=1:n
    ut(i)  =u1t(i);
    ut(i+n)=u2t(i);
  end
  ut=ut';
```

In each of the three routines, the basic requirement of computing the ODE derivative vector `ut` as an output from the input dependent variable `u` as an input is accomplished. You may wish to review the calculation of the derivative u_{xx} as explained in earlier listings of these routines with the same names. All of the preceding numerical solutions were computed with `pde_3` (for `mf=3`). Very similar solutions would have resulted using `pde_1` or `pde_2` (since $n = 401$ was determined to be an adequate number of grid points in x to achieve acceptable spatial convergence and `ode15s` provided acceptable accuracy in the t integration using the tolerances specified as input arguments).

REFERENCES

[1] Farlow, S. J. (1993), More on the D'Alembert Equation *Partial Differential Equations for Scientists and Engineers*, Dover Publications, New York, Chapter 18, pp. 137–145

[2] Cajori, F. (1961), *A History of Mathematics*, MacMillan, New York

[3] Kreyszig, E. (1993), *Advanced Engineering Mathematics – Seventh Edition*, Wiley, New York

[4] Leveque, R. J. (2002), *Finite Volume Methods for Hyperbolic Problems*, Cambridge University Press, Cambridge, UK

[5] Shu, C.-W. (1998), Essentially Non-Oscillatory and Weighted Essential Non-oscillatory Schemes for Hyperbolic Conservation Laws, In: B. Cockburn, C. Johnson, C.-W. Shu, and E. Tadmor (Eds.), *Advanced Numerical Approximation of Nonlinear Hyperbolic Equations*, Lecture Notes in Mathematics, vol. 1697, Springer, New York, pp. 325–432

[6] Strauss, W. A. (1992), *Partial Differential Equations: An Introduction*, Wiley, New York
[7] Wesseling, P. (2001), *Principles of Computational Fluid Dynamics*, Springer, New York
[8] Koren, B. (2006), Computational Fluid Dynamics: Science and Tool, In: *Centrum Wiskunde & Informatica (CWI), Modeling, Analysis and Simulation [MAS]*, Report E 0602, ISSN: 1386-3703, January 2006; available online at `ftp://ftp.cwi.nl/pub/CWIreports/MAS/MAS-E0602.pdf`

9

Maxwell's Equations

This partial differential equation (PDE) application introduces the following mathematical concepts and computational methods:

1. We start with a general PDE system in three dimensions (3D) that, with some simplifying assumptions, is reduced to a 1D linear PDE. Here is some terminology:
   (a) The starting point of this application is a classic PDE system, Maxwell's equations of electromagnetic (EM) field theory.
   (b) After reduction of these equations to 1D, followed by some additional simplifications, we arrive at the *damped wave equation* (DWE).
2. The use of coordinate-free PDEs that can then be specialized to a particular coordinate system; for the following analysis, this is Cartesian coordinates.
3. Spatial convergence of the DWE numerical solution by h- and p-refinement.
4. The effect of the method of lines (MOL) finite-difference (FD) approximations on the bandwidth and sparsity of the ordinary differential equation (ODE) Jacobian matrix.
5. A general method for the construction of PDE test problems is illustrated by a specific example for the DWE.
6. The physical significance of the DWE (which we can say without exaggeration is *profound*).

We start the analysis with the *differential form of Maxwell's equations for EM fields*:

$$\nabla \times \mathbf{H} = \mathbf{J} + \mathbf{J_d} = \mathbf{J} + \frac{\partial \mathbf{D}}{\partial t} \tag{9.1a}$$

$$\nabla \times \mathbf{E} = -\frac{\partial \mathbf{B}}{\partial t} \tag{9.1b}$$

$$\nabla \bullet \mathbf{D} = \rho \tag{9.1c}$$

$$\nabla \bullet \mathbf{B} = 0 \tag{9.1d}$$

$$\frac{\partial \rho}{\partial t} + \nabla \bullet \mathbf{J} = 0 \tag{9.1e}$$

Table 9.1. Variables, parameters, and operators in Eqs. (9.1a)–(9.1e)

H	Magnetic field intensity (amps/m)
E	Electric field intensity (volts/m)
D	Electric flux density (coulombs/m^2)
B	Magnetic flux density (webers/m^2)
J	Electric current density (amps/m^2)
$\mathbf{J}_d$	Displacement current density (amps/m^2)
ρ	Charge density (coulombs/m^3)
t	Time (s)
$\times$	Curl vector operator
$\bullet$	Vector dot product
∇	Del vector operator (1/m)

where boldface indicates a vector (except for the vector dot product $\bullet$, which we use for clarity in place of the less apparent dot $\cdot$). The variables, parameters, and operators in Eqs. (9.1a)–(9.1e), with a set of representative units, are given in Table 9.1.

Equation system (9.1a)–(9.1e) clearly has more dependent variables than equations. We therefore use some *constitutive equations* to provide the required additional relationships between the dependent variables:

$$\mathbf{D} = \epsilon \mathbf{E}, \tag{9.1f}$$

$$\mathbf{B} = \mu \mathbf{H}, \tag{9.1g}$$

$$\mathbf{J} = \sigma \mathbf{E} \tag{9.1h}$$

where

ϵ	capacitivity or permittivity (farads/m)
μ	inductivity or permeability (henrys/m)
σ	conductivity (mohs/m)

Equations (9.1a)–(9.1h) are a complete set of PDEs (number of dependent variables = number of equations). We now proceed to combine these equations and finally obtain a single equation in **E**.

If Eqs. (9.1f), (9.1g), and (9.1h) are substituted into Eq. (9.1a),

$$(1/\mu)\nabla \times \mathbf{B} = \sigma \mathbf{E} + \epsilon \frac{\partial \mathbf{E}}{\partial t} \tag{9.2}$$

Differentiation of Eq. (9.2) with respect to t (assuming a linear, homogeneous, isotropic medium) gives

$$(1/\mu)\nabla \times \frac{\partial \mathbf{B}}{\partial t} = \sigma \frac{\partial \mathbf{E}}{\partial t} + \epsilon \frac{\partial^2 \mathbf{E}}{\partial t^2} \tag{9.3}$$

where the order of the LHS differentiation with respect to t and ∇ (space) has been interchanged.

Substitution of Eq. (9.1b) into Eq. (9.3) gives

$$-\nabla \times (\nabla \times \mathbf{E}) = \mu\sigma\frac{\partial \mathbf{E}}{\partial t} + \mu\epsilon\frac{\partial^2 \mathbf{E}}{\partial t^2} \tag{9.4}$$

The identity

$$\nabla \times (\nabla \times \mathbf{J}) = \nabla(\nabla \bullet \mathbf{J}) - \nabla^2 \mathbf{J}$$

with $\nabla \bullet J = 0$ (constant charge density in Eq. (9.1e)) and Eq. (9.1h) gives

$$\nabla \times (\nabla \times \mathbf{E}) = -\nabla^2 \mathbf{E}$$

Substitution into Eq. (9.4) finally gives a single equation for **E**

$$\mu\epsilon\frac{\partial^2 \mathbf{E}}{\partial t^2} + \mu\sigma\frac{\partial \mathbf{E}}{\partial t} = \nabla^2 \mathbf{E} \tag{9.5}$$

Equation (9.5) is the *time-dependent Maxwell equation for the electric field*, **E**. It also applies to **H** and **J** in place of **E**. Equation (9.1h) represents *Ohm's law*. For the case of a nonconductor where $\sigma = 0$, Eq. (9.5) reduces to the *wave equation*.

Equation (9.5) is both *hyperbolic* (from $\partial^2\mathbf{E}/\partial t^2$ and $\nabla^2\mathbf{E}$) and *parabolic* (from $\partial\mathbf{E}/\partial t$ and $\nabla^2\mathbf{E}$). As expressed in terms of ∇, it is *coordinate independent*. If it is reduced to 1D in *Cartesian coordinates*, we have

$$\mu\epsilon\frac{\partial^2 \mathbf{E}}{\partial t^2} + \mu\sigma\frac{\partial \mathbf{E}}{\partial t} = \frac{\partial^2 \mathbf{E}}{\partial x^2} \tag{9.6}$$

Equation (9.6), a *linear, constant-coefficient PDE*, is the starting point for the MOL analysis. It is *second order in t and x*. We take as the two required initial conditions (ICs)

$$\mathbf{E}(x, t = 0) = \cos(\pi x), \tag{9.7a}$$

$$\frac{\partial \mathbf{E}(x, t = 0)}{\partial t} = 0 \tag{9.7b}$$

The two required boundary conditions (BCs) we take to be *homogeneous Neumann* are

$$\frac{\partial \mathbf{E}(x = 0, t)}{\partial x} = 0, \tag{9.8a}$$

$$\frac{\partial \mathbf{E}(x = 1, t)}{\partial x} = 0 \tag{9.8b}$$

Equations (9.6)–(9.8) are the complete specification of the PDE problem for the following MOL analysis.

A main program for the MOL solution of Eqs. (9.6)–(9.8) is given in Listing 9.1 (in accordance with the usual naming convention, the dependent variable is u rather than the E of Eqs. (9.6)–(9.8) in the subsequent programming).

```
%
% Clear previous files
  clear all
  clc
%
%
% Parameters shared with the ODE routine
  global  ncall ndss c1 c2 xl xu n
%
% Select case
%
%   Case 1 - second order FDs
%
%   Case 2 - fourth order FDs
%
%   Case 3 - sixth order FDs
%
  for ncase=1:3
%
% Problem parameters
  eps=1.0;
  mu=1.0;
  sigma=1.0;
  c1=sigma/eps;
  c2=1.0/(mu*eps);
%
% Boundaries, number of grid points
  xl=0.0;
  xu=1.0;
  n=101;
%
% Initial condition
  t0=0.0;
  dx=(xu-xl)/(n-1);
  for i=1:n
    x(i)=xl+(i-1)*dx;
    u0(i)=cos(pi*x(i));
    u0(i+n)=0.0;
  end
%
% Independent variable for ODE integration
  tf=1.0;
  tout=[t0:0.2:tf]';
  nout=6;
  ncall=0;
%
% ODE integration
```

```
  reltol=1.0e-04; abstol=1.0e-04;
  options=odeset('RelTol',reltol,'AbsTol',abstol);
  if    (ncase==1) ndss=42;
  elseif(ncase==2) ndss=44;
  elseif(ncase==3) ndss=46; end
%
% Implicit (sparse stiff) integration
  S=jpattern_num;
  pause
  options=odeset(options,'JPattern',S)
  [t,u]=ode15s(@pde_1,tout,u0,options);
%
% One vector to two vectors
  for it=1:nout
  for i=1:n
    u1(it,i)=u(it,i);
    u2(it,i)=u(it,i+n);
  end
  end
%
% Display selected output
  fprintf('\n  ncase = %2d  ndss = %2d  c1 = %4.2f
         c2 = %4.2f\n\n', ncase,ndss,c1,c2);
  for it=1:nout
    fprintf('     t        x          u(x,t)         ut(x,t)\n');
    for i=1:10:n
      fprintf('%6.2f%8.3f%15.6f%15.6f\n',t(it),x(i), ...
             u1(it,i),u2(it,i));
    end
    fprintf('\n');
  end
  fprintf('  ncall = %4d\n\n',ncall);
%
% Plot numerical solution
  figure(ncase)
  plot(x,u1,'-')
  xlabel('x')
  ylabel('u(x,t)')
  title('Maxwell‘s equation; t = 0, 0.2,... 1.0')
% print -deps -r300 pde.eps; print -dps -r300 pde.ps;
  print -dpng -r300 pde.png
%
% Next case
  end
```

Listing 9.1. Main program pde_1_main.m for the solution of Eqs. (9.6)–(9.8)

We can note the following points about this main program:

1. A *global* area is specified to share parameters with other routines. A `for` loop then steps through three cases for FDs of second, fourth, and sixth order. The parameters of Eq. (9.6) are set numerically and passed to the ODE routine as global parameters `c1` and `c2`. We will later return to the significance of these parameters.

```
%
% Clear previous files
  clear all
  clc
%
%
% Parameters shared with the ODE routine
  global  ncall ndss c1 c2 xl xu n
%
% Select case
%
%    Case 1 - second order FDs
%
%    Case 2 - fourth order FDs
%
%    Case 3 - sixth order FDs
%
  for ncase=1:3
%
% Problem parameters
  eps=1.0;
  mu=1.0;
  sigma=1.0;
  c1=sigma/eps;
  c2=1.0/(mu*eps);
```

2. The boundaries and number of grid points in x are set. Then the grid in x is used to define the ICs (Eqs. 9.7). Note that $u(x, t = 0)$ is stored in array `u0(i)` for $1 \leq i \leq n$ and the derivative $\partial u(x, t = 0)/\partial t$ is also stored in `u0(i)` for $n + 1 \leq i \leq 2n$, so that the 1D array (vector) of dependent variables is initialized as required by `ode15s`.

```
%
% Boundaries, number of grid points
  xl=0.0;
  xu=1.0;
```

```
  n=101;
%
% Initial condition
  t0=0.0;
  dx=(xu-xl)/(n-1);
  for i=1:n
    x(i)=xl+(i-1)*dx;
    u0(i)=cos(pi*x(i));
    u0(i+n)=0.0;
  end
```

3. The t interval is $0 \leq t \leq 1$ with an output interval of 0.2 so that six outputs are displayed (including the initial point $t = 0$). For the three cases ncase=1,2,3 (from the for loop at the beginning), ndss is passed as a global parameter to select one of the differentiation routines dss042, dss044, dss046, respectively (called by the ODE routine pde_1 discussed next). The sparse matrix version of ode15s integrates the n=(2)(101)=202 ODEs using the ODE routine pde_1.

```
%
% Independent variable for ODE integration
  tf=1.0;
  tout=[t0:0.2:tf]';
  nout=6;
  ncall-0;
%
% ODE integration
  reltol=1.0e-04; abstol=1.0e-04;
  options=odeset('RelTol',reltol,'AbsTol',abstol);
  if     (ncase==1) ndss=42;
  elseif(ncase==2) ndss=44;
  elseif(ncase==3) ndss=46; end
%
% Implicit (sparse stiff) integration
  S=jpattern_num;
  pause
  options=odeset(options,'JPattern',S)
  [t,u]=ode15s(@pde_1,tout,u0,options);
```

4. The array u returned by ode15s with the ODE solutions is transformed into two 1D arrays (with $u(x, t)$ and $\partial u(x, t)/\partial t$) to facilitate the handling of the output.

```
%
% One vector to two vectors
  for it=1:nout
  for i=1:n
    u1(it,i)=u(it,i);
    u2(it,i)=u(it,i+n);
  end
  end
```

5. Selected tabular output and plots of the solution are then produced.

```
%
% Display selected output
  fprintf('\n  ncase = %2d  ndss = %2d  c1 = %4.2f
         c2 = %4.2f\n\n', ncase,ndss,c1,c2);
  for it=1:nout
    fprintf('     t       x          u(x,t)          ut(x,t)\n');
    for i=1:10:n
      fprintf('%6.2f%8.3f%15.6f%15.6f\n',t(it),x(i),
           u1(it,i),u2(it,i));
    end
    fprintf('\n');
  end
  fprintf('  ncall = %4d\n\n',ncall);
%
% Plot numerical solution
  figure(ncase)
  plot(x,u1,'-')
  xlabel('x')
  ylabel('u(x,t)')
  title('Maxwell's equation; t = 0, 0.2,... 1.0')
% print -deps -r300 pde.eps; print -dps -r300 pde.ps;
  print -dpng -r300 pde.png
%
% Next case
  end
```

The final `end` statement terminates the `for` loop (for `ncase=1,2,3`).

We next review the output from `pde_1_main` before discussing the ODE routine `pde_1`. The abbreviated tabulated output is given in Table 9.2.

Table 9.2. Selected output from `pde_1_main` for `ncase=1,2,3`

```
options =

             AbsTol: 1.0000e-004
                BDF: []
             Events: []
        InitialStep: []
           Jacobian: []
          JConstant: []
           JPattern: [202x202 double]
               Mass: []
       MassConstant: []
       MassSingular: []
           MaxOrder: []
            MaxStep: []
        NormControl: []
          OutputFcn: []
          OutputSel: []
             Refine: []
             RelTol: 1.0000e-004
              Stats: []
         Vectorized: []
    MStateDependence: []
          MvPattern: []
       InitialSlope: []


 ncase =  1  ndss = 42  c1 = 1.00  c2 = 1.00

     t       x          u(x,t)         ut(x,t)
   0.00    0.000       1.000000        0.000000
   0.00    0.100       0.951057        0.000000
   0.00    0.200       0.809017        0.000000
   0.00    0.300       0.587785        0.000000
   0.00    0.400       0.309017        0.000000
   0.00    0.500       0.000000        0.000000
   0.00    0.600      -0.309017        0.000000
   0.00    0.700      -0.587785        0.000000
   0.00    0.800      -0.809017        0.000000
   0.00    0.900      -0.951057        0.000000
   0.00    1.000      -1.000000        0.000000

     t       x          u(x,t)         ut(x,t)
   0.20    0.000       0.820991       -1.673205
   0.20    0.100       0.780810       -1.591313
   0.20    0.200       0.664197       -1.353649
```

(continued)

Table 9.2 (*continued*)

```
0.20   0.300      0.482567     -0.983482
0.20   0.400      0.253701     -0.517047
0.20   0.500     -0.000000     -0.000000
0.20   0.600     -0.253701      0.517047
0.20   0.700     -0.482567      0.983482
0.20   0.800     -0.664197      1.353649
0.20   0.900     -0.780810      1.591313
0.20   1.000     -0.820991      1.673205
         .                         .
         .                         .
         .                         .

 Output for t = 0.4, 0.6, 0.8 is removed

         .                         .
         .                         .
         .                         .
   t        x         u(x,t)       ut(x,t)
1.00   0.000     -0.602024     -0.077062
1.00   0.100     -0.572559     -0.073288
1.00   0.200     -0.487048     -0.062321
1.00   0.300     -0.353861     -0.045258
1.00   0.400     -0.186035     -0.023821
1.00   0.500      0.000000     -0.000000
1.00   0.600      0.186035      0.023821
1.00   0.700      0.353861      0.045258
1.00   0.800      0.487048      0.062321
1.00   0.900      0.572559      0.073288
1.00   1.000      0.602024      0.077062

ncall =  244

         .                         .
         .                         .
         .                         .

  Options output from ode15s is removed

         .                         .
         .                         .
         .                         .

ncase =  2  ndss = 44  c1 = 1.00  c2 = 1.00

   t        x         u(x,t)       ut(x,t)
0.00   0.000      1.000000      0.000000
```

```
0.00   0.100      0.951057        0.000000
0.00   0.200      0.809017        0.000000
0.00   0.300      0.587785        0.000000
0.00   0.400      0.309017        0.000000
0.00   0.500      0.000000        0.000000
0.00   0.600     -0.309017        0.000000
0.00   0.700     -0.587785        0.000000
0.00   0.800     -0.809017        0.000000
0.00   0.900     -0.951057        0.000000
0.00   1.000     -1.000000        0.000000

   t       x         u(x,t)         ut(x,t)
0.20   0.000      0.820969       -1.673268
0.20   0.100      0.780788       -1.591373
0.20   0.200      0.664178       -1.353702
0.20   0.300      0.482553       -0.983522
0.20   0.400      0.253693       -0.517068
0.20   0.500      0.000000        0.000000
0.20   0.600     -0.253693        0.517068
0.20   0.700     -0.482553        0.983522
0.20   0.800     -0.664178        1.353702
0.20   0.900     -0.780788        1.591373
0.20   1.000     -0.820969        1.673268
          .                          .
          .                          .
          .                          .

 Output for t = 0.4, 0.6, 0.8 is removed

          .                          .
          .                          .
          .                          .
   t       x         u(x,t)         ut(x,t)
1.00   0.000     -0.601934       -0.076826
1.00   0.100     -0.572473       -0.073066
1.00   0.200     -0.486975       -0.062154
1.00   0.300     -0.353808       -0.045157
1.00   0.400     -0.186008       -0.023741
1.00   0.500     -0.000000       -0.000000
1.00   0.600      0.186008        0.023741
1.00   0.700      0.353808        0.045157
1.00   0.800      0.486975        0.062154
1.00   0.900      0.572473        0.073066
1.00   1.000      0.601934        0.076826

ncall =  246
```

(continued)

Table 9.2 *(continued)*

```
       .                      .
       .                      .
       .                      .

   Options output from ode15s is removed

       .                      .
       .                      .
       .                      .

ncase =  3  ndss = 46  c1 = 1.00  c2 = 1.00

   t        x         u(x,t)         ut(x,t)
 0.00    0.000       1.000000       0.000000
 0.00    0.100       0.951057       0.000000
 0.00    0.200       0.809017       0.000000
 0.00    0.300       0.587785       0.000000
 0.00    0.400       0.309017       0.000000
 0.00    0.500       0.000000       0.000000
 0.00    0.600      -0.309017       0.000000
 0.00    0.700      -0.587785       0.000000
 0.00    0.800      -0.809017       0.000000
 0.00    0.900      -0.951057       0.000000
 0.00    1.000      -1.000000       0.000000

   t        x         u(x,t)         ut(x,t)
 0.20    0.000       0.820969      -1.673268
 0.20    0.100       0.780788      -1.591373
 0.20    0.200       0.664178      -1.353702
 0.20    0.300       0.482553      -0.983522
 0.20    0.400       0.253693      -0.517068
 0.20    0.500      -0.000000       0.000000
 0.20    0.600      -0.253693       0.517068
 0.20    0.700      -0.482553       0.983522
 0.20    0.800      -0.664178       1.353702
 0.20    0.900      -0.780788       1.591373
 0.20    1.000      -0.820969       1.673268
       .                      .
       .                      .
       .                      .

  Output for t = 0.4, 0.6, 0.8 is removed

       .                      .
       .                      .
       .                      .
   t        x         u(x,t)         ut(x,t)
 1.00    0.000      -0.601934      -0.076826
```

```
1.00   0.100      -0.572473        -0.073066
1.00   0.200      -0.486975        -0.062154
1.00   0.300      -0.353808        -0.045157
1.00   0.400      -0.186008        -0.023741
1.00   0.500      -0.000000        -0.000000
1.00   0.600       0.186008         0.023741
1.00   0.700       0.353808         0.045157
1.00   0.800       0.486975         0.062154
1.00   0.900       0.572473         0.073066
1.00   1.000       0.601934         0.076826

ncall =  248
```

We can note the following details about this output:

1. The options selected before the ODE integrator, ode15s, is called are summarized. Note in particular the 202 × 202 ODE Jacobian matrix. The structure of this matrix is displayed graphically, as discussed subsequently.

```
options =

             AbsTol: 1.0000e-004
                BDF: []
             Events: []
        InitialStep: []
           Jacobian: []
          JConstant: []
           JPattern: [202x202 double]
               Mass: []
       MassConstant: []
       MassSingular: []
           MaxOrder: []
            MaxStep: []
        NormControl: []
          OutputFcn: []
          OutputSel: []
             Refine: []
             RelTol: 1.0000e-004
              Stats: []
         Vectorized: []
    MStateDependence: []
          MvPattern: []
       InitialSlope: []
```

2. The ICs, Eqs. (9.7a) and (9.7b), are verified. This is an important check (if the ICs are not correct, the solution will certainly not be correct).

```
ncase =  1  ndss = 42  c1 = 1.00  c2 = 1.00

   t       x          u(x,t)       ut(x,t)
0.00    0.000       1.000000      0.000000
0.00    0.100       0.951057      0.000000
0.00    0.200       0.809017      0.000000
0.00    0.300       0.587785      0.000000
0.00    0.400       0.309017      0.000000
0.00    0.500       0.000000      0.000000
0.00    0.600      -0.309017      0.000000
0.00    0.700      -0.587785      0.000000
0.00    0.800      -0.809017      0.000000
0.00    0.900      -0.951057      0.000000
0.00    1.000      -1.000000      0.000000
```

3. The solution evolves in a way that is not easily visualized from the tabulated output, and we will therefore rely on the subsequent plotted output to better understand the solution. However, we can check how the solution varies with the order of the FD approximations. For example, we have from `dss042` (second-order FDs), `dss044` (fourth-order FDs), and `dss046` (sixth-order FDs) the following output at $t = 0.2$.

```
dss042 (ncase = 1):

   t       x          u(x,t)       ut(x,t)
0.20    0.000       0.820991     -1.673205
0.20    0.100       0.780810     -1.591313
0.20    0.200       0.664197     -1.353649
0.20    0.300       0.482567     -0.983482
0.20    0.400       0.253701     -0.517047
0.20    0.500      -0.000000     -0.000000
0.20    0.600      -0.253701      0.517047
0.20    0.700      -0.482567      0.983482
0.20    0.800      -0.664197      1.353649
0.20    0.900      -0.780810      1.591313
0.20    1.000      -0.820991      1.673205

dss044 (ncase = 2):

   t       x          u(x,t)       ut(x,t)
0.20    0.000       0.820969     -1.673268
0.20    0.100       0.780788     -1.591373
```

```
0.20    0.200       0.664178      -1.353702
0.20    0.300       0.482553      -0.983522
0.20    0.400       0.253693      -0.517068
0.20    0.500       0.000000       0.000000
0.20    0.600      -0.253693       0.517068
0.20    0.700      -0.482553       0.983522
0.20    0.800      -0.664178       1.353702
0.20    0.900      -0.780788       1.591373
0.20    1.000      -0.820969       1.673268

dss046 (ncase = 3):

   t       x          u(x,t)         ut(x,t)
0.20    0.000       0.820969      -1.673268
0.20    0.100       0.780788      -1.591373
0.20    0.200       0.664178      -1.353702
0.20    0.300       0.482553      -0.983522
0.20    0.400       0.253693      -0.517068
0.20    0.500      -0.000000       0.000000
0.20    0.600      -0.253693       0.517068
0.20    0.700      -0.482553       0.983522
0.20    0.800      -0.664178       1.353702
0.20    0.900      -0.780788       1.591373
0.20    1.000      -0.820969       1.673268
```

We observe that through this *p-refinement* (change in the order of the FDs approximations),

(a) The agreement between the three solutions is better than four figures (and is six figures for dss044 compared to dss046). Thus, through this p-refinement, we have some confidence that the solutions are accurate to at least four figures (and we arrived at this conclusion without the benefit of an analytical solution).

(b) The symmetry around $x = 0.5$ is as expected (if this was not observed, this would be an indication of an error in the differentiation routines or the MOL solution). Of course, this symmetry also suggests that the solution need only be computed over the interval $0 \leq x \leq 0.5$, which would result in a twofold reduction in the number of ODEs.

(c) The numerical solution requires a modest computational effort.

```
dss042: ncall =  244

dss044: ncall =  246

dss046: ncall =  248
```

Also, at least for this problem, the use of higher-order FDs did not result in a substantially greater number of calls to `pde_1`. However, within each call, the number of arithmetic operations increases with the order of the FDs (the weighted sums in the FDs have more terms with increasing order).

(d) As an incidental point, if the cos in IC (9.7a) is changed to sin, the corresponding numerical solution becomes substantially more difficult to compute. This is because with this change, IC (9.7a) becomes inconsistent with BCs (9.8a) and (9.8b). In other words, the slope of $u(x, t = 0)$ at $x = 0, 1$ is not zero initially from IC (9.7a) (with sin), but changes discontinuously to zero from BCs (9.8a) and (9.8b) for $t > 0$. The discontinuities at the boundaries cause computational difficulties as reflected in the solution (which has some small spatial oscillation) and a substantial increase in `ncall`. Generally, *numerical methods do not like discontinuous functions* (except for those that are designed specifically for discontinuous functions).

The plotted output from `pde_1_main` follows. First, Figure 9.1a is a map of the Jacobian matrix produced by the sparse version of `ode15s` (the Jacobian matrix is required by the ODE integration algorithm in `ode15s` to provide stability for the numerical integration of stiff ODEs).

Figure 9.1a lists the number of the ODE dependent variable (out of 202 ODEs) along the bottom and the number of the ODE down the side. For example, if dependent variable 25 appears in ODE 26, a marker will appear at the coordinates `(25 (horizontal), 26 (vertical))`; thus, the map indicates which dependent variables appear in the RHS of which ODEs.

We can note the following details about the map in Figure 9.1a:

1. The map is sparse with mostly zeros. In fact, only 505 of the 202^2 =40,804 elements of the Jacobian matrix are nonzero (1.238% are nonzero). This degree of sparsity is not unusual in applications, and clearly demonstrates the utility of a sparse matrix integrator such as `ode15s` that works primarily with only the nonzero elements.
2. The nonzero elements appear in bands. The lower left band reflects the FDs of `dss042`; it is three elements wide because the `dss042` FDs use three values of the 202 dependent variables to compute numerical derivatives (they are three-point FD approximations).

The plotted solution is in Figure 9.2, which Figure 9.2 indicates that

1. The solution starts out with a half cosine wave according to IC (9.7a).
2. As expected, the solution is symmetric with respect to $x = 0.5$, so that, again, the solution only over half the interval $0 \le x \le 1$ is actually required and the use of symmetry could reduce the number of ODEs by one half.
3. The solution approaches $u(x, t = \infty) = 0$. This is expected because of the term $\mu\sigma\partial\mathbf{E}/\partial t$ in Eq. (9.6) that provides damping of the solution.

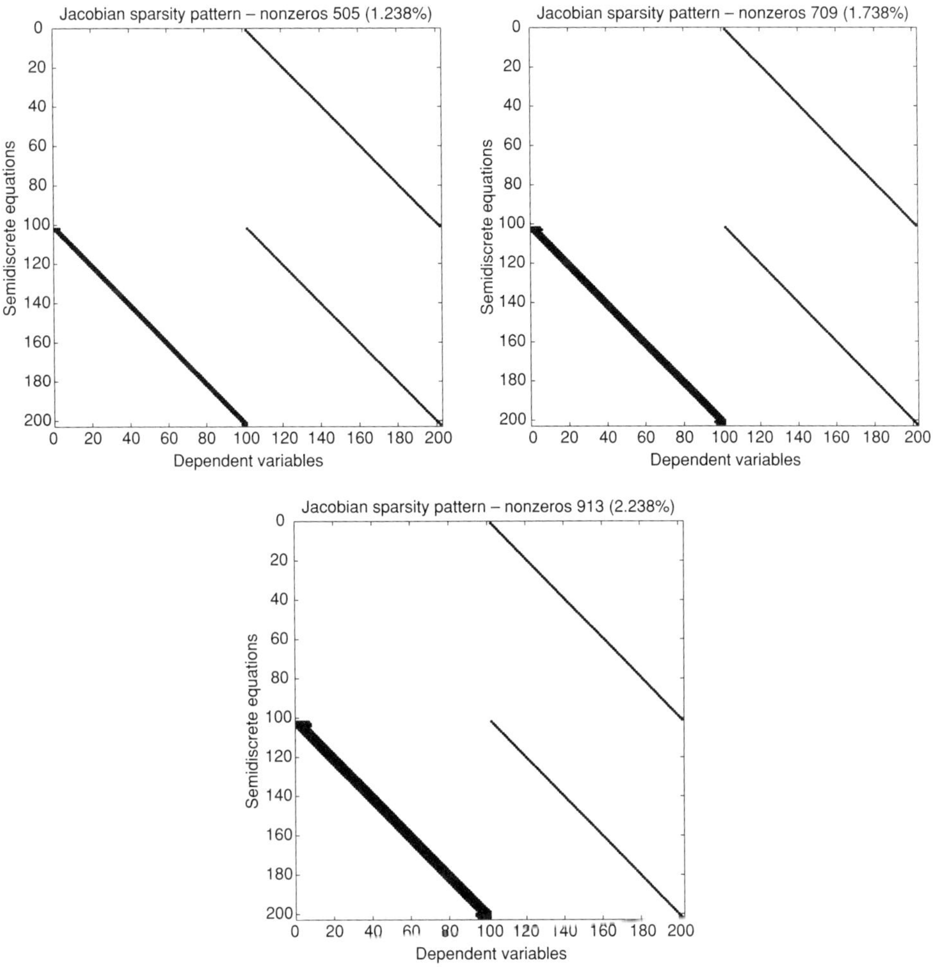

Figure 9.1. Jacobian matrix map for the MOL/ODE approximation of Eqs. (9.6)–(9.8), (a) ncase=1, (b) ncase=2, and (c) ncase=3

To complete the picture of the output from `pde_1_main`, we also include the Jacobian maps and plots for `ncase=2,3` (from `dss044, dss046`) in Figures 9.1b and 9.1c. These figures indicate that the width of the lower left band of the Jacobian map increases with the order of the FD approximation, which is to be expected since the higher order (of the FDs) is achieved by using more values of the dependent variable in calculating the numerical derivatives. Still, the Jacobian matrices are quite sparse.

The plots of the solutions for `ncase=2,3` are indistinguishable from Figure 9.2 for `ncase=1` and are therefore not presented here. However, to elucidate the numerical solution, we also include some 3D plots produced with the following code:

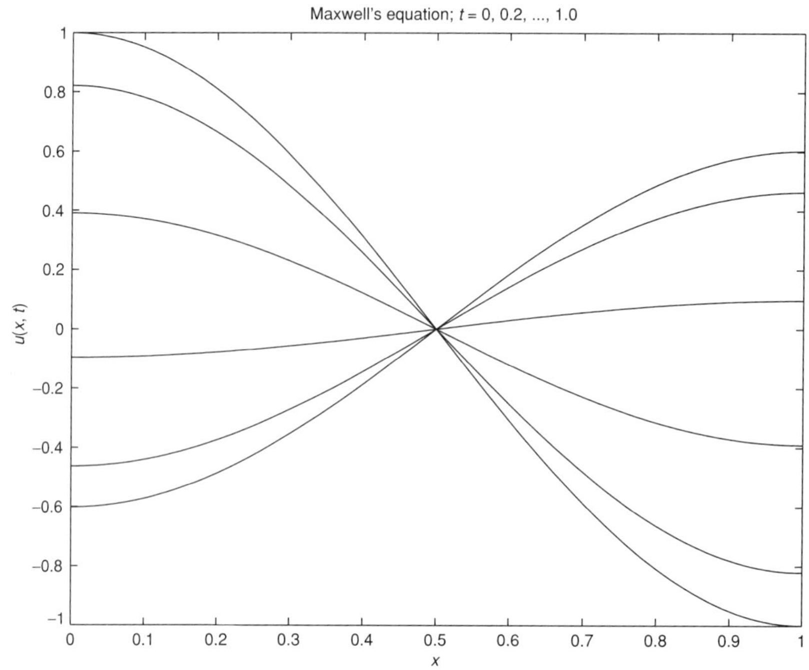

Figure 9.2. Plot of the numerical solution for Eqs. (9.6)–(9.8) from pde_1_main and pde_1 for ncase=1

```
% 3D Plots

figure(3);
surfl(x,t,u1); shading interp;
title('E(x,t)');
xlabel('x');
ylabel('t');
zlabel('E(x,t)');
axis tight
print -dpng -r300 fig3.png;

figure(4);
surfl(x,t,u2); shading interp;
title('\partial/\partialt E(x,t)');
xlabel('x');
ylabel('t');
zlabel('\partial/\partialt E(x,t)');
axis tight
rotate3d on
print -dpng -r300 fig4.png;
```

The resulting 3D plots are shown in Figures 9.3 and 9.4.

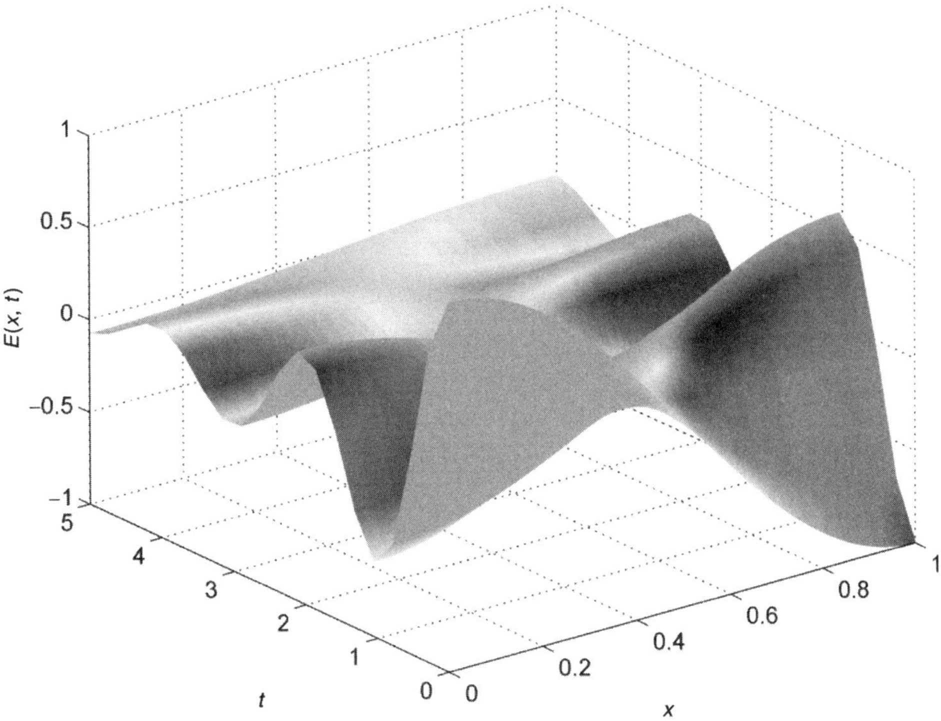

Figure 9.3. MOL/ODE solution of Eqs. (9.6)–(9.8), ncase=1, u1(x,t)

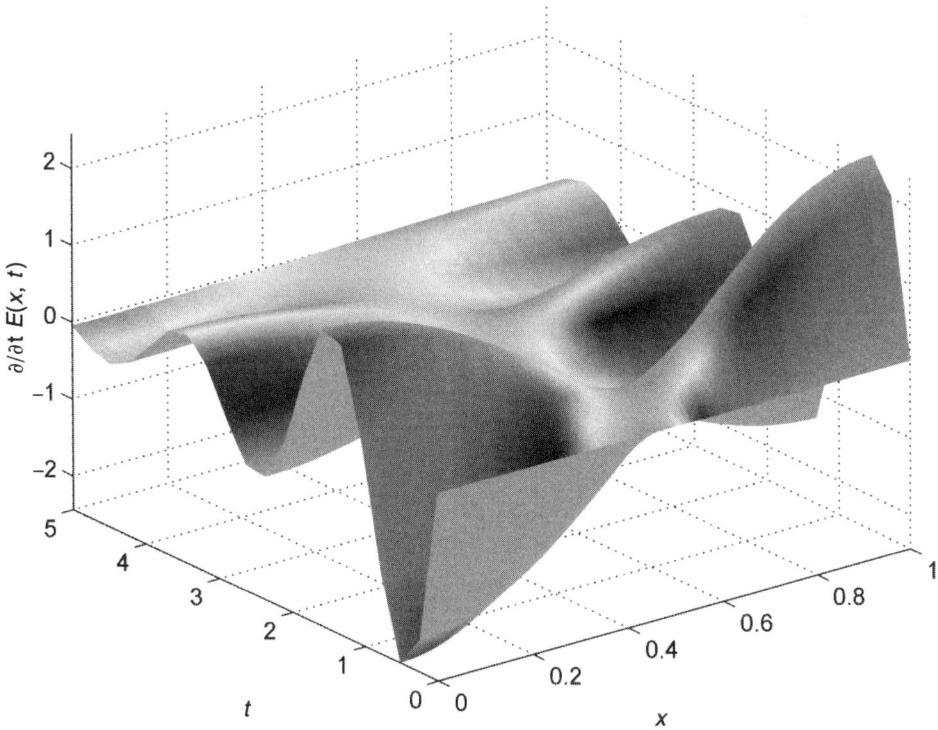

Figure 9.4. MOL/ODE solution of Eqs. (9.6)–(9.8), ncase=1, u2(x,t)

The MOL implementation of Eqs. (9.6) and (9.8) is in the ODE routine pde_1 (see Listing 9.2).

```
  function ut=pde_1(t,u)
%
% Function pde_1 computes the t derivative vector for
% Maxwell's equation
%
  global  ncall ndss c1 c2 xl xu n
%
% One vector to two vectors
  for i=1:n
    u1(i)=u(i);
    u2(i)=u(i+n);
  end
%
% Calculate u1xx with ux(1) = ux(n) = 0 as BCs
  u1x(1)=0.0;
  u1x(n)=0.0;
  nl=2;
  nu=2;
  if     (ndss==42) u1xx=dss042(xl,xu,n,u1,u1x,nl,nu);
  % second order
  elseif(ndss==44) u1xx=dss044(xl,xu,n,u1,u1x,nl,nu);
  % fourth order
  elseif(ndss==46) u1xx=dss046(xl,xu,n,u1,u1x,nl,nu);
  % sixth order
  elseif(ndss==48) u1xx=dss048(xl,xu,n,u1,u1x,nl,nu);
  % eighth order
  elseif(ndss==50) u1xx=dss050(xl,xu,n,u1,u1x,nl,nu);
  % tenth order
  end
%
% PDE
  for i=1:n
    u1t(i)=u2(i);
    u2t(i)=-c1*u1t(i)+c2*u1xx(i);
  end
%
% Two vectors to one vector
  for i=1:n
    ut(i)   =u1t(i);
    ut(i+n)=u2t(i);
  end
```

```
  ut=ut';
%
% Increment calls to pde_1
  ncall=ncall+1;
```

Listing 9.2. pde_1 for the MOL solution of Eqs. (9.6)–(9.8)

We can note the following details about pde_1:

1. After the function is defined and a *global* area is set up to share parameters with the main program of Listing 9.1, the ODE dependent-variable vector u is divided into two vectors for $u(x,t)$ = u1) and $\partial u(x,t)/\partial t$ (= u2), respectively.

```
  function ut=pde_1(t,u)
%
% Function pde_1 computes the t derivative vector for
% Maxwell's equation
%
  global  ncall ndss c1 c2 xl xu n
%
% One vector to two vectors
  for i=1:n
    u1(i)=u(i);
    u2(i)=u(i+n);
  end
```

   Equation (9.6), which is second order in t, is expressed as two PDEs first order in t (u = u1, $\partial u/\partial t$ = u2). This is a general procedure required for ODE/PDEs higher than first order in the initial value t since library ODEs integrators such as ode15s can accommodate only first-order ODEs; this procedure for replacing a ODE/PDE nth order in t with n ODE/PDEs first order in t is almost always possible and is easy to implement, so the limitation of library ODEs to first-order equations is not really a significant restriction.
2. The second derivative $\partial^2 u(x,t)/\partial x^2$ (= u1xx) is calculated by a call to one of the differentiation routines dss042, dss044, dss046 (for ncase=1,2,3, respectively). As this is done, BCs (9.8a) and (9.8b) are included.

```
%
% Calculate u1xx with ux(1) = ux(n) = 0 as BCs
  u1x(1)=0.0;
  u1x(n)=0.0;
  nl=2;
  nu=2;
```

```
if     (ndss==42) u1xx=dss042(xl,xu,n,u1,u1x,nl,nu);
% second order
elseif(ndss==44) u1xx=dss044(xl,xu,n,u1,u1x,nl,nu);
% fourth order
elseif(ndss==46) u1xx=dss046(xl,xu,n,u1,u1x,nl,nu);
% sixth order
elseif(ndss==48) u1xx=dss048(xl,xu,n,u1,u1x,nl,nu);
% eighth order
elseif(ndss==50) u1xx=dss050(xl,xu,n,u1,u1x,nl,nu);
% tenth order
end
```

3. The two first-order (in t) PDEs representing Eq. (9.6) are programmed.

```
%
% PDE
  for i=1:n
    u1t(i)=u2(i);
    u2t(i)=-c1*u1t(i)+c2*u1xx(i);
  end
  ut=ut';
%
% Increment calls to pde_1
  ncall=ncall+1;
```

Note that $\partial u/\partial t$ = `u1t` = `u2`, $\partial^2 u/\partial t^2$ = `u2t`.

4. The usual transpose is then included and the number of calls, `ncall`, to `pde_1` is incremented.

Finally, `S=jpattern_num` is listed in Listing 9.3 without a detailed discussion.

```
  function S=jpattern_num
%
% Set global variables
     global n
%
% Sparsity pattern of the Jacobian matrix based on a
% numerical evaluation
%
% Set independent, dependent variables for the calculation
% of the sparsity pattern
  tbase=0;
  for i=1:n
```

```
    ybase(i)=0.5;
    ybase(i+n)=0.5;
  end
  ybase=ybase';
%
% Compute the corresponding derivative vector
  ytbase=pde_1(tbase,ybase);
  fac=[];
  thresh=1e-16;
  vectorized='on';
  [Jac,fac]=numjac(@pde_1,tbase,ybase,ytbase,thresh,fac, ...
                   vectorized);
%
% Replace nonzero elements by "1" (so as to create a "0-1"
% map of the Jacobian matrix)
  S=spones(sparse(Jac));
%
% Plot the map
  figure
  spy(S);
  xlabel('dependent variables');
  ylabel('semi-discrete equations');
%
% Compute the percentage of non-zero elements
  [njac,mjac]=size(S);
  ntotjac=njac*mjac;
  non_zero=nnz(S);
  non_zero_percent=non_zero/ntotjac*100;
  stat=sprintf('Jacobian sparsity pattern - nonzeros
             %d (%.3f%%)', non_zero,non_zero_percent);
  title(stat);
```

Listing 9.3. S=jpattern_num Called by the Sparse Option of ode15s

We note a few points about S=jpattern_num:

1. It is set up for two PDEs (Eq. (9.6) expressed as two first-order PDEs, each approximated over n grid points), but can be readily extended to other numbers of PDEs.

```
  for i=1:n
    ybase(i)=0.5;
    ybase(i+n)=0.5;
  end
```

2. The Jacobian matrix is mapped using Matlab routine `numjac` that calls the ODE routine `pde_1` of Listing 9.2.

```
[Jac,fac]=numjac(@pde_1,tbase,ybase,ytbase,thresh, ...
                 fac,vectorized);
```

3. The remainder of `S=jpattern_num` plots the Jacobian map and computes and displays the number of nonzero elements (as in Figures 9.1a–9.1c).

In summary, our intention in presenting this chapter is to indicate how:

1. A particular PDE, in this case Maxwell's equation for the time-dependent electric field, $E(x, t)$, can be obtained as a special case of a more general PDE system, in this case the Maxwell's field equations.
2. The use of the differentiation routines facilitates changing the order of the FD approximations; all that is required is to change the number of the routine since the arguments remain the same. Since the order of numerical approximations is often given the symbol p, this procedure of changing the order of the FDs is termed *p-refinement*.
3. Additionally, the number of grid points can be changed (in `pde_1_main` of Listing 9.1). Since the grid spacing in numerical approximations is often given the symbol h, changing the number of grid points is termed *h-refinement*.
4. The effect of h- and p-refinement on the numerical solution can be observed through repeated solutions. If the solution remains essentially unchanged, for example, as in Table 9.2, then reasonable *spatial convergence* can be inferred. Note that this procedure does not require knowledge of an analytical solution (but, of course, it is not a mathematical proof of convergence).

We conclude with two further ideas:

1. We did not use an analytical solution to evaluate the numerical solution in the preceding discussion; an analytical solution to Eqs. (9.6)–(9.8) would not be difficult to derive. However, we consider just briefly how analytical solutions might generally be derived that can be useful for testing numerical solutions of PDEs. This is done through an example application for Eqs. (9.6)–(9.8).
   (a) We begin by assuming an analytical solution that satisfies as much of the PDE problem as we can include a priori. For example, for Eq. (9.6), we might assume a solution of the form

$$E(x, t) = e^{at} \cos bt \cos \pi x \tag{9.9}$$

   where a and b are constants to be determined. The choice of the form of Eq. (9.9) is dictated by
   (i) e^{at}, $\cos bt$, and $\cos \pi x$, which when substituted in Eq. (9.6) differentiate to e^{at}, $\cos bt$, $\sin bt$, and $\cos \pi x$ that can then be manipulated so as to evaluate a and b in Eq. (9.9) (this is explained later).
   (ii) $\cos \pi x$ satisfies BCs (9.8a) and (9.8b) (and therefore Eq. (9.9) satisfies BCs (9.8a) and (9.8b) because they are homogeneous).
   (iii) $\mathbf{E}(x, t)$ from Eq. (9.9) satisfies IC (9.7a).

(b) Substitution of Eq. (9.9) into Eq. (9.6) gives

Term from Eq. (9.6)	Term from Eq. (9.9)
$\mu\epsilon\dfrac{\partial^2\mathbf{E}}{\partial t^2}$	$\mu\epsilon(-b^2\, e^{at}\, \cos bt - ab\, e^{at}\, \sin bt$ $-ab\, e^{at}\, \cos bt + a^2\, e^{at}\, \cos bt) \cos \pi x$
$\mu\sigma\dfrac{\partial\mathbf{E}}{\partial t}$	$\mu\sigma(-b\, e^{at}\, \sin bt + a\, e^{at}\, \cos bt) \cos \pi x$
$-\dfrac{\partial^2\mathbf{E}}{\partial x^2}$	$-\pi^2(e^{at}\, \cos bt) \cos \pi x$
$\sum = 0$	$\sum = 0(?)$

Collecting terms in $\sin bt$, we have

$$\mu\epsilon(-ab) + \mu\sigma(-b) = 0$$

or (assuming $b \neq 0$)

$$a = -\sigma/\epsilon \tag{9.10a}$$

Collecting terms in $\cos bt$, we have

$$\mu\epsilon(-b^2 - ab + a^2) + \mu\sigma(a) - \pi^2 = 0$$

or rearranging as a quadratic in b,

$$\mu\epsilon b^2 + a\mu\epsilon b - (\mu\epsilon a^2 - a\mu\sigma + \pi^2) = 0 \tag{9.10b}$$

Thus, Eqs. (9.10a) and (9.10b) can be used to solve for a and b in Eq. (9.9).

(c) So far, Eq. (9.9) satisfies Eqs. (9.6), (9.7a), and (9.8). Unfortunately, it does not satisfy Eq. (9.7b), and in fact, from Eq. (9.9)

$$\frac{\partial\mathbf{E}(x, t = 0)}{\partial t} = a \cos \pi x \tag{9.7c}$$

Thus, Eq. (9.9) is a solution to Eqs. (9.6), (9.7a), (9.7c), and (9.8) (and is therefore not a solution to the original problem, Eqs. (9.6), (9.7a), (9.7b), and (9.8). However, the aforementioned process is useful as it does provide an analytical solution to this related problem. Thus, it can be used as an alternative problem to test a candidate numerical algorithm prior to using it to solve the original problem. If the candidate algorithm provides an accurate solution to the related problem, it is likely to provide an accurate solution to the original problem – although this is not guaranteed.

(d) More generally, this approach to an analytical solution involves starting with a proposed analytical solution (e.g., Eq. (9.9)) and then differentiating it to arrive at a PDE (e.g., Eq. (9.6)) and associated set of ICs and BCs (e.g., Eqs. (9.7a), (9.7c), and (9.8)). This approach has been used extensively to generate analytical solutions that can be used to test numerical PDE solutions.

2. To conclude, the second idea begins with the special case $\sigma = 0$, for which Eq. (9.6) reduces to the *wave equation*.

$$\frac{\partial^2 \mathbf{E}}{\partial t^2} = \frac{1}{\mu\epsilon}\frac{\partial^2 \mathbf{E}}{\partial x^2} \tag{9.11}$$

But the coefficient of the derivative $\partial^2\mathbf{E}/\partial x^2$ is the square of the velocity, c, of the wave described by Eq. (9.11), that is,

$$c = \frac{1}{\sqrt{\mu\epsilon}} \tag{9.12}$$

In other words, EM waves travel at the velocity c given by Eq. (9.12) (here we have considered the electric field $\mathbf{E}$, but Eq. (9.6) also applies to the magnetic field $\mathbf{H}$, or in other words, Eq. (9.6) applies to EM fields). If we use values for μ and ϵ for free space, $\mu = 4\pi \times 10^{-7}$ H/m, $\epsilon = 10^{-9}/(36\pi)$ F/m, we have from Eq. (9.12)

$$c = \frac{1}{\sqrt{(4\pi \times 10^{-7})(10^{-9}/(36\pi))}} = 3 \times 10^8$$

which is the *speed of light in free space* in m/s. Thus, we come to the remarkable conclusion that Maxwell's equations predict that *the speed of EM waves is equal to the speed of light*; in fact, *light waves are EM waves*.

To bring these last two ideas together, for $\sigma = 0$, $a = 0$ and $b = \pi c$ from Eqs. (9.10a) and (9.10b), we obtain velocity c from Eq. (9.12). Thus, IC (9.7b) and IC (9.7c) are the same (for $a = 0$) and Eq. (9.9) becomes an analytical solution for the original problem, Eqs. (9.6), (9.7a), (9.7b), and (9.8). In other words, this analytical solution can be used to test the numerical solution from `pde_1_main` (with `sigma=0.0;`), `pde_1` and `S=jpattern_num`. This numerical solution will oscillate indefinitely, since for $\sigma = 0$, there is no damping in Eq. (9.6).

10

Elliptic Partial Differential Equations: Laplace's Equation

This partial differential equation (PDE) application introduces the following mathematical concepts and computational methods:

1. Laplace's equation is
   (a) a classical PDE with many physical applications;
   (b) a PDE in two dimensions (2D), and therefore the following analysis demonstrates the method of lines (MOL) solution of 2D PDEs;
   (c) an elliptic PDE so that the following analysis demonstrates the application of the MOL to elliptic PDEs;
   (d) a basic PDE with a variety of extensions, for example, Poisson's equation and Helmholtz's equation.
2. Dirichlet and Neumann boundary conditions implemented on a 2D domain.
3. Evaluation of the accuracy of the numerical solution, including a comparison between the numerical and analytical solutions.
4. Application of h- and p-refinement to achieve spatial convergence of the numerical solution.
5. Introduction to the concept of continuation for the solution of elliptic PDEs and other important classes of problems.

Laplace's equation is

$$\frac{\partial^2 u}{\partial x^2} + \frac{\partial^2 u}{\partial y^2} = 0 \tag{10.1}$$

where x and y are boundary-value (spatial) independent variables. In subscript notation, Eq. (10.1) is

$$u_{xx} + u_{yy} = 0$$

Equation (10.1) is classified as an *elliptic* PDE; it is a *boundary-value problem only* (since it does not have an initial-value independent variable).

Since the MOL generally involves the integration of a system of initial-value ODEs, it cannot be applied directly to Eq. (10.1) (again, Eq. (10.1) does not have an initial-value independent variable). Thus, to perform a MOL solution of Eq. (10.1), we add a derivative with respect to the initial-value variable t to form a *parabolic PDE*:

$$\frac{\partial u}{\partial t} = \frac{\partial^2 u}{\partial x^2} + \frac{\partial^2 u}{\partial y^2} \tag{10.2}$$

Equation (10.2) is the *heat conduction equation*. It can be integrated to a *steady state* or *equilibrium condition*, corresponding to $t \rightarrow \infty$ for which $\partial u/\partial t \approx 0$, and then Eq. (10.2) reverts to Eq. (10.1).

Equation (10.2) requires one *"pseudo" initial condition* (IC) in t (since it is first order in t) and two boundary conditions (BCs) in x and y (since it is second order in x and y). We use the term "pseudo" because t in Eq. (10.2) is not part of the original problem of Eq. (10.1). Thus, we select the IC arbitrarily with the expectation that for $t \rightarrow \infty$, this IC will not determine the final equilibrium solution of Eq. (10.1); multiple solutions may be possible for different ICs, but we choose the IC as close as possible to the final solution of interest (admittedly, a rather imprecise procedure that can be tested by observing how the solution approaches the final equilibrium solution). For this purpose, we choose just a piecewise constant function, as explained subsequently.

$$u(x, y, t = 0) = u_0 \tag{10.3}$$

For the BCs in x, we use the *homogeneous Neumann conditions*

$$\frac{\partial u(x = 0, y, t)}{\partial x} = 0 \tag{10.4}$$

$$\frac{\partial u(x = 1, y, t)}{\partial x} = 0 \tag{10.5}$$

For the BCs in y, we use the *Dirichlet conditions*

$$u(x, y = 0, t) = 0 \tag{10.6}$$

$$u(x, y = 1, t) = 1 \tag{10.7}$$

Equations (10.2)–(10.7) for $t \rightarrow \infty$ have the analytical solution

$$u(x, y, t = \infty) = y \tag{10.8}$$

which will be used to evaluate the accuracy of the solution to Eq. (10.1) (subject to BCs (10.4)–(10.7).

A main program for the MOL solution of Eqs. (10.2)–(10.7) is given in Listing 10.1.

```
%
% Clear previous files
  clear all
  clc
%
% Parameters shared with the ODE routine
  global ncall nx ny ndss mmf

% Initial condition (which also is consistent with the
% boundary conditions)
  nx=11;
  ny=11;
  for i=1:nx
  for j=1:ny
    if(j==1)     u0(i,j)=0.0;
    elseif(j==ny)u0(i,j)=1.0;
    else         u0(i,j)=0.5;
    end
  end
  end
%
% Matrix conversion method flag
%
%   mmf = 1 - explicit subscripting for matrix conversion
%
%   mmf = 2 - Matlab reshape function
%
  mmf=2;
%
% 2D to 1D matrix conversion
%
  if(mmf==1)
    for i=1:nx
    for j=1:ny
      y0((i-1)*ny+j)=u0(i,j);
    end
    end
  end
%
  if(mmf==2)
    y0=reshape(u0',1,nx*ny);
  end
%
% Independent variable for ODE integration
  t0=0.0;
```

```
  tf=0.15;
  tout=[t0:0.03:tf]';
  nout=6;
  ncall=0;
%
% ODE integration
  mf=2;
  reltol=1.0e-04; abstol=1.0e-04;
  options=odeset('RelTol',reltol,'AbsTol',abstol);
  if(mf==1) % explicit FDs
    [t,y]=ode15s(@pde_1,tout,y0,options); end
  if(mf==2) ndss=4; % ndss = 2, 4, 6, 8 or 10 required
    [t,y]=ode15s(@pde_2,tout,y0,options); end
  if(mf==3) ndss=44; % ndss = 42, 44, 46, 48 or 50 required
    [t,y]=ode15s(@pde_3,tout,y0,options); end
%
% 1D to 2D matrix conversion (plus t), with 3D plotting at
% t = 0, 0.03, ..., 0.15
%
% Composite 3D plots showing evolution of solution
  figure(1);
  for it=1:nout
%
% 2D to 3D matrix conversion
    u(it,:,:)=reshape(y(it,:),ny,nx)';
    subplot(3,2,it)
    uview(:,:)=u(it,:,:);
    surf(uview);
  end
%
% 3D plot of initial condition
  figure(2);
  uview(:,:)=u(1,:,:);
  surf(uview);
  xlabel('y grid number');
  ylabel('x grid number');
  zlabel('u(x,y)');
  s1=sprintf('Laplace Equation - MOL Solution');
  s2=sprintf('Parametric plot at t=0 - initial condition');
  title([{s1}, {s2}], 'fontsize', 12);
  rotate3d on;
%
% 3D plot of final solution
  figure(3);
  uview(:,:)=u(nout,:,:);
  surf(uview);
  xlabel('y grid number');
```

```
  ylabel('x grid number');
  zlabel('u(x,y)');
  s1=sprintf('Laplace Equation - MOL Solution');
  s2=sprintf('Parametric plot at t=%4.2f - final solution', ...
             tout(nout));
  title([{s1}, {s2}], 'fontsize', 12);
  rotate3d on;
%
% Display selected output
  for it=1:nout
    fprintf('\n\n t = %5.2f',t(it));
    fprintf('\n x across (x=0,0.2,...,1.0)');
    fprintf('\n y  down  (y=1.0,0.8,...,0)');
  for j=ny:-2:1
    fprintf('\n%10.3f%10.3f%10.3f%10.3f%10.3f%10.3f', ...
            u(it,1:2:nx,j));
  end
  end
  fprintf('\n\n ncall = %4d\n',ncall);
```

Listing 10.1. Main program pde_1_main.m

We can note the following points about this program:

1. After some variables and parameters are declared *global* so that they can be shared with the MOL ordinary differential equation (ODE) routine, IC Eq. (10.3) is coded as $u(x, y, t = 0) = 0.5$ over a 11×11 grid in x and y; note, however, that along the boundary $y = 0$, $u(x, y = 0, t) = 0$, and along the boundary $y = 1$, $u(x, y = 1, t) = 1$, which is consistent with the BCs (Eqs. (10.6) and (10.7). In other words, we program the IC to be consistent with the BCs; the reason for this will be explained in the subsequent discussion.

```
%
% Clear previous files
  clear all
  clc
%
% Parameters shared with the ODE routine
  global ncall nx ny ndss mmf

% Initial condition (which also is consistent with the
% boundary conditions)
  nx=11;
  ny=11;
  for i=1:nx
```

```
    for j=1:ny
      if(j==1)       u0(i,j)=0.0;
      elseif(j==ny)u0(i,j)=1.0;
      else           u0(i,j)=0.5;
      end
    end
    end
```

The choice of nx=11 and ny=11 and grid points in x and y was made rather arbitrarily. Thus we have the requirement to determine if these numbers are large enough to give sufficient accuracy in the MOL solution of Eq. (10.1). Of course, we should not choose these numbers of grid points to be larger than necessary to achieve acceptable accuracy since this would merely result in longer computer runs to achieve unnecessary accuracy in the numerical solution of Eq. (10.1). We can ascertain the adequacy of nx=11 and ny=1 in at least three ways:

(a) The numerical solution can be compared with the analytical solution (Eq. (10.8)) (although in general we will not have an analytical solution to evaluate the numerical solution).
(b) The numbers of grid points can be changed and the effect on the numerical solution can be observed (since the grid spacing in numerical approximations is often given the symbol "h," this is referred to as *h-refinement*).
(c) The order of the approximations of the derivatives in x and y in Eqs. (10.1) and (10.2) can be changed and the effect on the numerical solution observed (since the order of numerical approximations is often given the symbol "p," this is referred to as *p-refinement*).

2. Since library ODE integrators, such as the Matlab integrators, generally require all of the dependent variables to be arranged in a single 1D vector, we must convert the 2D IC matrix u0(i,j) ($= u(x, y, t = 0)$) of Eq. (10.2) into a 1D vector, in this case y0. This is done in two equivalent ways: (a) the explicit programming of the matrix subscripting (mmf=1) and (b) the use of a Matlab utility, reshape, (mmf=2).

```
%
% Matrix conversion method flag
%
%   mmf = 1 - explicit subscripting for matrix conversion
%
%   mmf = 2 - Matlab reshape function
%
  mmf=2;
%
% 2D to 1D matrix conversion
%
  if(mmf==1)
```

```
    for i=1:nx
    for j=1:ny
      y0((i-1)*ny+j)=u0(i,j);
    end
    end
  end
%
  if(mmf==2)
    y0=reshape(u0',1,nx*ny);
  end
```

The details of this conversion in a pair of nested `for` loops (`mmf=1`) should be studied since this is an essential operation for the solution of 2D PDEs (with straightforward extensions to 3D PDEs). This can be done by considering the successive passes through the two `for` loops: for example, for `i=1, j=1,2,...,ny`; `i=2, j=1,2,...,ny`; `...;i=nx, j=1,2,...,ny`, so that there are $nx \times ny = (11)(11) = 121$ passes through the two `for` loops. In other words, for each increasing value of `i` in the outer `for` loop, `ny` values are assigned to `y0` through the inner `for` loop; thus, for `nx` passes through the outer `for` loop, 121 values are assigned to `y0`.

The Matlab `reshape` function allows the user to manipulate an array in order to change the number of rows and columns, but without changing the overall number of elements. Here we use the `reshape` function to change the 2D matrix `u0` to the 1D matrix (or vector) `y0`, which is then an input to the Matlab integrator `ode15s` (as discussed subsequently).

The relative merits of the two approaches for the 2D to 1D conversion are as follows:

(a) For `mmf=1`, the approach is general in the sense that it illustrates the basic operations for the 2D to 1D conversion, and can therefore be used in any procedural programming language. Also, this coding clarifies what takes place in `reshape`.

(b) For `mmf=1`, the execution is modestly more efficient (faster) than for `mmf=2`. This becomes clear when running the two cases (`mmf=1,2`).

(c) For `mmf=2`, the code is more compact (one line vs. six lines for `mmf=1`).

3. The total interval in t is defined ($0 \le t \le 0.15$). The solution is to be displayed six times, so the output interval is 0.03.

```
%
% Independent variable for ODE integration
  t0=0.0;
  tf=0.15;
  tout=[t0:0.03:tf]';
  nout=6;
  ncall=0;
```

4. The ODE integration (a total of 11 × 11 ODEs) is integrated by Matlab integrator ode15s; we chose a stiff integrator only because the MOL approximating ODEs are frequently stiff (although we did not test this idea by, for example, also using a nonstiff integrator and comparing the performance through the final value of ncall).

```
%
% ODE integration
  mf=2;
  reltol=1.0e-04; abstol=1.0e-04;
  options=odeset('RelTol',reltol,'AbsTol',abstol);
  if(mf==1) % explicit FDs
    [t,y]=ode15s(@pde_1,tout,y0,options); end
  if(mf==2) ndss=4; % ndss = 2, 4, 6, 8 or 10 required
    [t,y]=ode15s(@pde_2,tout,y0,options); end
  if(mf==3) ndss=44; % ndss = 42, 44, 46, 48 or 50 required
    [t,y]=ode15s(@pde_3,tout,y0,options); end
```

Three cases are programmed:

(a) mf = 1: Routine pde_1 is called by ode15s with the explicit programming of the finite-difference (FDs) approximations of the derivatives $\partial^2 u/\partial x^2$ and $\partial^2 u/\partial y^2$ in Eqs. (10.1) and (10.2). The details of pde_1 are discussed subsequently.

(b) mf = 2: Routine pde_2 is called by ode15s with the calculation of FD approximations of the derivatives $\partial^2 u/\partial x^2$ and $\partial^2 u/\partial y^2$ in Eqs. (10.1) and (10.2) by a library routine for first-order derivatives, for example, DSS004. The second-order derivatives are computed by successive calculation of first-order derivatives (termed *stagewise differentiation*). The details of pde_2 are discussed subsequently.

(c) mf = 3: Routine pde_3 is called by ode15s with the calculation of FD approximations of the derivatives $\partial^2 u/\partial x^2$ and $\partial^2 u/\partial y^2$ in Eqs. (10.1) and (10.2) by a library routine for second-order derivatives, for example, DSS044. The details of pde_3 are discussed subsequently.

5. After integration of the ODEs by ode15s, the solution is returned as a 2D matrix y; the two dimensions of y are for t with a first dimension of 6 and a combination of x and y with a second dimension of 121. This matrix is then expressed as a 2D matrix in x and y at a series of nout=6 values of t through the index it, so that a parametric series of surface plots can be produced through the Matlab routine surf.

```
%
% 1D to 2D matrix conversion (plus t), with 3D plotting at
% t = 0, 0.03, ..., 0.15
```

```
%
% Composite 3D plots showing evolution of solution
  figure(1);
  for it=1:nout
%
% 2D to 3D matrix conversion
    u(it,:,:)=reshape(y(it,:),ny,nx)';
    subplot(3,2,it)
    uview(:,:)=u(it,:,:);
    surf(uview);
  end
%
% 3D plot of initial condition
  figure(2);
  uview(:,:)=u(1,:,:);
  surf(uview);
  xlabel('y grid number');
  ylabel('x grid number');
  zlabel('u(x,y)');
  s1=sprintf('Laplace Equation - MOL Solution');
  s2=sprintf('Parametric plot at t=0 - initial condition');
  title([{s1}, {s2}], 'fontsize', 12);
  rotate3d on;
%
% 3D plot of final solution
  figure(3);
  uview(:,:)=u(nout,:,:);
  surf(uview);
  xlabel('y grid number');
  ylabel('x grid number');
  zlabel('u(x,y)');
  s1=sprintf('Laplace Equation - MOL Solution');
  s2=sprintf('Parametric plot at t=%4.2f-final solution', ...
             tout(nout));
  title([{s1}, {s2}], 'fontsize', 12);
  rotate3d on;
```

6. Additionally, tabular output is displayed so that the numerical solution can be compared with the analytical solution (Eq. (10.8)).

```
%
% Display selected output
  for it=1:nout
    fprintf('\n\n t = %5.2f',t(it));
```

```
    fprintf('\n x across (x=0,0.2,...,1.0)');
    fprintf('\n y  down  (y=1.0,0.8,...,0)');
  for j=ny:-2:1
    fprintf('\n%10.3f%10.3f%10.3f%10.3f%10.3f%10.3f',...
            u(it,1:2:nx,j));
  end
  end
  fprintf('\n\n ncall = %4d\n',ncall);
```

The three ODE routines called by ode15s, pde_1.m, pde_2.m, pde_3.m, are now discussed. A listing of pde_1.m is given in Listing 10.2.

```
  function yt=pde_1(t,y)
%
% Function pde_1 computes the temporal derivative in the
% pseudo transient solution of Laplace's equation by explicit
% finite differences
%
% Problem parameters
  global ncall nx ny mmf
  xl=0.0;
  xu=1.0;
  yl=0.0;
  yu=1.0;
%
% Initially zero derivatives in x, y, t
  uxx=zeros(nx,ny); uyy=zeros(nx,ny); ut=zeros(nx,ny);
%
% 1D to 2D matrix conversion
  if(mmf==1)
    for i=1:nx
    for j=1:ny
      u(i,j)=y((i-1)*ny+j);
    end
    end
  end
  if(mmf==2)
    u=reshape(y,ny,nx)';
  end
%
% PDE
  dx2=((xu-xl)/(nx-1))^2;
  dy2=((yu-yl)/(ny-1))^2;
  for i=1:nx
```

```
  for j=1:ny
%
%   uxx
    if(i==1)      uxx(i,j)=2.0*(u(i+1,j)-u(i,j))/dx2;
    elseif(i==nx) uxx(i,j)=2.0*(u(i-1,j)-u(i,j))/dx2;
    else          uxx(i,j)=(u(i+1,j)-2.0*u(i,j)+u(i-1,j))/dx2;
    end
%
%   uyy
    if(j~=1)&(j~=ny)
       uyy(i,j)=(u(i,j+1)-2.0*u(i,j)+u(i,j-1))/dy2;
    end
%
%   ut = uxx + uyy
    if(j==1)|(j==ny)ut(i,j)=0.0;
    else ut(i,j)=uxx(i,j)+uyy(i,j);
    end
  end
  end
%
% 2D to 1D matrix conversion
  if(mmf==1)
    for i=1:nx
    for j=1:ny
      yt((i-1)*ny+j)=ut(i,j);
    end
    end
    yt=yt';
  end
  if(mmf==2)
    yt=reshape(ut',nx*ny,1);
  end
%
% Increment calls to pde_1
  ncall=ncall+1;
```

Listing 10.2. ODE routine pde_1.m

We can note the following points about pde_1.m:

1. The definition of pde_1, function yt=pde_1(t,y), indicates two important points:
   (a) The input arguments, t,y, the ODE system independent variable, and dependent-variable vector, t and y respectively, are available for the programming in pde_1.
   (b) The output argument, yt, is the vector of dependent-variable derivatives (with respect to t) that must be defined numerically before the execution of pde_1 is completed. Thus, if y is of length n, then all of the n elements of

yt must be defined (failing to evaluate even one of these derivatives will generally produce an incorrect solution; this may seem obvious, but since in this case there are $n = 121$ dependent variables, failure to evaluate all of the derivatives in yt can easily happen). Some problem parameters are then declared as *global* so that they can be shared with pde_1_main in Listing 10.1, or other routines. In this case, the dimensions of the 2D $x - y$ domain are defined. We also set all of the PDE derivatives to zero (via the Matlab utility zeros) to minimize the effect of failing to set any of these derivatives.

```
  function yt=pde_1(t,y)
%
% Function pde_1 computes the temporal derivative in the
% pseudo transient solution of Laplace's equation by explicit
% finite differences
%
% Problem parameters
  global ncall nx ny mmf
  xl=0.0;
  xu=1.0;
  yl=0.0;
  yu=1.0;
%
% Initially zero derivatives in x, y, t
  uxx=zeros(nx,ny); uyy=zeros(nx,ny); ut=zeros(nx,ny);
```

2. A conversion of the 1D vector, y, to the 2D matrix u is made so that the subsequent programming can be done in terms of the problem-oriented variable u in Eqs. (10.1) and (10.2). Although this conversion is, in principle, not required, programming in terms of problem-oriented variables, for example, u, is a major convenience. Again, we follow two approaches: (a) explicit programming of the conversion and (b) use of the Matlab utility reshape, depending on the value of mmf set in the main program of Listing 10.1.

```
%
% 1D to 2D matrix conversion
  if(mmf==1)
    for i=1:nx
    for j=1:ny
      u(i,j)=y((i-1)*ny+j);
    end
    end
```

```
      end
      if(mmf==2)
        u=reshape(y,ny,nx)';
      end
```

3. The programming of the MOL ODEs is next, starting with the square of the increments for the FD approximations in x and y.

```
%
% PDE
      dx2=((xu-xl)/(nx-1))^2;
      dy2=((yu-yl)/(ny-1))^2;
```

4. We then move throughout the $x - y$ domain using a pair of nested `for` loops. Each pass through this pair of `for` loops executes the programming for another ODE (a total of 121 passes or ODEs).

```
      for i=1:nx
      for j=1:ny
%
%       uxx
        if(i==1)       uxx(i,j)=2.0*(u(i+1,j)-u(i,j))/dx2;
        elseif(i==nx)  uxx(i,j)=2.0*(u(i-1,j)-u(i,j))/dx2;
        else           uxx(i,j)=(u(i+1,j)-2.0*u(i,j)+u(i-1,j))/dx2;
        end
%
%       uyy
        if(j~=1)&(j~=ny)
           uyy(i,j)=(u(i,j+1)-2.0*u(i,j)+u(i,j-1))/dy2;
        end
%
%       ut = uxx + uyy
        if(j==1)|(j==ny)ut(i,j)=0.0;
        else ut(i,j)=uxx(i,j)+uyy(i,j);
        end
      end
      end
```

The coding of the FD approximations requires some explanation. For the derivative $\partial^2 u/\partial x^2$ = uxx, three cases are required:

(a) For $x = 0$, BC (10.4) is applied, so the FD approximation of $\partial^2 u(x = 0, y)/\partial x^2$ is

```
if(i==1)uxx(i,j)=2.0*(u(i+1,j)-u(i,j))/dx2;
```

In other words, the BC is applied as `u(0,j) = u(2,j)` to eliminate the fictitious value `u(0,j)`.

(b) For $x = 1$, BC (10.5) is applied, so the FD approximation of $\partial^2 u(x = 1, y)/\partial x^2$ is

```
elseif(i==nx)uxx(i,j)=2.0*(u(i-1,j)-u(i,j))/dx2;
```

The BC is applied as `u(nx+1,j) = u(nx-1,j)` to eliminate the fictitious value `u(nx+1,j)`.

(c) For the interior points `i = 2` to `i = nx-1`, the FD approximation of the second derivative $\partial^2 u(x, y)/\partial x^2$ is

```
else uxx(i,j)=(u(i+1,j)-2.0*u(i,j)+u(i-1,j))/dx2;
```

Similarly, the coding of the FD approximation of $\partial^2 u/\partial y^2$ = `uyy` requires some explanation.

(a) For $\partial^2 u/\partial y^2$ at $y \neq 0, 1$, the coding is

```
if(j~=1)&(j~=ny)uyy(i,j)=(u(i,j+1)-2.0*u(i,j)+u(i,j-1))/dy2; end
```

that is, the usual FD approximation for a second-order derivative.

(b) For $y = 0, 1$, BCs (10.6) and (10.7) apply. The constant values $u(x, y = 0, t) = 0$, $u(x, y = 1, t) = 1$ are implemented as $\partial u(x, y = 0, t)/\partial t = 0$, $\partial u(x, y = 1, t)/\partial t = 0$ or `if(j==1)|(j==ny)ut(i,j)=0.0;`

(c) Thus, the values of $u(x, y = 0, t) = 0$, $u(x, y = 1, t) = 1$ from BCs (10.6) and (10.7) set in the coding of IC (10.3) (see this coding in `pde_1_main` in Listing 10.1) remain at their prescribed values in `pde_1` (as reflected in the zero derivatives in t). This is the rationalization for making the ICs and BCs consistent. To explain a bit further, ODE integrator `ode15s` does not permit setting the dependent variables in `pde_1`, for example,

```
if(j==1) u(i,j)=0.0;
if(j==ny)u(i,j)=1.0;
```

If this coding were used, *it would have no effect*. In other words, the *dependent variables can be defined only in terms of their derivatives* in `pde_1`, for example, `if(j==1)|(j==ny)ut(i,j)=0.0;`

(d) For the remaining grid points for which special conditions are not imposed, Eq. (10.2) is programmed as `else ut(i,j)=uxx(i,j)+uyy(i,j);`. The resemblance of this code to Eq. (10.2) is clear.

5. Finally, the 2D matrix `ut` is converted to a 1D vector required by `ode15s`, and since `ode15s` requires a column derivative vector, a transpose of `ut` is added at the end for `mmf=1`. The counter for the calls to `pde_1`, `ncall` is incremented (the final value of this counter at the end of the solution is displayed by `pde_1_main` and gives an indication of the computational effort required to produce the solution).

```
%
% 2D to 1D matrix conversion
  if(mmf==1)
    for i=1:nx
    for j=1:ny
      yt((i-1)*ny+j)=ut(i,j);
    end
    end
    yt=yt';
  end
  if(mmf==2)
    yt=reshape(ut',nx*ny,1);
  end
%
% Increment calls to pde_1
  ncall=ncall+1;
```

The numerical output from `pde_1_main` of Listing 10.1 and `pde_1` of Listing 10.2 for `mf=1`, mmf=1 is given in Table 10.1. The approach to the analytical solution of Eq. (10.8) for large t is clear (if the code runs to a larger final time (`tf > 0.15`), the numerical solution agrees with Eq. (10.8) to four figures).

Figure 10.1 produced by `surf` indicates how the solution evolves with each iteration. The intermediate figures at $t = 0.03, 0.06, \ldots, 0.012$ show clearly the transition from IC (10.3) (the piecewise constant function) to the final solution of Eq. (10.8) as the integration proceeds.

In addition, for clarity, the first and last figures produced by `surf` are indicated in Figure 10.2 (corresponding to `t = 0` and `t = 0.15`). The IC of Figure 10.2a programmed in `pde_1_main` of Listing 10.1, is in Figure 10.2a and the final solution of Eq. (10.8) is in Figure 10.2b.

Table 10.1. Numerical output for the solution of Eqs. (10.2)–(10.7) from pde_1, with mf=1, mmf=1

```
t =  0.00
x across (x=0,0.2,...,1.0)
y  down  (y=1.0,0.8,...,0)
    1.000     1.000     1.000     1.000     1.000     1.000
    0.500     0.500     0.500     0.500     0.500     0.500
    0.500     0.500     0.500     0.500     0.500     0.500
    0.500     0.500     0.500     0.500     0.500     0.500
    0.500     0.500     0.500     0.500     0.500     0.500
    0.000     0.000     0.000     0.000     0.000     0.000

t =  0.03
x across (x=0,0.2,...,1.0)
y  down  (y=1.0,0.8,...,0)
    1.000     1.000     1.000     1.000     1.000     1.000
    0.706     0.706     0.706     0.706     0.706     0.706
    0.545     0.545     0.545     0.545     0.545     0.545
    0.455     0.455     0.455     0.455     0.455     0.455
    0.294     0.294     0.294     0.294     0.294     0.294
    0.000     0.000     0.000     0.000     0.000     0.000

t =  0.06
x across (x=0,0.2,...,1.0)
y  down  (y=1.0,0.8,...,0)
    1.000     1.000     1.000     1.000     1.000     1.000
    0.770     0.770     0.770     0.770     0.770     0.770
    0.582     0.582     0.582     0.582     0.582     0.582
    0.418     0.418     0.418     0.418     0.418     0.418
    0.230     0.230     0.230     0.230     0.230     0.230
    0.000     0.000     0.000     0.000     0.000     0.000

t =  0.09
x across (x=0,0.2,...,1.0)
y  down  (y=1.0,0.8,...,0)
    1.000     1.000     1.000     1.000     1.000     1.000
    0.791     0.791     0.791     0.791     0.791     0.791
    0.594     0.594     0.594     0.594     0.594     0.594
    0.406     0.406     0.406     0.406     0.406     0.406
    0.209     0.209     0.209     0.209     0.209     0.209
    0.000     0.000     0.000     0.000     0.000     0.000

t =  0.12
x across (x=0,0.2,...,1.0)
y  down  (y=1.0,0.8,...,0)
    1.000     1.000     1.000     1.000     1.000     1.000
```

```
    0.797      0.797      0.797      0.797      0.797      0.797
    0.598      0.598      0.598      0.598      0.598      0.598
    0.402      0.402      0.402      0.402      0.402      0.402
    0.203      0.203      0.203      0.203      0.203      0.203
    0.000      0.000      0.000      0.000      0.000      0.000

t =  0.15
x across (x=0,0.2,...,1.0)
y  down  (y=1.0,0.8,...,0)
    1.000      1.000      1.000      1.000      1.000      1.000
    0.799      0.799      0.799      0.799      0.799      0.799
    0.599      0.599      0.599      0.599      0.599      0.599
    0.401      0.401      0.401      0.401      0.401      0.401
    0.201      0.201      0.201      0.201      0.201      0.201
    0.000      0.000      0.000      0.000      0.000      0.000

ncall =  203
```

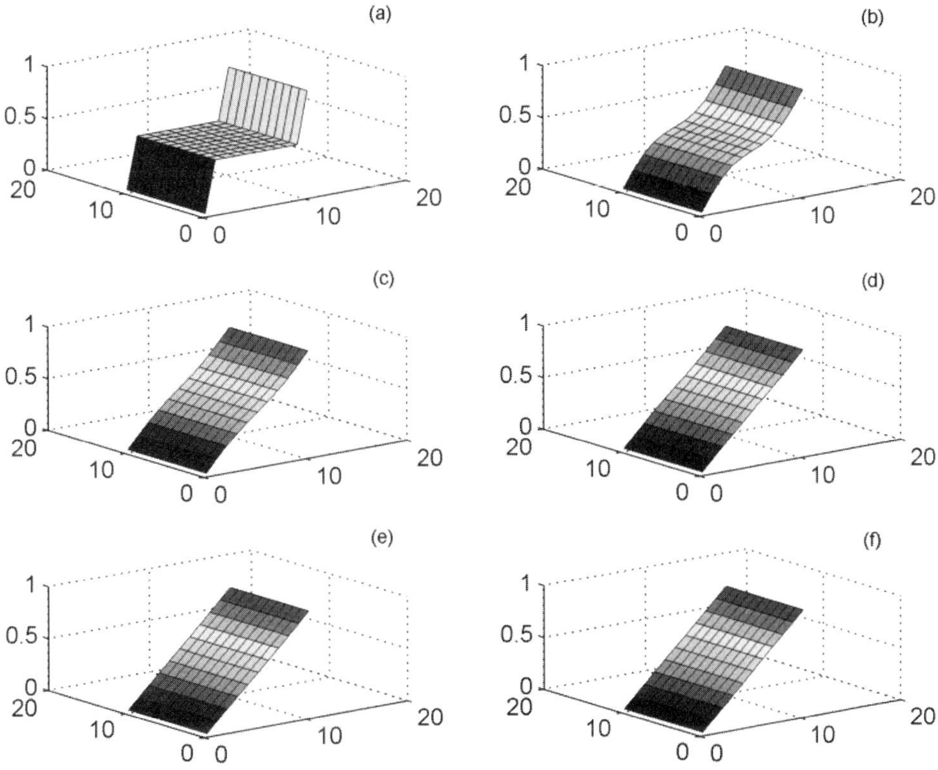

Figure 10.1. Output of pde_1_main and pde1_1 as the solution evolves. The individual plots correspond to (a) $t = 0$; (b) $t = 0.03$; (c) $t = 0.06$; (d) $t = 0.09$; (e) $t = 0.0.12$; (f) $t = 0.15$

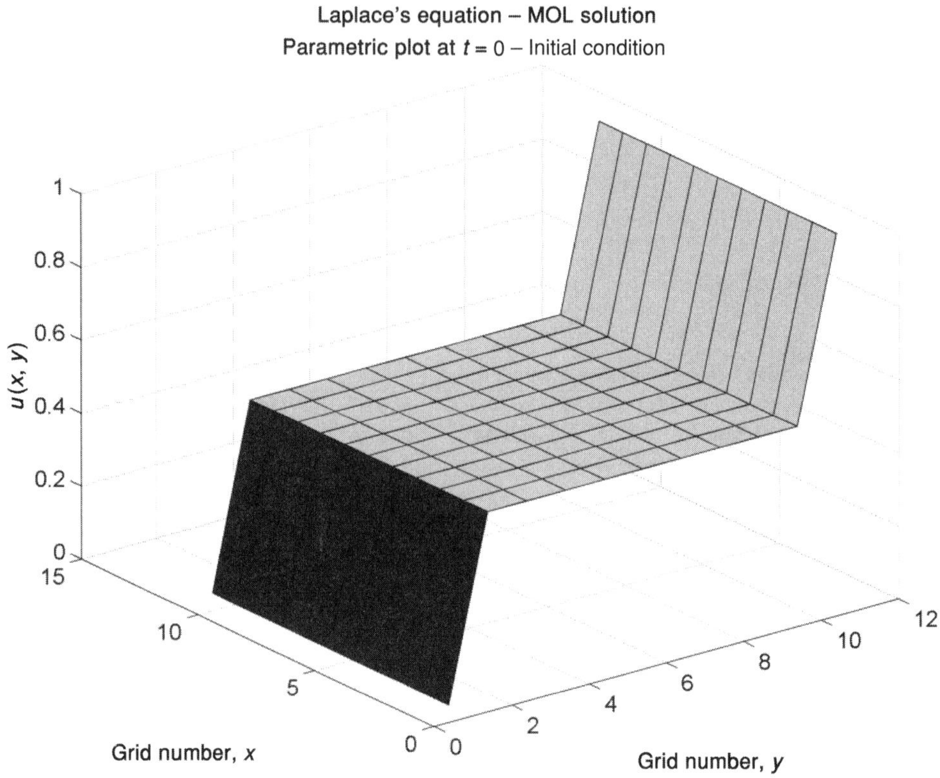

(a)

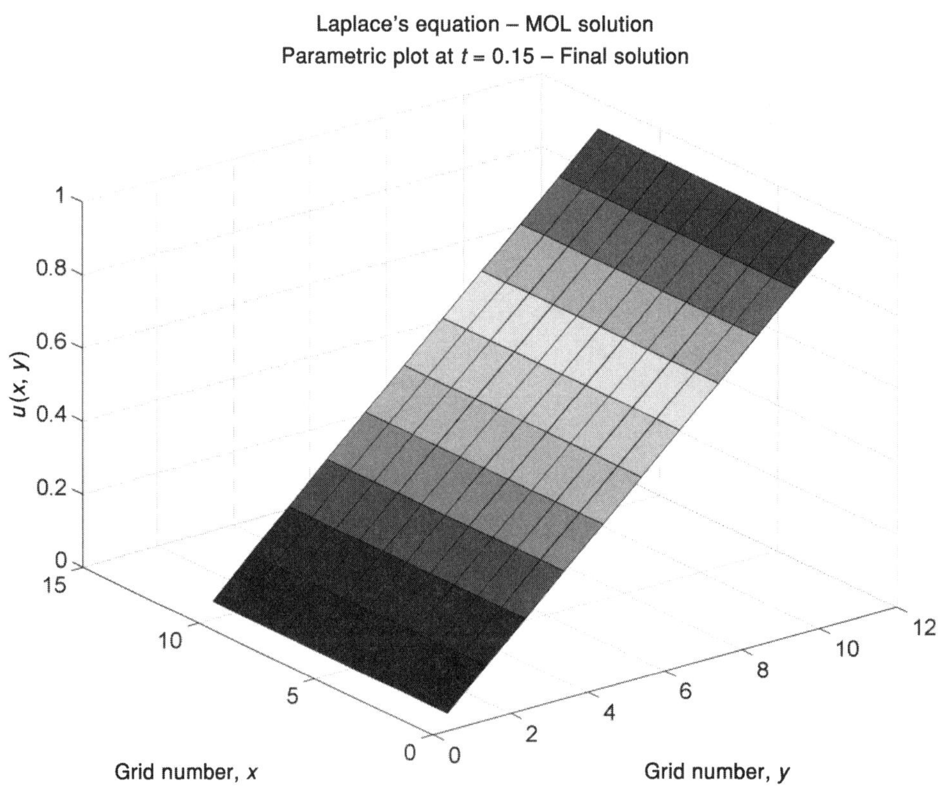

(b)

Figure 10.2. (a) Output of pde_1_main and pde1_1 at $t = 0$; (b) output of pde_1_main and pde1_1 at $t = 0.15$

For mf=2 in pde_1_main, pde_2 is the ODE routine called by ode15s (See Listing 10.3).

```
  function yt=pde_2(t,y)
%
% Function pde_2 computes the temporal derivative in the
% pseudo transient solution of Laplace's equation by
% stagewise differentiation
%
%
% Problem parameters
  global ncall nx ny ndss mmf
  xl=0.0;
  xu=1.0;
  yl=0.0;
  yu=1.0;
%
% 1D to 2D matrix conversion
  if(mmf==1)
    for i=1:nx
    for j=1:ny
      u(i,j)=y((i-1)*ny+j);
    end
    end
  end
  if(mmf==2)
    u=reshape(y,ny,nx)';
  end
%
% PDE
%
% ux
  for j=1:ny
    u1d=u(:,j);
    if     (ndss== 2) ux1d=dss002(xl,xu,nx,u1d); % second order
    elseif(ndss== 4) ux1d=dss004(xl,xu,nx,u1d); % fourth order
    elseif(ndss== 6) ux1d=dss006(xl,xu,nx,u1d); % sixth order
    elseif(ndss== 8) ux1d=dss008(xl,xu,nx,u1d); % eighth order
    elseif(ndss==10) ux1d=dss010(xl,xu,nx,u1d); % tenth order
    end
%
% uxx
    ux1d(1) =0.0;
    ux1d(nx)=0.0;
    if     (ndss== 2) uxx1d=dss002(xl,xu,nx,ux1d); % second order
    elseif(ndss== 4) uxx1d=dss004(xl,xu,nx,ux1d); % fourth order
```

```
    elseif(ndss== 6) uxx1d=dss006(xl,xu,nx,ux1d); % sixth order
    elseif(ndss== 8) uxx1d=dss008(xl,xu,nx,ux1d); % eighth order
    elseif(ndss==10) uxx1d=dss010(xl,xu,nx,ux1d); % tenth order
    end
%
% 1D to 2D
    uxx(:,j)=uxx1d(:);
  end
%
% uy
  for i=1:nx
    u1d=u(i,:);
    if     (ndss== 2) uy1d=dss002(yl,yu,ny,u1d); % second order
    elseif(ndss== 4) uy1d=dss004(yl,yu,ny,u1d); % fourth order
    elseif(ndss== 6) uy1d=dss006(yl,yu,ny,u1d); % sixth order
    elseif(ndss== 8) uy1d=dss008(yl,yu,ny,u1d); % eighth order
    elseif(ndss==10) uy1d=dss010(yl,yu,ny,u1d); % tenth order
    end
%
% uyy
    if     (ndss== 2) uyy1d=dss002(yl,yu,ny,uy1d); % second order
    elseif(ndss== 4) uyy1d=dss004(yl,yu,ny,uy1d); % fourth order
    elseif(ndss== 6) uyy1d=dss006(yl,yu,ny,uy1d); % sixth order
    elseif(ndss== 8) uyy1d=dss008(yl,yu,ny,uy1d); % eighth order
    elseif(ndss==10) uyy1d=dss010(yl,yu,ny,uy1d); % tenth order
    end
%
% 1D to 2D
    uyy(i,:)=uyy1d(:);
  end
%
% ut = uxx + uyy
  ut=uxx+uyy;
  ut(:,1) =0.0;
  ut(:,ny)=0.0;
%
% 2D to 1D matrix conversion
  if(mmf==1)
    for i=1:nx
    for j=1:ny
      yt((i-1)*ny+j)=ut(i,j);
    end
    end
    yt=yt';
  end
  if(mmf==2)
```

```
    yt=reshape(ut',nx*ny,1);
  end
%
% Increment calls to pde_2
  ncall=ncall+1;
```

Listing 10.3. ODE routine pde_2.m

We can note the following details about pde_2:

1. The function is defined, selected parameters and variables are declared as global, the spatial dimensions are defined, and the conversion of the 1D matrix y to the 2D matrix u is made so that programming in terms of problem-oriented variables is set up.

```
  function yt=pde_2(t,y)
%
% Function pde_2 computes the temporal derivative in the
% pseudo transient solution of Laplace's equation by stagewise
% differentiation
%
%
% Problem parameters
  global ncall nx ny ndss mmf
  xl=0.0;
  xu=1.0;
  yl=0.0;
  yu=1.0;
```

2. A 1D to 2D conversion allows programming in terms of the problem-oriented dependent variable, u.

```
%
% 1D to 2D matrix conversion
  if(mmf==1)
    for i=1:nx
    for j=1:ny
      u(i,j)=y((i-1)*ny+j);
    end
    end
  end
  if(mmf==2)
    u=reshape(y,ny,nx)';
  end
```

3. The calculation of the partial derivative in x, u_{xx}, is computed by calling one of a group of five differentiation in space (DSS) routines (which compute first-order derivatives). In this case, dss004 is selected by setting ndss=4 in pde_1_main (with mf=2).

```
%
% PDE
%
% ux
  for j=1:ny
    u1d=u(:,j);
    if     (ndss== 2) ux1d=dss002(xl,xu,nx,u1d); % second order
    elseif(ndss== 4) ux1d=dss004(xl,xu,nx,u1d); % fourth order
    elseif(ndss== 6) ux1d=dss006(xl,xu,nx,u1d); % sixth order
    elseif(ndss== 8) ux1d=dss008(xl,xu,nx,u1d); % eighth order
    elseif(ndss==10) ux1d=dss010(xl,xu,nx,u1d); % tenth order
    end
%
% uxx
    ux1d(1) =0.0;
    ux1d(nx)=0.0;
    if     (ndss== 2) uxx1d=dss002(xl,xu,nx,ux1d); % second order
    elseif(ndss== 4) uxx1d=dss004(xl,xu,nx,ux1d); % fourth order
    elseif(ndss== 6) uxx1d=dss006(xl,xu,nx,ux1d); % sixth order
    elseif(ndss== 8) uxx1d=dss008(xl,xu,nx,ux1d); % eighth order
    elseif(ndss==10) uxx1d=dss010(xl,xu,nx,ux1d); % tenth order
    end
%
% 1D to 2D
    uxx(:,j)=uxx1d(:);
  end
```

Since these DSS routines compute numerical derivatives of 1D arrays, it is necessary before calling them to temporarily store the 2D array u in a 1D array, u1d, which is easily done using the subscripting utilities of Matlab (in this case, the : operator). After the first derivative of u1d is computed (by a call to dss004), the boundary values are reset in accordance with Eqs. (10.4) and (10.5).

Then a second call to dss004 computes the second derivative uxx1d, again a 1D array. uxx1d is returned to a 2D array, uxx, for subsequent MOL programming of Eq. (10.2). Note in particular the use of *stagewise differentiation* using $u \rightarrow u_x \rightarrow u_{xx}$ by two successive calls to dss004.

4. The same procedure is then repeated for the derivative in y, u_{yy}, which eventually is stored in uyy.

```
%
% uy
  for i=1:nx
    u1d=u(i,:);
    if     (ndss== 2) uy1d=dss002(yl,yu,ny,u1d); % second order
    elseif(ndss== 4) uy1d=dss004(yl,yu,ny,u1d); % fourth order
    elseif(ndss== 6) uy1d=dss006(yl,yu,ny,u1d); % sixth order
    elseif(ndss== 8) uy1d=dss008(yl,yu,ny,u1d); % eighth order
    elseif(ndss==10) uy1d=dss010(yl,yu,ny,u1d); % tenth order
    end
%
% uyy
    if     (ndss== 2) uyy1d=dss002(yl,yu,ny,uy1d); % second order
    elseif(ndss== 4) uyy1d=dss004(yl,yu,ny,uy1d); % fourth order
    elseif(ndss== 6) uyy1d=dss006(yl,yu,ny,uy1d); % sixth order
    elseif(ndss== 8) uyy1d=dss008(yl,yu,ny,uy1d); % eighth order
    elseif(ndss==10) uyy1d=dss010(yl,yu,ny,uy1d); % tenth order
    end
%
% 1D to 2D
    uyy(i,:)=uyy1d(:);
  end
```

Table 10.2. Numerical output for the solution of Eqs. (10.2)–(10.7) from pde_1.m and pde_2, mf=2, mmf=1

```
t =  0.00
x across (x=0,0.2,...,1.0)
y  down  (y=1.0,0.8,...,0)
    1.000    1.000    1.000    1.000    1.000    1.000
    0.500    0.500    0.500    0.500    0.500    0.500
    0.500    0.500    0.500    0.500    0.500    0.500
    0.500    0.500    0.500    0.500    0.500    0.500
    0.500    0.500    0.500    0.500    0.500    0.500
    0.000    0.000    0.000    0.000    0.000    0.000

t =  0.03
x across (x=0,0.2,...,1.0)
y  down  (y=1.0,0.8,...,0)
    1.000    1.000    1.000    1.000    1.000    1.000
    0.706    0.706    0.706    0.706    0.706    0.706
    0.544    0.544    0.544    0.544    0.544    0.544
```

(continued)

Table 10.2 (*continued*)

```
     0.456      0.456      0.456      0.456      0.456      0.456
     0.294      0.294      0.294      0.294      0.294      0.294
     0.000      0.000      0.000      0.000      0.000      0.000

t =  0.06
x across (x=0,0.2,...,1.0)
y  down  (y=1.0,0.8,...,0)
     1.000      1.000      1.000      1.000      1.000      1.000
     0.772      0.772      0.772      0.772      0.772      0.772
     0.582      0.582      0.582      0.582      0.582      0.582
     0.418      0.418      0.418      0.418      0.418      0.418
     0.228      0.228      0.228      0.228      0.228      0.228
     0.000      0.000      0.000      0.000      0.000      0.000

t =  0.09
x across (x=0,0.2,...,1.0)
y  down  (y=1.0,0.8,...,0)
     1.000      1.000      1.000      1.000      1.000      1.000
     0.791      0.791      0.791      0.791      0.791      0.791
     0.595      0.595      0.595      0.595      0.595      0.595
     0.405      0.405      0.405      0.405      0.405      0.405
     0.209      0.209      0.209      0.209      0.209      0.209
     0.000      0.000      0.000      0.000      0.000      0.000

t =  0.12
x across (x=0,0.2,...,1.0)
y  down  (y=1.0,0.8,...,0)
     1.000      1.000      1.000      1.000      1.000      1.000
     0.797      0.797      0.797      0.797      0.797      0.797
     0.598      0.598      0.598      0.598      0.598      0.598
     0.402      0.402      0.402      0.402      0.402      0.402
     0.203      0.203      0.203      0.203      0.203      0.203
     0.000      0.000      0.000      0.000      0.000      0.000

t =  0.15
x across (x=0,0.2,...,1.0)
y  down  (y=1.0,0.8,...,0)
     1.000      1.000      1.000      1.000      1.000      1.000
     0.799      0.799      0.799      0.799      0.799      0.799
     0.599      0.599      0.599      0.599      0.599      0.599
     0.401      0.401      0.401      0.401      0.401      0.401
     0.201      0.201      0.201      0.201      0.201      0.201
     0.000      0.000      0.000      0.000      0.000      0.000

ncall =  170
```

5. With the two partial derivatives in the RHS of Eq. (10.2) now available, this PDE can be programmed.

```
%
% ut = uxx + uyy
  ut=uxx+uyy;
  ut(:,1) =0.0;
  ut(:,ny)=0.0;
```

Note the application of the BCs in y (Eqs. (10.6) and (10.7)).

6. Finally, a 2D to 1D conversion produces the derivative vector yt, which is then transposed as required by ode15s. The counter for calls to pde_1 is incremented at the end of pde_1.

```
%
% 2D to 1D matrix conversion
  if(mmf==1)
    for i=1:nx
    for j=1:ny
      yt((i-1)*ny+j)=ut(i,j);
    end
    end
    yt=yt';
  end
  if(mmf==2)
    yt=reshape(ut',nx*ny,1);
  end
%
% Increment calls to pde_2
  ncall=ncall+1;
```

The numerical output from pde_1_main and pde_2 with mf=2, mmf=1 is given in Table 10.2. This output is very similar to that for mf=1 in Table 10.1. Again, the agreement with the analytical solution, Eq. (10.8), is evident. Also, any possible advantage of using the fourth-order FDs in dss004 (for mf=2) over the second-order FDs in pde_1 (for mf=1) is not evident since the final solution is so smooth; specifically, the solution is linear in y and constant in x, so that second- and fourth-order FDs are exact. However, in general, using higher-order FDs for the same gridding produces more accurate solutions (as demonstrated in other PDE chapters).

For mf=3 in pde_1_main, pde_3 is the ODE routine called by ode15s (See Listing 10.4).

```
  function yt=pde_3(t,y)
%
% Function pde_3 computes the temporal derivative in the
% pseudo transient solution of Laplace's equation by library
% routines for the second derivative
%
% Problem parameters
  global ncall nx ny ndss mmf
  xl=0.0;
  xu=1.0;
  yl=0.0;
  yu=1.0;
%
% 1D to 2D matrix conversion
  if(mmf==1)
    for i=1:nx
    for j=1:ny
      u(i,j)=y((i-1)*ny+j);
    end
    end
  end
  if(mmf==2)
    u=reshape(y,ny,nx)';
  end
%
% PDE
%
% uxx
  for j=1:ny
    u1d=u(:,j);
    nl=2; % Neumann
    nu=2; % Neumann
    ux1d(1) =0.0;
    ux1d(nx)=0.0;
    if    (ndss==42) uxx1d=dss042(xl,xu,nx,u1d,ux1d,nl,nu);
    % second order
    elseif(ndss==44) uxx1d=dss044(xl,xu,nx,u1d,ux1d,nl,nu);
    % fourth order
    elseif(ndss==46) uxx1d=dss046(xl,xu,nx,u1d,ux1d,nl,nu);
    % sixth order
    elseif(ndss==48) uxx1d=dss048(xl,xu,nx,u1d,ux1d,nl,nu);
    % eighth order
    elseif(ndss==50) uxx1d=dss050(xl,xu,nx,u1d,ux1d,nl,nu);
    % tenth order
    end
%
% 1D to 2D
    uxx(:,j)=uxx1d(:);
  end
```

```
%
% uyy
  for i=1:nx
    u1d=u(i,:);
    nl=1; % Dirichlet
    nu=1; % Dirichlet
    uy1d(:)=0.0;
    if    (ndss==42) uyy1d=dss042(yl,yu,ny,u1d,uy1d,nl,nu);
    % second order
    elseif(ndss==44) uyy1d=dss044(yl,yu,ny,u1d,uy1d,nl,nu);
    % fourth order
    elseif(ndss==46) uyy1d=dss046(yl,yu,ny,u1d,uy1d,nl,nu);
    % sixth order
    elseif(ndss==48) uyy1d=dss048(yl,yu,ny,u1d,uy1d,nl,nu);
    % eighth order
    elseif(ndss==50) uyy1d=dss050(yl,yu,ny,u1d,uy1d,nl,nu);
    % tenth order
    end
%
% 1D to 2D
    uyy(i,:)=uyy1d(:);
  end
%
% ut = uxx + uyy
  ut=uxx+uyy;
  ut(:,1) =0.0;
  ut(:,ny)=0.0;
%
% 2D to 1D matrix conversion
  if(mmf==1)
    for i=1:nx
    for j=1:ny
      yt((i-1)*ny+j)=ut(i,j);
    end
    end
    yt=yt';
  end
  if(mmf==2)
    yt=reshape(ut',nx*ny,1);
  end
%
% Increment calls to pde_3
  ncall=ncall+1;
```

Listing 10.4. ODE routine pde_3.m

We can note the following points about pde_3:

1. The beginning code is to declare variables as global, set the dimensions of the $x - y$ grid, and convert the 1D matrix (vector) y to the 2D matrix u is the same as in pde_1 and pde_2.

```
  function yt=pde_3(t,y)
%
% Function pde_3 computes the temporal derivative in the
% pseudo transient solution of Laplace's equation by library
% routines for the second derivative
%
% Problem parameters
  global ncall nx ny ndss mmf
  xl=0.0;
  xu=1.0;
  yl=0.0;
  yu=1.0;
%
% 1D to 2D matrix conversion
  if(mmf==1)
    for i=1:nx
    for j=1:ny
      u(i,j)=y((i-1)*ny+j);
    end
    end
  end
  if(mmf==2)
    u=reshape(y,ny,nx)';
  end
```

2. The spatial derivative u_{xx} is now calculated by dss044 (ndss=44 with mf=3 in pde_1_main).

```
%
% PDE
%
% uxx
  for j=1:ny
    u1d=u(:,j);
    nl=2; % Neumann
    nu=2; % Neumann
```

```
      ux1d(1) =0.0;
      ux1d(nx)=0.0;
      if     (ndss==42) uxx1d=dss042(xl,xu,nx,u1d,ux1d,nl,nu);
      % second order
      elseif(ndss==44) uxx1d=dss044(xl,xu,nx,u1d,ux1d,nl,nu);
      % fourth order
      elseif(ndss==46) uxx1d=dss046(xl,xu,nx,u1d,ux1d,nl,nu);
      % sixth order
      elseif(ndss==48) uxx1d=dss048(xl,xu,nx,u1d,ux1d,nl,nu);
      % eighth order
      elseif(ndss==50) uxx1d=dss050(xl,xu,nx,u1d,ux1d,nl,nu);
      % tenth order
      end
%
% 1D to 2D
    uxx(:,j)=uxx1d(:);
  end
```

Note that since BCs (10.4) and (10.5) are Neumann, `nl=nu=2` are inputs to `dss044` corresponding to $x = 0, 1$, respectively. Also, BCs (10.4) and (10.5) are set as the boundary derivatives $u_x(x = 0, t) = u_x(x = 1, t) = 0$ as inputs to `dss044`. Then `dss044` returns the second derivative directly in `uxx1d`.

3. The code is then essentially the same for u_{yy}. The exception are the Dirichlet BCs at $y = 0, 1$ for which `nl=nu=1`.

```
%
% uyy
  for i=1:nx
    u1d=u(i,:);
    nl=1; % Dirichlet
    nu=1; % Dirichlet
    uy1d(:)=0.0;
    if     (ndss==42) uyy1d=dss042(yl,yu,ny,u1d,uy1d,nl,nu);
    % second order
    elseif(ndss==44) uyy1d=dss044(yl,yu,ny,u1d,uy1d,nl,nu);
    % fourth order
    elseif(ndss==46) uyy1d=dss046(yl,yu,ny,u1d,uy1d,nl,nu);
    % sixth order
    elseif(ndss==48) uyy1d=dss048(yl,yu,ny,u1d,uy1d,nl,nu);
    % eighth order
    elseif(ndss==50) uyy1d=dss050(yl,yu,ny,u1d,uy1d,nl,nu);
    % tenth order
    end
```

```
%
% 1D to 2D
    uyy(i,:)=uyy1d(:);
  end
```

Also, since Matlab requires that input arguments to a function be assigned a value, we use `uy1d(:)=0.0;` to meet this requirement, even though the first derivative `uy1d` is not used by `dss044` (since the BCs in y are Dirichlet, not Neumann as in x).

4. Finally, with u_{xx} and u_{yy} available, Eq. (10.2) can be programmed (as in `pde_2`).

```
%
% ut = uxx + uyy
  ut=uxx+uyy;
  ut(:,1) =0.0;
  ut(:,ny)=0.0;
%
% 2D to 1D matrix conversion
  if(mmf==1)
    for i=1:nx
    for j=1:ny
      yt((i-1)*ny+j)=ut(i,j);
    end
    end
    yt=yt';
  end
  if(mmf==2)
    yt=reshape(ut',nx*ny,1);
  end
%
% Increment calls to pde_3
  ncall=ncall+1;
```

The numerical solution produced by `pde_1_main` and `pde_3` is essentially the same as that from `pde_1` and `pde_2` in Tables 10.1 and 10.2, so it will not be listed here.

We conclude this discussion of the solution of 2D PDEs, as illustrated with Laplace's equation (Eq. (10.1)), with the following points:

1. This approach to elliptic PDEs by converting them to parabolic PDEs (essentially, by adding an initial-value derivative to each PDE) is a general method for the solution of elliptic problems, which is usually termed the method of

pseudo transients or *false transients* [1, 2]. These names stem from the solution of the parabolic problem with respect to the parameter t; in other words, the solution appears to be transient as it approaches the steady-state solution of the elliptic problem.

2. t in the previous example is a form of a *continuation parameter* that is used to continue a known, initial solution (such as Eq. (10.3)) to the final, desired solution (for which the derivatives in t are essentially zero in the preceding example). A variety of ways to *embed a continuation parameter* in the problem of interest (other than as a derivative in t) are widely used to continue a problem from a known solution to a final, desired solution [3]. We will not discuss further this very important and useful method to the solution of complex problems.
3. When the parameter is embedded, care is required in doing this in such a way that the modified problem is stable. For example, if t is embedded in Eq. (10.1) as $u_t = -u_{xx} - u_{yy}$, that is, the sign of the derivative u_t is inverted, the resulting PDE is actually unstable, so continuation to a final solution for which $u_t \approx 0$ is not possible.
4. To repeat, the embedding method previously illustrated for the solution of Eq. (10.1) is quite general. For example, the previous analysis can easily be extended to

$$\frac{\partial^2 u}{\partial x^2} + \frac{\partial^2 u}{\partial y^2} = f(x, y) \tag{10.9}$$

$$\frac{\partial^2 u}{\partial x^2} + \frac{\partial^2 u}{\partial y^2} = u \tag{10.10}$$

$$\frac{\partial^2 u}{\partial x^2} + \frac{\partial^2 u}{\partial y^2} = g(u) \tag{10.11}$$

   Equations (10.9) and (10.10) are known as *Poisson's equation* and *Helmholtz's equation*, respectively. If $g(u)$ in Eq. (10.11) is nonlinear, for example, $e^{(-1/u)}$, a full range of nonlinear effects can be investigated. All that is required in the solution of Eqs. (10.9)–(10.11) is to add programming for $f(x, y)$, $u(x, y)$, or $g(u)$ to subroutines `pde_1`, `pde_2`, or `pde_3`. For example, for Eq. (10.10), the programming `ut=uxx+uyy;` could be replaced by `ut=uxx+uyy+u`.
5. However, we should keep in mind that the equilibrium solution to which the PDE problem in t converges can be determined by the assumed IC (for $t = 0$), especially for nonlinear problems such as Eq. (10.11).
6. The generality and ease of use of the MOL solution of elliptic problems as illustrated by the preceding example is due in part to the use of library ODE integrators for the integration with respect to the embedded parameter t (e.g., `ode15s`). In other words, we can take advantage of quality initial-value ODE integrators that are widely available.
7. One possible choice of an integrator is the sparse option of `ode15s`, particularly if the ODE system has a *sparse Jacobian matrix*. In the present case,

121 ODEs is a rather modest problem and the sparse option may not provide a significant computational advantage (of course, this could be investigated). Another possibility is to directly use the Matlab sparse matrix utilities; we have investigated this approach and have provided a Matlab code that is available as part of the supplemental software library.

8. Since MOL can be applied to parabolic problems (such as the heat equation $u_t = u_{xx}$) and hyperbolic problems (such as the wave equation $u_{tt} = u_{xx}$), MOL can be applied to *all three major classes of PDEs, elliptic, parabolic, and hyperbolic*. Also, since MOL can be applied to systems of equations of mixed type, such as hyperbolic–parabolic PDEs, it is a general framework for the numerical solution of PDE systems.

REFERENCES

[1] Schiesser, W. E. (1991), *The Numerical Method of Lines*, Academic Press, San Diego, CA

[2] Schiesser, W. E. (1994), *Computational Mathematics in Engineering and Applied Science: ODEs, DAEs and PDEs*, CRC Press, Boca Raton, FL

[3] Lee, E. S. (1968), *Quasilinearization and Invariant Imbedding: With Applications to Chemical Engineering and Adaptive Control*, Academic Press, New York

11

Three-Dimensional Partial Differential Equation

This partial differential equation (PDE) application introduces the following mathematical concepts and computational methods:

1. A PDE in three dimensions (3D), and therefore the following analysis, illustrates the method of lines (MOL) solution of 3D PDEs.
2. An elliptic PDE so that the following analysis demonstrates the application of the MOL to elliptic PDEs.
3. The implementation of Dirichlet, Neumann, and third-type (mixed, Robin) boundary conditions (BCs).
4. A basic PDE with a variety of extensions, for example, inhomogeneous, PDEs with linear and nonlinear source terms.
5. Evaluation of the accuracy of the numerical solution, including a comparison between the numerical and analytical solutions.
6. Application of h- and p-refinement to achieve spatial convergence of the numerical solution.
7. The concept of continuation for the solution of elliptic PDEs and other important classes of problems.

The 3D PDE is

$$0 = \frac{\partial^2 u}{\partial x^2} + \frac{\partial^2 u}{\partial y^2} + \frac{\partial^2 u}{\partial z^2}$$

where x, y, and z are boundary-value (spatial) independent variables. In subscript notation,

$$0 = u_{xx} + u_{yy} + u_{yy} \tag{11.1}$$

Equation (11.1) is classified as an *elliptic* PDE; it is a *boundary-value problem only* (since it does not have an initial-value independent variable). We follow the general approach of false transients, as discussed in Chapter 10.

Since the MOL generally involves the integration of a system of initial-value ordinary differential equations (ODEs), it cannot be applied directly to Eq. (11.1) (again, Eq. (11.1) does not have an initial-value independent variable). Thus, to perform a MOL solution of Eq. (11.1), we add a derivative with respect to the initial-value variable t to form a *parabolic PDE*:

$$u_t = u_{xx} + u_{yy} + u_{yy} \tag{11.2}$$

Equation (11.2) can be integrated to a *steady state* or *equilibrium condition*, corresponding to $t \to \infty$ for which $u_t \approx 0$, and then Eq. (11.2) reverts to Eq. (11.1).

For the BCs in x, we use the *nonhomogeneous or inhomogeneous (nonzero) Dirichlet conditions*

$$u(x = 0, y, z, t) = u(x = 1, y, z, t) = 1 \qquad (11.3)(11.4)$$

For the BCs in y, we use the *homogeneous (zero) Neumann conditions*

$$u_y(x, y = 0, z, t) = u_y(x, y = 1, z, t) = 0 \qquad (11.5)(11.6)$$

For the BCs in z, we use the *third-type (mixed, Robin) conditions*

$$u_z(x, y, z = 0, t) + u(x, y, z = 0, t) = 1 \tag{11.7}$$

$$u_z(x, y, z = 1, t) + u(x, y, z = 1, t) = 1 \tag{11.8}$$

Equation (11.2) requires one *"pseudo" initial condition* (IC) in t (since it is first order in t). We use the term "pseudo" because t in Eq. (11.2) is not part of the original problem of Eq. (11.1). Thus, we select the IC arbitrarily with the expectation that for $t \to \infty$, this IC will not determine the final equilibrium solution of Eq. (11.1); multiple solutions may be possible for different ICs, but we choose the IC as close as possible to the final solution of interest (admittedly a rather imprecise procedure that can be tested by observing how the solution approaches the final equilibrium solution). For this purpose, we choose just a piecewise constant function, as explained subsequently:

$$u(x = 0, y, z, t = 0) = u(x = 1, y, z, t = 0) = 1 \tag{11.9a}$$

Otherwise

$$u(x, y, z, t = 0) = 0 \quad (x \neq 0, 1) \tag{11.9b}$$

Equations (11.2)–(11.9) for $t \to \infty$ have the analytical solution

$$u(x, y, z, t \to \infty) = 1 \tag{11.10}$$

which will be used to evaluate the accuracy of the solution to Eq. (11.1) (subject to BCs (11.3)–(11.8)).

A main program for the solution of Eqs. (11.2)–(11.9) is given in Listing 11.1.

```
%
% Clear previous files
  clear all
  clc
%
% Parameters shared with the ODE routine
```

```
  global ncall nx ny nz ndss

% Initial condition (which also is consistent with the boundary
% conditions)
  nx=11;
  ny=11;
  nz=11;
  for i=1:nx
  for j=1:ny
  for k=1:nz
    if(i==1)     u0(i,j,k)=1.0;
    elseif(i==nx)u0(i,j,k)=1.0;
    else         u0(i,j,k)=0.0;
    end
  end
  end
  end
%
% 3D to 1D matrix conversion
  for i=1:nx
  for j=1:ny
  for k=1:nz
     y0((i-1)*ny*nz+(j-1)*nz+k)=u0(i,j,k);
  end
  end
  end
%
% Independent variable for ODE integration
  t0=0.0;
  tf=1.0;
  tout=[t0:0.1:tf]';
  nout=11;
  ncall=0;
%
% ODE integration
  mf=3;
  reltol=1.0e-04; abstol=1.0e-04;
  options=odeset('RelTol',reltol,'AbsTol',abstol);
  if(mf==1) % explicit FDs
    [t,y]=ode15s(@pde_1,tout,y0,options); end
  if(mf==2) ndss=4; % ndss = 2, 4, 6, 8 or 10 required
    [t,y]=ode15s(@pde_2,tout,y0,options); end
  if(mf==3) ndss=44; % ndss = 42, 44, 46, 48 or 50 required
    [t,y]=ode15s(@pde_3,tout,y0,options); end
%
% 1D to 3D matrix conversion (plus t)
  for it=1:nout
```

```
  for i=1:nx
  for j=1:ny
  for k=1:nz
    u(it,i,j,k)=y(it,(i-1)*ny*nz+(j-1)*nz+k);
  end
  end
  end
  end
%
% Display selected output as t,
%   u(t,x=0,y=0.5,z=0.5)
%   u(t,x=1,y=0.5,z=0.5)
%   u(t,x=0.5,y=0,z=0.5)
%   u(t,x=0.5,y=1,z=0.5)
%   u(t,x=0.5,y=0.5,z=0)
%   u(t,x=0.5,y=0.5,z=1)
%   u(t,x=0.5,y=0.5,z=0.5)
  fprintf('\n mf = %2d\n',mf);
  for it=1:2:nout
    fprintf('\n t = %5.3f\n%10.3f%10.3f\n%10.3f%10.3f
            \n%10.3f%10.3f\n%10.3f\n',t(it), ...
            u(it,1,6,6),u(it,nx,6,6), ...
            u(it,6,1,6),u(it,6,ny,6),u(it,6,6,1), ...
            u(it,6,6,nz),u(it,6,6,6));
  end
  fprintf('\n ncall = %4d\n',ncall);
```

Listing 11.1. Main program pde_1_main.m

We can note the following points about this program:

1. After some variables and parameters are declared *global* so that they can be shared with the MOL ODE routine, IC Eqs. (11.9a) and (11.9b) are coded over a $11 \times 11 \times 11 = 1{,}331$-point grid in x, y, and z; note that along the boundaries $x = 0$ and $x = 1$, $u(x, y, z, t = 0) = 1$, and everywhere else $u(x, y, z, t = 0) = 0$. This consistency between IC (11.9) and BCs (11.3) and (11.4) facilitates the numerical solution (by avoiding an abrupt change at these boundaries for $t > 0$); in other words, choosing the IC to provide a smooth transition from $t = 0$ to $t > 0$ as much as possible minimizes numerical problems.

```
%
% Clear previous files
  clear all
  clc
%
```

```
% Parameters shared with the ODE routine
  global ncall nx ny nz ndss

% Initial condition (which also is consistent with the boundary
% conditions)
  nx=11;
  ny=11;
  nz=11;
  for i=1:nx
  for j=1:ny
  for k=1:nz
    if(i==1)      u0(i,j,k)=1.0;
    elseif(i==nx)u0(i,j,k)=1.0;
    else          u0(i,j,k)=0.0;
    end
  end
  end
  end
```

The choice of nx=11, ny=11, nz=11 grid points in x, y, and z was made rather arbitrarily. Thus, we have the requirement to determine if these numbers are large enough to give sufficient accuracy in the MOL solution of Eq. (11.1). Of course, we should not choose these numbers of grid points to be larger than necessary to achieve acceptable accuracy since this would merely result in longer computer runs to achieve unnecessary accuracy in the numerical solution of Eq. (11.1). This is especially true for 3D problems for which the total number of grid points increases very rapidly with increases in grid points in each spatial dimension (the so-called *curse of dimensionality*).

2. Since library ODE integrators, such as the Matlab integrators, generally require all of the dependent variables to be arranged in a single ID vector, we must convert the 3D IC matrix u0(i,j,k) ($= u(x, y, z, t = 0)$) of Eq. (11.9) into a 1D vector, in this case, y0.

```
%
% 3D to 1D matrix conversion
  for i=1:nx
  for j=1:ny
  for k=1:nz
     y0((i-1)*ny*nz+(j-1)*nz+k)=u0(i,j,k);
  end
  end
  end
```

The details of this conversion in three nested for loops should be studied since this is an essential operation for the solution of 3D PDEs. We choose here to explicitly code the conversion to clarify what has actually been done. This conversion could also be coded more compactly using the vector/matrix operations in Matlab, as illustrated in Chapter 10 where the `reshape` function has been used.

3. The total interval in t is defined ($0 \le t \le 1$). The solution is to be displayed 11 times, so the output interval is 0.1.

```
%
% Independent variable for ODE integration
  t0=0.0;
  tf=1.0;
  tout=[t0:0.1:tf]';
  nout=11;
  ncall=0;
```

4. The ODE integration (for a total of 1,331 ODEs) is integrated by Matlab integrator `ode15s`; we chose a stiff integrator only because the MOL approximating ODEs are frequently stiff (although we did not test this idea by, for example, also using a nonstiff integrator and comparing the performance through the final value of `ncall`). We did try the sparse matrix form of `ode15s` since the Jacobian matrix of this 1,331-ODE system is no doubt sparse, but this did not offer a significant reduction in computer run times.

```
%
% ODE integration
  mf=3;
  reltol=1.0e-04; abstol=1.0e-04;
  options=odeset('RelTol',reltol,'AbsTol',abstol);
  if(mf==1) % explicit FDs
    [t,y]=ode15s(@pde_1,tout,y0,options); end
  if(mf==2) ndss=4; % ndss = 2, 4, 6, 8 or 10 required
    [t,y]=ode15s(@pde_2,tout,y0,options); end
  if(mf==3) ndss=44; % ndss = 42, 44, 46, 48 or 50 required
    [t,y]=ode15s(@pde_3,tout,y0,options); end
```

Three cases are programmed:

(a) `mf = 1`: Routine `pde_1` is called by `ode15s` with the explicit programming of the finite-difference (FD) approximations of the derivatives u_{xx}, u_{yy}, u_{zz} in Eqs. (11.1) and (11.2). The details of `pde_1` are discussed subsequently.

(b) `mf = 2`: Routine `pde_2` is called by `ode15s` with the calculation of FD approximations of the derivatives u_{xx}, u_{yy}, u_{zz} in Eqs. (11.1) and (11.2) by a library routine for first-order derivatives, for example, `DSS004`. The second-order derivatives are computed by successive differentiation of first-order derivatives (termed *stagewise differentiation*). The details of pde_2 are discussed subsequently.

(c) mf = 3: Routine pde_3 is called by ode15s with the calculation of FD approximations of the derivatives u_{xx}, u_{yy}, u_{zz} in Eqs. (11.1) and (11.2) by a library routine for second-order derivatives, for example, DSS044. The details of pde_3 are discussed subsequently.

5. After integration of the ODEs by ode15s, the solution is returned as a 2D matrix y; the two dimensions of y are for t with a first dimension of 11 and a combination of x, y, and z with a second dimension of 1,331. This matrix is then expressed as a 4D matrix in x, y, and z (with indices i,j,k, respectively) at a series of nout=11 values of t through the index it.

```
%
% 1D to 3D matrix conversion (plus t)
  for it=1:nout
  for i=1:nx
  for j=1:ny
  for k=1:nz
    u(it,i,j,k)=y(it,(i-1)*ny*nz+(j-1)*nz+k);
  end
  end
  end
  end
```

6. Tabular output is displayed so that the numerical solution can be compared with the analytical solution (Eq. 11.10.)

```
%
% Display selected output as t,
%   u(t,x=0,y=0.5,z=0.5)
%   u(t,x=1,y=0.5,z=0.5)
%   u(t,x=0.5,y=0,z=0.5)
%   u(t,x=0.5,y=1,z=0.5)
%   u(t,x=0.5,y=0.5,z=0)
%   u(t,x=0.5,y=0.5,z=1)
%   u(t,x=0.5,y=0.5,z=0.5)
  fprintf('\n mf = %2d\n',mf);
  for it=1:2:nout
    fprintf('\n t = %5.3f\n%10.3f%10.3f\n%10.3f%10.3f
           \n%10.3f%10.3f\n%10.3f\n',t(it),u(it,1,6,6), ...
            u(it,nx,6,6),u(it,6,1,6),u(it,6,ny,6), ...
            u(it,6,6,1),u(it,6,6,nz),u(it,6,6,6));
  end
  fprintf('\n ncall = %4d\n',ncall);
```

The three ODE routines called by `ode15s`, `pde_1.m`, `pde_2.m`, `pde_3.m`, are now discussed. Refer to Listing 11.2 for `pde_1.m`.

```
  function yt=pde_1(t,y)
%
% Problem parameters
  global ncall nx ny nz
  xl=0.0;
  xu=1.0;
  yl=0.0;
  yu=1.0;
  zl=0.0;
  zu=1.0;
%
% 1D to 3D matrix conversion
  for i=1:nx
  for j=1:ny
  for k=1:nz
    u(i,j,k)=y((i-1)*ny*nz+(j-1)*nz+k);
  end
  end
  end
%
% PDE
  dx=(xu-xl)/(nx-1);
  dy=(yu-yl)/(ny-1);
  dz=(zu-zl)/(nz-1);
  dx2=dx^2;
  dy2=dy^2;
  dz2=dz^2;
  for i=1:nx
  for j=1:ny
  for k=1:nz
%
%   uxx
    if(i~=1)&(i~=nx)uxx(i,j,k)=(u(i+1,j,k)-2.0*u(i,j,k) ...
                              +u(i-1,j,k))/dx2; end
%
%   uyy
    if(j==1)          uyy(i,j,k)=2.0*(u(i,j+1,k)-u(i,j,k))/dy2;
    elseif(j==ny)     uyy(i,j,k)=2.0*(u(i,j-1,k)-u(i,j,k))/dy2;
    else              uyy(i,j,k)=(u(i,j+1,k)-2.0*u(i,j,k)...
                                 +u(i,j-1,k))/dy2;

    end
%
%   uzz
```

```
      if(k==1)        u0=u(i,j,k+1)-2.0*dz*(1.0-u(i,j,k));
                      uzz(i,j,k)=(u(i,j,k+1)-2.0*u(i,j,k)+u0)/dz2;
      elseif(k==nz)   u1=u(i,j,k-1)+2.0*dz*(1.0-u(i,j,k));
                      uzz(i,j,k)=(u1-2.0*u(i,j,k)+u(i,j,k-1))/dz2;
      else            uzz(i,j,k)=(u(i,j,k+1)-2.0*u(i,j,k) ...
                                 +u(i,j,k-1))/dz2;
      end
%
%     ut = uxx + uyy + uzz
      if(i==1)|(i==nx)ut(i,j,k)=0.0;
      else ut(i,j,k)=uxx(i,j,k)+uyy(i,j,k)+uzz(i,j,k);
      end
    end
    end
    end
%
% 3D to 1D matrix conversion
    for i=1:nx
    for j=1:ny
    for k=1:nz
      yt((i-1)*ny*nz+(j-1)*nz+k)=ut(i,j,k);
    end
    end
    end
    yt=yt';
%
% increment calls to pde_1
    ncall=ncall+1;
```

Listing 11.2. ODE routine pde_1.m

We can note the following points about pde_1.m:

1. The definition of pde_1, function yt=pde_1(t,y), indicates two important points:
   (a) The input arguments, t,y, the ODE system independent variable, and dependent-variable vector, t and y, respectively, are available for the programming in pde_1.
   (b) The output argument, yt, is the vector of dependent-variable derivatives (with respect to t) that must be defined numerically before the execution of pde_1 is completed. Thus, if y is of length n, then all of the n elements of yt must be defined (failing to evaluate even one of these derivatives will generally produce an incorrect solution; this may seem obvious, but since in this case there are $n = 1{,}331$ dependent variables, failure to evaluate all of the derivatives in ut can easily happen).

```
  function yt=pde_1(t,y)
%
% Problem parameters
  global ncall nx ny nz
  xl=0.0;
  xu=1.0;
  yl=0.0;
  yu=1.0;
  zl=0.0;
  zu=1.0;
```

Some problem parameters are then declared as *global* so that they can be shared with pde_1_main in Listing 11.1 or other routines. In this case, the dimensions of the 3D $x - y - z$ domain are defined (and, of course, these dimensions can be different in the three directions).

2. A conversion of the 1D vector, y, to the 3D matrix u is made so that the subsequent programming can be done in terms of the problem-oriented variable u in Eqs. (11.1)–(11.9). Although this conversion is, in principle, not required, programming in terms of problem-oriented variables, for example, u, is a major convenience.

```
%
% 1D to 3D matrix conversion
  for i=1:nx
  for j=1:ny
  for k=1:nz
    u(i,j,k)=y((i-1)*ny*nz+(j-1)*nz+k);
  end
  end
  end
```

3. The programming of the MOL ODEs is given next, starting with the square of the increments for the FD approximations in x, y, and z.

```
%
% PDE
  dx=(xu-xl)/(nx-1);
  dy=(yu-yl)/(ny-1);
  dz=(zu-zl)/(nz-1);
  dx2=dx^2;
  dy2=dy^2;
  dz2=dz^2;
```

4. We then move throughout the $x - y - z$ domain using three nested `for` loops. Each pass through this set of `for` loops executes the programming for another ODE (a total of 1,331 passes or ODEs).

```
      for i=1:nx
      for j=1:ny
      for k=1:nz
%
%       uxx
        if(i~=1)&(i~=nx)uxx(i,j,k)=(u(i+1,j,k)-2.0*u(i,j,k) ...
                                    +u(i-1,j,k))/dx2; end
%
%       uyy
        if(j==1)          uyy(i,j,k)=2.0*(u(i,j+1,k)-u(i,j,k))/dy2;
        elseif(j==ny)     uyy(i,j,k)=2.0*(u(i,j-1,k)-u(i,j,k))/dy2;
        else              uyy(i,j,k)=(u(i,j+1,k)-2.0*u(i,j,k) ...
                                    +u(i,j-1,k))/dy2;
        end
%
%       uzz
        if(k==1)          u0=u(i,j,k+1)-2.0*dz*(1.0-u(i,j,k));
                          uzz(i,j,k)=(u(i,j,k+1)-2.0*u(i,j,k)+u0)/dz2;
        elseif(k==nz)     u1=u(i,j,k-1)+2.0*dz*(1.0-u(i,j,k));
                          uzz(i,j,k)=(u1-2.0*u(i,j,k)+u(i,j,k-1))/dz2;
        else              uzz(i,j,k)=(u(i,j,k+1)-2.0*u(i,j,k) ...
                                    +u(i,j,k-1))/dz2;
        end
%
%       ut = uxx + uyy + uzz
        if(i==1)|(i==nx)ut(i,j,k)=0.0;
        else ut(i,j,k)=uxx(i,j,k)+uyy(i,j,k)+uzz(i,j,k);
        end
      end
      end
      end
```

The coding of the FD approximations requires some explanation. For the derivative u_{xx}, two cases are required:
(a) For $x \neq 0, 1$, $u_{xx}(x, y, z, t)$ is programmed as the FD approximation

```
%
%       uxx
        if(i~=1)&(i~=nx)uxx(i,j,k)=(u(i+1,j,k)-2.0*u(i,j,k) ...
                                    +u(i-1,j,k))/dx2; end
```

which is based on

$$u_{xx} \approx \frac{u(x+\Delta x, y, z, t) - 2u(x, y, z, t) + u(x-\Delta x, y, z, t)}{\Delta x^2} \tag{11.11}$$

(b) For $x = 0, 1$, BCs (11.3) and (11.4) are programmed toward the end of pde_1 as

```
      if(i==1)|(i==nx)ut(i,j,k)=0.0;
```

In other words, since the boundary values of Eqs. (11.3) and (11.4) are constant, their derivatives in t are zero.

For the derivative u_{yy}, two cases are again required:

(a) For BCs (11.5) and (11.6),

```
%
%     uyy
      if(j==1)         uyy(i,j,k)=2.0*(u(i,j+1,k)-u(i,j,k))/dy2;
      elseif(j==ny)    uyy(i,j,k)=2.0*(u(i,j-1,k)-u(i,j,k))/dy2;
```

This coding is based on the approximation of a second derivative (see Eq. (11.11)), including the approximation of homogeneous Neumann BCs (11.5) and (11.6) (i.e., $u(x, y = 0 - \Delta y, z, t) = u(x, y = 0 + \Delta y, z, t)$, $u(x, y = 1 + \Delta y, z, t) = u(x, y = 1 - \Delta y, z, t)$):

$$u_{yy}(x, y = 0, z, t) \approx 2\frac{u(x, y = 0 + \Delta y, z, t) - u(x, y = 0, z, t)}{\Delta y^2} \tag{11.12a}$$

$$u_{yy}(x, y = 1, z, t) \approx 2\frac{u(x, y = 1 - \Delta y, z, t) - u(x, y = 1, z, t)}{\Delta y^2} \tag{11.12b}$$

(b) For $y \neq 0, 1$, the approximation of uyy is programmed as

```
      else   uyy(i,j,k)=(u(i,j+1,k)-2.0*u(i,j,k)+u(i,j-1,k))/dy2;
```

corresponding to the FD

$$u_{yy} \approx \frac{u(x, y+\Delta y, z, t) - 2u(x, y, z, t) + u(x, y-\Delta y, z, t)}{\Delta y^2} \tag{11.12c}$$

For the derivative u_{zz}, two cases are also required:

(a) For BCs (11.7) and (11.8),

```
if(k==1)          u0=u(i,j,k+1)-2.0*dz*(1.0-u(i,j,k));
                  uzz(i,j,k)=(u(i,j,k+1)-2.0*u(i,j,k)+u0)/dz2;
elseif(k==nz)     u1=u(i,j,k-1)+2.0*dz*(1.0-u(i,j,k));
                  uzz(i,j,k)=(u1-2.0*u(i,j,k)+u(i,j,k-1))/dz2;
```

This coding is based on the approximation of a second derivative (see Eq. (11.11)), including the approximation of third-type BCs (11.7) and (11.8). For Eq. (11.7) we have (with the fictitious point u_0)

$$u_z(x, y, z = 0, t) \approx \frac{u(x, y, z = \Delta z, t) - u_0}{2\Delta z} = 1 - u(x, y, z = 0, t)$$

or

$$u_0 = (u(x, y, z = \Delta z, t) - 2\Delta z(1 - u(x, y, z = 0, t)) \tag{11.13a}$$

u_0 from Eq. (11.13a) can then be used in the FD approximation of u_{zz} (see Eq. (11.11)).

$$u_{zz} = \frac{u(x, y, z = \Delta z, t) - 2u(x, y, z = 0, t) + u_0}{\Delta z^2} \tag{11.13b}$$

(b) Similarly, for BC (11.8) (with fictitious point u_1),

$$u_z(x, y, z = 1, t) \approx \frac{u_1 - u(x, y, z = 1 - \Delta z, t)}{2\Delta z} = 1 - u(x, y, z = 1, t)$$

or

$$u_1 = (u(x, y, z = 1 - \Delta z, t) + 2\Delta z(1 - u(x, y, z = 1, t)) \tag{11.13c}$$

u_1 from Eq. (11.13c) can then be used in the FD approximation of u_{zz} (see Eq. (11.11)).

$$u_{zz} = \frac{u_1 - 2u(x, y, z = 1, t) + u(x, y, z = 1 - \Delta z, t)}{\Delta z^2} \tag{11.13d}$$

(c) For $z \neq 0, 1$, the approximation of u_{zz} is

```
else uzz(i,j,k)=(u(i,j,k+1)-2.0*u(i,j,k)+u(i,j,k-1))/dz2;
```

corresponding to the FD

$$u_{zz} \approx \frac{u(x, y, z + \Delta z, t) - 2u(x, y, z, t) + u(x, y, z - \Delta z, t)}{\Delta z^2} \tag{11.13e}$$

With the three derivatives u_{xx}, u_{yy}, u_{zz} computed, PDE (11.2) is programmed as

```
if(i==1)|(i==nx)ut(i,j,k)=0.0;
else ut(i,j,k)=uxx(i,j,k)+uyy(i,j,k)+uzz(i,j,k);
```

The numerical output from pde_1_main of Listing 11.1 and pde_1 of Listing 11.2 for mf=1 is given in Table 11.1.

Table 11.1. Numerical output for the solution of Eqs. (11.2)–(11.9) from pde_1_main and pde_1, mf=1

```
mf =   1

t = 0.000
       1.000        1.000
       0.000        0.000
       0.000        0.000
       0.000

t = 0.200
       1.000        1.000
       0.804        0.804
       0.692        0.874
       0.804

t = 0.400
       1.000        1.000
       0.967        0.967
       0.945        0.980
       0.967

t = 0.600
       1.000        1.000
       0.994        0.994
       0.990        0.996
       0.994

t = 0.800
       1.000        1.000
       0.999        0.999
       0.998        0.999
       0.999

t = 1.000
       1.000        1.000
       1.000        1.000
       1.000        1.000
       1.000

ncall = 1422
```

We can note the following points:

1. This output (which is quite abbreviated since the solution at 1,331 points is available) indicates the approach of the numerical solution to the analytical solution of Eq. (11.10). A more complete picture of the solution might be available from a series of 3D plots, although this would be rather challenging since there are four independent variables, x, y, z, t. Some subplots might be a better way to proceed, for example, plots at specific values of x, y, z as a function of t. We did not attempt this approach to displaying the solution graphically in the preceding Matlab; some further discussion is given in Appendix 6.
2. IC (11.9) introduces discontinuities at $x = 0, 1$ for which the numerical solution appears to remain quite smooth, which is a consequence of parabolic PDE (11.2); that is, parabolic PDEs tend to "diffuse" or smooth numerical solutions (the same is not true for hyperbolic PDEs that tend to propagate discontinuities and subsequently produce severe numerical distortions).
3. The computational effort is rather substantial (1,422 calls to `pde_1`, which is rather typical for 3D problems).

For `mf=2` in `pde_1_main`, `pde_2` is the ODE routine called by `ode15s` (see Listing 11.3).

```
  function yt=pde_2(t,y)
%
% Problem parameters
  global ncall nx ny nz ndss
  xl=0.0;
  xu=1.0;
  yl=0.0;
  yu=1.0;
  zl=0.0;
  zu=1.0;
%
% 1D to 3D matrix conversion
  for i=1:nx
  for j=1:ny
  for k=1:nz
    u(i,j,k)=y((i-1)*ny*nz+(j-1)*nz+k);
  end
  end
  end
%
% PDE
%
% ux
  for j=1:ny
  for k=1:nz
    u1d=u(:,j,k);
```

```
      if    (ndss== 2) ux1d=dss002(xl,xu,nx,u1d); % second order
      elseif(ndss== 4) ux1d=dss004(xl,xu,nx,u1d); % fourth order
      elseif(ndss== 6) ux1d=dss006(xl,xu,nx,u1d); % sixth order
      elseif(ndss== 8) ux1d=dss008(xl,xu,nx,u1d); % eighth order
      elseif(ndss==10) ux1d=dss010(xl,xu,nx,u1d); % tenth order
      end
  %
  % uxx
      if    (ndss== 2) uxx1d=dss002(xl,xu,nx,ux1d);
      % second order
      elseif(ndss== 4) uxx1d=dss004(xl,xu,nx,ux1d);
      % fourth order
      elseif(ndss== 6) uxx1d=dss006(xl,xu,nx,ux1d);
      % sixth order
      elseif(ndss== 8) uxx1d=dss008(xl,xu,nx,ux1d);
      % eighth order
      elseif(ndss==10) uxx1d=dss010(xl,xu,nx,ux1d);
      % tenth order
      end
  %
  % 1D to 3D
      uxx(:,j,k)=uxx1d(:);
    end
    end
  %
  % uy
    for i=1:nx
    for k=1:nz
      u1d=u(i,:,k);
      if    (ndss== 2) uy1d=dss002(yl,yu,ny,u1d); % second order
      elseif(ndss== 4) uy1d=dss004(yl,yu,ny,u1d); % fourth order
      elseif(ndss== 6) uy1d=dss006(yl,yu,ny,u1d); % sixth order
      elseif(ndss== 8) uy1d=dss008(yl,yu,ny,u1d); % eighth order
      elseif(ndss==10) uy1d=dss010(yl,yu,ny,u1d); % tenth order
      end
  %
  % uyy
      uy1d(1) =0.0;
      uy1d(ny)=0.0;
      if    (ndss== 2) uyy1d=dss002(yl,yu,ny,uy1d);
      % second order
      elseif(ndss== 4) uyy1d=dss004(yl,yu,ny,uy1d);
      % fourth order
      elseif(ndss== 6) uyy1d=dss006(yl,yu,ny,uy1d);
      % sixth order
      elseif(ndss== 8) uyy1d=dss008(yl,yu,ny,uy1d);
      % eighth order
```

```
    elseif(ndss==10) uyy1d=dss010(yl,yu,ny,uy1d);
    % tenth order
    end
%
% 1D to 3D
    uyy(i,:,k)=uyy1d(:);
  end
  end
%
% uz
  for i=1:nx
  for j=1:ny
    u1d=u(i,j,:);
    if    (ndss== 2) uz1d=dss002(zl,zu,nz,u1d); % second order
    elseif(ndss== 4) uz1d=dss004(zl,zu,nz,u1d); % fourth order
    elseif(ndss== 6) uz1d=dss006(zl,zu,nz,u1d); % sixth order
    elseif(ndss== 8) uz1d=dss008(zl,zu,nz,u1d); % eighth order
    elseif(ndss==10) uz1d=dss010(zl,zu,nz,u1d); % tenth order
    end
%
% uzz
    uz1d(1) =1.0-u1d(1);
    uz1d(nz)=1.0-u1d(nz);
    if    (ndss== 2) uzz1d=dss002(zl,zu,nz,uz1d);
    % second order
    elseif(ndss== 4) uzz1d=dss004(zl,zu,nz,uz1d);
    % fourth order
    elseif(ndss== 6) uzz1d=dss006(zl,zu,nz,uz1d);
    % sixth order
    elseif(ndss== 8) uzz1d=dss008(zl,zu,nz,uz1d);
    % eighth order
    elseif(ndss==10) uzz1d=dss010(zl,zu,nz,uz1d);
    % tenth order
    end
%
% 1D to 3D
    uzz(i,j,:)=uzz1d(:);
  end
  end
%
% ut = uxx + uyy + uzz
  ut=uxx+uyy+uzz;
  ut(1,:,:) =0.0;
  ut(nx,:,:)=0.0;
%
% 3D to 1D matrix conversion
  for i=1:nx
```

```
  for j=1:ny
  for k=1:nz
    yt((i-1)*ny*nz+(j-1)*nz+k)=ut(i,j,k);
  end
  end
  end
  yt=yt';
%
% increment calls to pde_2
  ncall=ncall+1;
```

Listing 11.3. ODE routine pde_2.m

We can note the following details about pde_2:

1. The function is defined, selected parameters and variables are declared as global, the spatial dimensions are defined, and the conversion of the 1D matrix y to the 3D matrix u is made so that programming in terms of problem-oriented variables is set up.

```
  function yt=pde_2(t,y)
%
% Problem parameters
  global ncall nx ny nz ndss
  xl=0.0;
  xu=1.0;
  yl=0.0;
  yu=1.0;
  zl=0.0;
  zu=1.0;
%
% 1D to 3D matrix conversion
  for i=1:nx
  for j=1:ny
  for k=1:nz
    u(i,j,k)=y((i-1)*ny*nz+(j-1)*nz+k);
  end
  end
  end
```

2. The calculation of the partial derivative in x, u_{xx}, is computed by calling one of a group of five differentiation in space (DSS) routines (which compute first-order derivatives). In this case, dss004 is selected by setting ndss=4 in pde_1_main (with mf=2).

```
%
% PDE
%
% ux
  for j=1:ny
  for k=1:nz
    u1d=u(:,j,k);
    if     (ndss== 2) ux1d=dss002(xl,xu,nx,u1d); % second order
    elseif(ndss== 4) ux1d=dss004(xl,xu,nx,u1d); % fourth order
    elseif(ndss== 6) ux1d=dss006(xl,xu,nx,u1d); % sixth order
    elseif(ndss== 8) ux1d=dss008(xl,xu,nx,u1d); % eighth order
    elseif(ndss==10) ux1d=dss010(xl,xu,nx,u1d); % tenth order
    end
%
% uxx
    if     (ndss== 2) uxx1d=dss002(xl,xu,nx,ux1d);
    % second order
    elseif(ndss== 4) uxx1d=dss004(xl,xu,nx,ux1d);
    % fourth order
    elseif(ndss== 6) uxx1d=dss006(xl,xu,nx,ux1d);
    % sixth order
    elseif(ndss== 8) uxx1d=dss008(xl,xu,nx,ux1d);
    % eighth order
    elseif(ndss==10) uxx1d=dss010(xl,xu,nx,ux1d);
    % tenth order
    end
%
% 1D to 3D
    uxx(:,j,k)=uxx1d(:);
  end
  end
```

Since these DSS routines compute numerical derivatives of 1D arrays, it is necessary before calling them to temporarily store the 3D array u in a 1D array, u1d, which is easily done using the subscripting utilities of Matlab (in this case, the : operator).

Then a second call to dss004 computes the second derivative uxx1d, again a 1D array. uxx1d is returned to a 3D array, uxx, for subsequent MOL programming of Eq. (11.2). Note, in particular, the use of *stagewise differentiation* using $u \rightarrow u_x \rightarrow u_{xx}$ by two successive calls to dss004.

Also, Dirichlet BCs (11.3) and (11.4) are not used in the calculation of u_{xx} at this point. In other words, it would seem reasonable that these BCs would be programmed before the first call to dss004 using something like

```
u1d(1 ,j,k)=1.0;
u1d(nx,j,k)=1.0;
```

but these statements would have no effect since dependent variables cannot be set in the ODE routine (a property of the Matlab integrators such as `ode15s`). Rather, these boundary values are set in the IC of Eq. (11.9) in `pde_1_main` and passed to `pde_1` through its second argument `y`. The derivatives of these boundary values are set to zero later in `pde_1` to ensure the constant boundary values (of Eqs. (11.3) and (11.4)) are maintained.

3. The same procedure is then repeated for the derivative in y, u_{yy}, which eventually is stored in uyy.

```
%
% uy
  for i=1:nx
  for k=1:nz
    u1d=u(i,:,k);
    if     (ndss== 2) uy1d=dss002(yl,yu,ny,u1d); % second order
    elseif(ndss== 4) uy1d=dss004(yl,yu,ny,u1d); % fourth order
    elseif(ndss== 6) uy1d=dss006(yl,yu,ny,u1d); % sixth order
    elseif(ndss== 8) uy1d=dss008(yl,yu,ny,u1d); % eighth order
    elseif(ndss==10) uy1d=dss010(yl,yu,ny,u1d); % tenth order
    end
%
% uyy
    uy1d(1) =0.0;
    uy1d(ny)=0.0;
    if     (ndss== 2) uyy1d=dss002(yl,yu,ny,uy1d);
    % second order
    elseif(ndss== 4) uyy1d=dss004(yl,yu,ny,uy1d);
    % fourth order
    elseif(ndss== 6) uyy1d=dss006(yl,yu,ny,uy1d);
    % sixth order
    elseif(ndss== 8) uyy1d=dss008(yl,yu,ny,uy1d);
    % eighth order
    elseif(ndss==10) uyy1d=dss010(yl,yu,ny,uy1d);
    % tenth order
    end
%
```

```
% 1D to 3D
    uyy(i,:,k)=uyy1d(:);
  end
  end
```

Note that Neumann BCs (11.5) and (11.6) are implemented as

```
    uy1d(1) =0.0;
    uy1d(ny)=0.0;
```

4. The derivative in z, u_{zz}, is computed and eventually is stored in uzz.

```
%
% uz
  for i=1:nx
  for j=1:ny
    u1d=u(i,j,:);
    if     (ndss== 2) uz1d=dss002(zl,zu,nz,u1d); % second order
    elseif(ndss== 4) uz1d=dss004(zl,zu,nz,u1d); % fourth order
    elseif(ndss== 6) uz1d=dss006(zl,zu,nz,u1d); % sixth order
    elseif(ndss== 8) uz1d=dss008(zl,zu,nz,u1d); % eighth order
    elseif(ndss==10) uz1d=dss010(zl,zu,nz,u1d); % tenth order
    end
%
% uzz
    uz1d(1) =1.0-u1d(1);
    uz1d(nz)=1.0-u1d(nz);
    if     (ndss== 2) uzz1d=dss002(zl,zu,nz,uz1d);
    % second order
    elseif(ndss== 4) uzz1d=dss004(zl,zu,nz,uz1d);
    % fourth order
    elseif(ndss== 6) uzz1d=dss006(zl,zu,nz,uz1d);
    % sixth order
    elseif(ndss== 8) uzz1d=dss008(zl,zu,nz,uz1d);
    % eighth order
    elseif(ndss==10) uzz1d=dss010(zl,zu,nz,uz1d);
    % tenth order
    end
%
```

```
% 1D to 3D
    uzz(i,j,:)=uzz1d(:);
  end
  end
```

Note that the third-type BCs (11.7) and (11.8) are implemented as

```
    uz1d(1) =1.0-u1d(1);
    uz1d(nz)=1.0-u1d(nz);
```

5. With the three partial derivatives in the RHS of Eq. (11.2) now available, this PDE can be programmed.

```
%
% ut = uxx + uyy + uzz
  ut=uxx+uyy+uzz;
  ut(1,:,:) =0.0;
  ut(nx,:,:)=0.0;
```

   Note the application of the BCs in x, Eqs. (11.3) and (11.4), by setting the derivatives in t to zero to maintain these constant values.

6. Finally, a 3D to 1D conversion produces the derivative vector yt, which is then transposed as required by ode15s. The counter for calls to pde_1 is incremented at the end of pde_1.

```
%
% 3D to 1D matrix conversion
  for i=1:nx
  for j=1:ny
  for k=1:nz
    yt((i-1)*ny*nz+(j-1)*nz+k)=ut(i,j,k);
  end
  end
  end
  yt=yt';
%
% increment calls to pde_2
  ncall=ncall+1;
```

The numerical output from pde_1_main and pde_2 is given in Table 11.2.

Table 11.2. Numerical output for the solution of Eqs. (11.2)-(11.9) from `pde_1_main` and `pde_2`, `mf=2`

```
mf =  2

t = 0.000
     1.000       1.000
     0.000       0.000
     0.000       0.000
     0.000

t = 0.200
     1.000       1.000
     0.806       0.806
     0.693       0.875
     0.806

t = 0.400
     1.000       1.000
     0.967       0.967
     0.946       0.980
     0.967

t = 0.600
     1.000       1.000
     0.994       0.994
     0.991       0.997
     0.994

t = 0.800
     1.000       1.000
     0.999       0.999
     0.998       0.999
     0.999

t = 1.000
     1.000       1.000
     1.000       1.000
     1.000       1.000
     1.000

ncall = 1420
```

This output is very similar to that for `mf=1` in Table 11.1 Again, the agreement with the analytical solution, Eq. (11.10), is evident. Also, any possible advantage of using the fourth-order FDs in `dss004` (for `mf=2`) over the second-order FDs in `pde_1` (for `mf=1`) is not evident since the final solution, Eq. (11.10), is so smooth; specifically, the final solution is constant so that second- and fourth-order FDs are exact. However, in general, using higher-order FDs for the same gridding produces more accurate solutions (as demonstrated in other PDE chapters).

For `mf=3` in `pde_1_main`, `pde_3` is the ODE routine called by `ode15s` (see Listing 11.4).

```
  function yt=pde_3(t,y)
%
% Problem parameters
  global ncall nx ny nz ndss
  xl=0.0;
  xu=1.0;
  yl=0.0;
  yu=1.0;
  zl=0.0;
  zu=1.0;
%
% 1D to 3D matrix conversion
  for i=1:nx
  for j=1:ny
  for k=1:nz
    u(i,j,k)=y((i-1)*ny*nz+(j-1)*nz+k);
  end
  end
  end
%
% PDE
%
% uxx
  for j=1:ny
  for k=1:nz
    u1d=u(:,j,k);
    nl=1; % Dirichlet
    nu=1; % Dirichlet
    ux1d(:)=0.0;
    if     (ndss==42) uxx1d=dss042(xl,xu,nx,u1d,ux1d,nl,nu);
    % second order
    elseif(ndss==44) uxx1d=dss044(xl,xu,nx,u1d,ux1d,nl,nu);
    % fourth order
    elseif(ndss==46) uxx1d=dss046(xl,xu,nx,u1d,ux1d,nl,nu);
    % sixth order
```

```
    elseif(ndss==48) uxx1d=dss048(xl,xu,nx,u1d,ux1d,nl,nu);
    % eighth order
    elseif(ndss==50) uxx1d=dss050(xl,xu,nx,u1d,ux1d,nl,nu);
    % tenth order
    end
%
% 1D to 3D
    uxx(:,j,k)=uxx1d(:);
  end
  end
%
% uyy
  for i=1:nx
  for k=1:nz
    u1d=u(i,:,k);
    nl=2; % Neumann
    nu=2; % Neumann
    uy1d(1) =0.0;
    uy1d(ny)=0.0;
    if    (ndss==42) uyy1d=dss042(yl,yu,ny,u1d,uy1d,nl,nu);
    % second order
    elseif(ndss==44) uyy1d=dss044(yl,yu,ny,u1d,uy1d,nl,nu);
    % fourth order
    elseif(ndss==46) uyy1d=dss046(yl,yu,ny,u1d,uy1d,nl,nu);
    % sixth order
    elseif(ndss==48) uyy1d=dss048(yl,yu,ny,u1d,uy1d,nl,nu);
    % eighth order
    elseif(ndss==50) uyy1d=dss050(yl,yu,ny,u1d,uy1d,nl,nu);
    % tenth order
    end
%
% 1D to 3D
    uyy(i,:,k)=uyy1d(:);
  end
  end
%
% uzz
  for i=1:nx
  for j=1:ny
    u1d=u(i,j,:);
    nl=2; % Neumann
    nu=2; % Neumann
    uz1d(1) =1.0-u1d(1);
    uz1d(nz)=1.0-u1d(nz);
    if    (ndss==42) uzz1d=dss042(zl,zu,nz,u1d,uz1d,nl,nu);
    % second order
    elseif(ndss==44) uzz1d=dss044(zl,zu,nz,u1d,uz1d,nl,nu);
```

```
    % fourth order
    elseif(ndss==46) uzz1d=dss046(zl,zu,nz,u1d,uz1d,nl,nu);
    % sixth order
    elseif(ndss==48) uzz1d=dss048(zl,zu,nz,u1d,uz1d,nl,nu);
    % eighth order
    elseif(ndss==50) uzz1d=dss050(zl,zu,nz,u1d,uz1d,nl,nu);
    % tenth order
    end
%
% 1D to 3D
    uzz(i,j,:)=uzz1d(:);
  end
  end
%
% ut = uxx + uyy + uzz
  ut=uxx+uyy+uzz;
  ut(1,:,:) =0.0;
  ut(nx,:,:)=0.0;
%
% 3D to 1D matrix conversion
  for i=1:nx
  for j=1:ny
  for k=1:nz
    yt((i-1)*ny*nz+(j-1)*nz+k)=ut(i,j,k);
  end
  end
  end
  yt=yt';
%
% increment calls to pde_3
  ncall=ncall+1;
```

Listing 11.4. ODE routine pde_3.m

pde_3 closely parallels pde_2 (Listing 11.3). The essential differences are as follows:

1. The use of dss044 for the direct calculation of the derivatives u_{xx}, u_{yy}, u_{zz} rather than dss004 for the calculation of these second derivatives via the corresponding first derivatives (through stagewise differentiation).
2. The programming of the Dirichlet BCs of Eqs. (11.3) and (11.4) is

```
    nl=1; % Dirichlet
    nu=1; % Dirichlet
    ux1d(:)=0.0;
```

The setting of the first derivative `ux1d` is required, even though this first derivative is not used, because it is an input argument of `dss044`; Matlab has the requirement that each input argument of a function must have at least one assigned value.

The programming of the Neumann BCs of Eqs. (11.5) and (11.6), that is, the derivatives $u_y(x, y = 0, z, t)$ = `uy1d(1)` and $u_y(x, y = 1, z, t)$ = `uy1d(ny)`, is

```
nl=2; % Neumann
nu=2; % Neumann
uy1d(1) =0.0;
uy1d(ny)=0.0;
```

3. The programming of the third-type BCs of Eqs. (11.7) and (11.8) is done as two Neumann BCs by defining the derivatives $u_z(x, y, z = 0, t)$ = `uz1d(1)` and $u_z(x, y, z = 1, t)$ = `uz1d(nz)`:

```
nl=2; % Neumann
nu=2; % Neumann
uz1d(1) =1.0-u1d(1);
uz1d(nz)=1.0-u1d(nz);
```

The output from `pde_1_main` and `tt pde_3` is given in Table 11.3. This output is very similar to the output in Tables 11.1 and 11.2 for `pde_1` and `pde_2`.

We conclude this discussion of the solution of 3D PDEs, as illustrated by Eqs. (11.1) and (11.2), with the following points:

1. We could investigate the accuracy of the numerical solution (in addition to comparing it with the analytical solution, Eq. (11.10)) by
   (a) Varying the number of spatial grid points, but with modest increases because of the increase in total ODEs, which is particularly limiting for 3D problems. If the solution remains essentially unchanged to a reasonable number of figures, for example, 3–4, with changes in the number of grid points, we have some assurance that the numerical solution is accurate to that number of figures. Since the grid spacing is often given the symbol h in the numerical analysis literature, varying the number of grid points is generally termed *h-refinement*.
   (b) Varying the order of the FD approximations. This is easily done in the present case by changing the calls to the differentiation routines, for example, `dss004` to `dss006` or `dss044` to `dss046`. Only the name of the routine changes; the arguments can remain the same. Then we can observe the effect of changing the order of the FD approximations on the numerical solution. Since the order of the FD approximations is often given the symbol p, this is termed *p-refinement*.

Table 11.3. Numerical output for the solution of Eqs. (11.2)–(11.9) from `pde_1_main` and `pde_3`, `mf=3`

```
mf =  3

t = 0.000
     1.000      1.000
     0.000      0.000
     0.000      0.000
     0.000

t = 0.200
     1.000      1.000
     0.805      0.805
     0.693      0.875
     0.805

t = 0.400
     1.000      1.000
     0.967      0.967
     0.946      0.980
     0.967

t = 0.600
     1.000      1.000
     0.994      0.994
     0.991      0.997
     0.994

t = 0.800
     1.000      1.000
     0.999      0.999
     0.998      0.999
     0.999

t = 1.000
     1.000      1.000
     1.000      1.000
     1.000      1.000
     1.000
ncall = 1427
```

   (c) Using the third commonly used method to assess solution accuracy, which we have not used here, that is, to adaptively refine the grid. This is usually termed *adaptive mesh refinement (AMR)* or *r-refinement*.

2. The preceding approach to elliptic PDEs by converting them to parabolic PDEs (essentially, by adding an initial-value derivative to each PDE) is a general method for the solution of elliptic problems, which is usually termed the method of *pseudo transients* or *false transients*. These names stem from the solution of the parabolic problem with respect to the parameter t; in other words, the solution appears to be transient as it approaches the steady-state solution of the elliptic problem.
3. t in the previous example is a form of a *continuation parameter* that is used to continue a known, initial solution (such as Eq. (11.9)) to the final, desired solution (for which the derivatives in t are essentially zero in the preceding example). A variety of ways to *embed a continuation parameter* in the problem of interest (other than as a derivative in t) are widely used to continue a problem from a known solution to a final, desired solution. We will not discuss further this very important and useful method to the solution of complex problems.
4. When the parameter is embedded, care is required in doing this in such a way that the modified problem is stable. For example, if t is embedded in Eq. (11.1) as $u_t = -u_{xx} - u_{yy} - u_{yy}$, that is, the sign of the derivative u_t is inverted, the resulting PDE is actually unstable, so continuation to a final solution for which $u_t \approx 0$ is not possible.
5. To repeat, the embedding method previously illustrated for the solution of Eq. (11.1) is quite general. For example, the previous analysis can easily be extended to

$$\frac{\partial^2 u}{\partial x^2} + \frac{\partial^2 u}{\partial y^2} + \frac{\partial^2 u}{\partial z^2} = f(x, y, z) \tag{11.14a}$$

$$\frac{\partial^2 u}{\partial x^2} + \frac{\partial^2 u}{\partial y^2} + \frac{\partial^2 u}{\partial z^2} = u \tag{11.14b}$$

$$\frac{\partial^2 u}{\partial x^2} + \frac{\partial^2 u}{\partial y^2} + \frac{\partial^2 u}{\partial z^2} = g(u) \tag{11.14c}$$

   If $g(u)$ in Eq. (11.14c) is nonlinear, for example, $e^{(-1/u)}$, a full range of nonlinear effects can be investigated. All that is required in the solution of Eqs. (11.14a)–(11.14c) is to add programming for $f(x, y, z)$, $u(x, y, z)$, or $g(u)$ to subroutines `pde_1`, `pde_2`, or `pde_3`. For example, for Eq. (11.14b), the programming `ut=uxx+uyy+yzz;` could be replaced by `ut=uxx+uyy+uzz+u`.
6. Nonlinear BCs can also easily be included. For example, if the third-type BCs of Eqs. (11.7) and (11.8) have the nonlinear form

$$u_z(x, y, z = 0, t) + u^4(x, y, z = 0, t) = 1 \tag{11.15a}$$

$$u_z(x, y, z = 1, t) + u^4(x, y, z = 1, t) = 1 \tag{11.15b}$$

the preceding code for BCs (11.7) and (11.8) changes simply to

```
nl=2; % Neumann
nu=2; % Neumann
uz1d(1) =1.0-u1d(1)^4;
uz1d(nz)=1.0-u1d(nz)^4;
```

This discussion of Eqs. (11.14) and (11.15) demonstrates one of the important advantages of the numerical approach to PDEs (rather than analytical), namely, the ease with which nonlinearities can be included and how easily the form of the nonlinearities can be changed.

7. The generality and ease of use of the MOL solution of elliptic problems as illustrated by the preceding example is due in part to the use of library ODE integrators for the integration with respect to the embedded parameter t (e.g., `ode15s`). In other words, we can take advantage of quality initial-value ODE integrators that are widely available.
8. Since MOL can be applied to parabolic problems (such as the heat equation $u_t = u_{xx}$) and hyperbolic problems (such as the wave equation $u_{tt} = u_{xx}$), MOL can, in principle, be applied to *all three major classes of PDEs, elliptic, parabolic, and hyperbolic*. Also, since MOL can be applied to systems of equations of mixed type, such as hyperbolic–parabolic PDEs, it is a general framework for the numerical solution of PDE systems.
9. Additionally, this generality for PDEs can be readily extended to systems of ODEs and 1D, 2D, and 3D PDEs (essentially by replicating the methods discussed here for each of the equations). Thus, we can, in principle, investigate numerically a broad class of differential equation systems.

12

Partial Differential Equation with a Mixed Partial Derivative

This partial differential equation (PDE) application introduces the following mathematical concepts and computational methods:

1. A PDE with an exact solution that can be used to assess the accuracy of a numerical method of lines (MOL) solution.
2. MOL solution of a PDE with a mixed partial derivative.
3. Implicit ordinary differential equations (ODEs) and their origin in a MOL solution.
4. Matrix formulation of implicit ODE solutions.
5. Differential-algebraic (DAE) systems.

PDEs with mixed partial derivatives are commonplace in the physical sciences. Therefore, we consider the computation of numerical solutions to such PDEs within the MOL framework.

We first consider some examples of PDEs with mixed partials ([1], section 3.5). A two-dimensional (2D) example is

$$\frac{\partial^2 u}{\partial x\,\partial y} = f(u)$$

where x and y are boundary-value (spatial) independent variables. This PDE can therefore be considered *elliptic* since it does not have an initial-value variable (see Chapter 10 for a discussion of this geometric classification). Special cases include ([1], p. 268)

$$\frac{\partial^2 u}{\partial x\,\partial y} = a\,\sinh u$$

the *sinh-Gordon equation*, and

$$\frac{\partial^2 u}{\partial x\,\partial y} = a\,\sin u$$

the *sine-Gordon equation*.

Another 2D elliptic PDE with a mixed partial derivative is the Monge-Ampère equation, which has applications in differential geometry, gas dynamics, and meteorology ([1], p. 451):

$$\left(\frac{\partial^2 u}{\partial x\,\partial y}\right)^2 - \frac{\partial^2 u}{\partial x^2}\frac{\partial^2 u}{\partial y^2} = f(x, y)$$

An initial-value variable, t, can also be included in the PDE. For example, a nonlinear form of the 1D *Calogero equation* ([1], chapter 7)

$$\frac{\partial^2 u}{\partial x\,\partial t} = u\frac{\partial^2 u}{\partial x^2} + a\left(\frac{\partial u}{\partial x}\right)^2$$

and the 2D *Khokhlov-Zabolotskaya equation*, which describes the propagation of sound in a nonlinear medium

$$\frac{\partial^2 u}{\partial x\,\partial t} = u\frac{\partial^2 u}{\partial x^2} + \left(\frac{\partial u}{\partial x}\right)^2 + \frac{\partial^2 u}{\partial y^2}$$

have t in the mixed partial derivative.

The origin of PDEs with mixed partial derivatives is illustrated in Appendix 1, where an *anisotropic diffusion equation* is derived in three dimensions in *cylindrical* and *spherical* coordinates.

To illustrate the numerical integration of a PDE with a mixed partial derivative, we consider the following 1D PDE:

$$\frac{\partial u}{\partial t} = \frac{\partial^3 u}{\partial x^2\,\partial t} + u$$

or in subscript notation,

$$u_t = u_{xxt} + u \tag{12.1}$$

where

u dependent variable
x boundary-value (spatial) independent variable
t initial-value independent variable

Equation (12.1) is first order in t and second order in x (through the mixed partial $\partial^3 u/\partial x^2 \partial t$). It therefore requires one *initial condition* (IC) and two *boundary conditions* (BCs). The IC is taken as

$$u(x, t = 0) = \sin(\pi x/L) \tag{12.2}$$

and the two BCs as

$$u(x = 0, t) = u(x = L, t) = 0 \qquad (12.3)(12.4)$$

The analytical solution to Eqs. (12.1)–(12.4) is

$$u(x, t) = \sin(\pi x/L)e^{at}; \quad a = \frac{1}{1 - (\pi/L)^2} \tag{12.5}$$

Equations (12.1)–(12.4) constitute the complete PDE problem. The solution to this problem, Eq. (12.5), is used in the subsequent programming and analysis to evaluate the numerical MOL solution.

We now consider some Matlab routines for a numerical MOL solution of Eqs. (12.1)–(12.4) with the analytical solution, Eq. (12.5), included. A main program, pde_1_main, is given in Listing 12.1.

```
%
%Clear previous files
  clear all
  clc
%
% Parameters shared with the ODE routine
  global  a     x    xl    xu    dx    cm     n  ncall
%
% Boundaries, number of grid points
  xl=0.0;
  xu=1.0;
  n=49;
%
% Initial condition
  t0=0.0;
  u0=inital_1(t0);
%
% Independent variable for ODE integration
  tf=20.0;
  tout=[t0:4.0:tf]';
  nout=6;
  ncall=0;
%
% Coefficient matrix
  for i=1:n
  for j=1:n
    if(i==j)cm(i,j)=(1.0-2.0/dx^2);
    elseif(abs(i-j)==1)cm(i,j)=1.0/dx^2;
    else cm(i,j)=0.0;
    end
  end
  end
%
% ODE integration
  reltol=1.0e-06; abstol=1.0e-06;
  options=odeset('RelTol',reltol,'AbsTol',abstol);
%
```

```
% Explicit (nonstiff) integration
  [t,u]=ode45(@pde_1,tout,u0,options);
%
% Analytical solution and difference between the numerical and
% analytical solutions at selected points
  for it=1:nout
    fprintf('\n\n   t     x     u(x,t)    u(x,t)       err\n')
    fprintf('                 num       anal            \n')
    for i=1:n
      u_anal(it,i)=ua(t(it),x(i));
      err(it,i)=u(it,i)-u_anal(it,i);
    end
    for i=1:5:n
      fprintf('%6.2f%8.3f%12.6f%12.6f%12.6f\n',...
              t(it),x(i),u(it,i),u_anal(it,i),err(it,i));
    end
  end
  fprintf('\n  ncall = %4d\n\n',ncall);
%
% Plot numerical and analytical solutions
  figure(1)
  x=[0.0 x 1.0];
  for it=1:nout
    uplot(it,1)=0.0;
    uplot(it,2:n+1)=u(it,1:n);
    uplot(it,n+2)=0.0;
    uaplot(it,1)=0.0;
    uaplot(it,2:n+1)=u_anal(it,1:n);
    uaplot(it,n+2)=0.0;
  end
  plot(x,uplot,'o',x,uaplot,'-')
  xlabel('x')
  ylabel('u(x,t)')
  title('Mixed partial PDE; t = 0, 5,..., 20;
      o - numerical; solid - analytical')
```

Listing 12.1. Main program pde_1_main

We can note the following points about this program:

1. After specifying some *global* variables, the program sets the parameters for the spatial grid for the interval $0 \leq x \leq 1$ with 49 *interior points* (grid points at x=xl,xu are not included in the grid). Note that these parameters are global so that they can be shared with other routines. IC (12.2) is then defined through a call to inital_1 (discussed subsequently).

```
%
% Clear previous files
  clear all
  clc
%
% Parameters shared with the ODE routine
  global  a     x    xl    xu    dx    cm     n  ncall
%
% Boundaries, number of grid points
  xl=0.0;
  xu=1.0;
  n=49;
%
% Initial condition
  t0=0.0;
  u0=inital_1(t0);
```

2. The interval in t is defined as $0 \le t \le 20$ with an output interval of 4 so that the solution will be displayed six times (counting $t = 0$). The counter for the ODE routine, `ncall`, is also initialized.

```
%
% Independent variable for ODE integration
  tf=20.0;
  tout=[t0:4.0:tf]';
  nout=6;
  ncall=0;
```

3. A coefficient matrix, `cm`, (also termed a *coupling matrix* or a *mass matrix*) is defined that originates with the MOL approximation of the mixed partial derivative $\partial^3 u/\partial x^2\,\partial t$ in Eq. (12.1).

```
%
% Coefficient matrix
  for i=1:n
  for j=1:n
    if(i==j)cm(i,j)=(1.0-2.0/dx^2);
    elseif(abs(i-j)==1)cm(i,j)=1.0/dx^2;
    else cm(i,j)=0.0;
    end
  end
  end
```

The origin and use of this matrix is considered after this discussion of the main program in Listing 12.1 is concluded.

4. The MOL ODEs are integrated by `ode45`, which, although a nonstiff integrator, is quite efficient for this application, as demonstrated by the output discussed subsequently. The ODE routine called by `ode45` is `pde_1`.

```
%
% ODE integration
  reltol=1.0e-06; abstol=1.0e-06;
  options=odeset('RelTol',reltol,'AbsTol',abstol);
%
% Explicit (nonstiff) integration
  [t,u]=ode45(@pde_1,tout,u0,options);
```

5. The analytical solution of Eq. (12.5) is evaluated by `ua` (discussed subsequently) and the numerical and analytical solutions and their difference are then displayed. The counter `ncall` for the calls to `pde_1` is then displayed.

```
%
% Analytical solution and difference between the numerical and
% analytical solutions at selected points
  for it=1:nout
    fprintf('\n\n   t      x     u(x,t)     u(x,t)          err\n')
    fprintf('                           num        anal               \n')
    for i=1:n
      u_anal(it,i)=ua(t(it),x(i));
      err(it,i)=u(it,i)-u_anal(it,i);
    end
    for i=1:5:n
      fprintf('%6.2f%8.3f%12.6f%12.6f%12.6f\n', ...
              t(it),x(i),u(it,i),u_anal(it,i),err(it,i));
    end
  end
  fprintf('\n  ncall = %4d\n\n',ncall);
```

6. Finally, the numerical and analytical solutions are plotted together so that they may be compared. The `for` loop merely fills in BCs (12.3) and (12.4) (at `x=xl,xu`) since the spatial grid does not include these boundary points; this is done to improve slightly the appearance of the plotted output.

```
%
% Plot numerical and analytical solutions
  figure(1)
  x=[0.0 x 1.0];
```

```
for it=1:nout
  uplot(it,1)=0.0;
  uplot(it,2:n+1)=u(it,1:n);
  uplot(it,n+2)=0.0;
  uaplot(it,1)=0.0;
  uaplot(it,2:n+1)=u_anal(it,1:n);
  uaplot(it,n+2)=0.0;
end
plot(x,uplot,'o',x,uaplot,'-')
xlabel('x')
ylabel('u(x,t)')
title('Mixed partial PDE; t = 0, 5,..., 20;
     o - numerical; solid - analytical')
```

The subordinate routines `inital_1, ua, pde_1` called by `ode45` and the output from the main program `pde_1_main` are discussed after the following development of the MOL analysis of Eqs. (12.1)–(12.4). Equation (12.1) can be approximated as a system of ODEs if the x differentiation is approximated with the three-point finite difference (FD)

$$u_{xx} \approx \frac{u((i+1)\Delta x) - 2u(\Delta xi) + u((i-1)\Delta x)}{\Delta x^2}, \qquad \Delta x = L/(n+1),\ i = 1, 2, \ldots, n \tag{12.6}$$

where $u(0, t) = u((n+1)\Delta x, t) = 0$ from BCs (12.3) and (12.4). Note, again, that the index i spans the *interior* points of the grid and does not include the boundary points at $x = 0, L$.

Application of Eq. (12.6) to Eq. (12.1) gives

$$u_t(i\Delta x, t) = -\frac{u_t((i+1)\Delta x, t) - 2u_t(i\Delta x, t) + u_t((i-1)\Delta x, t)}{\Delta x^2} + u(i, t) \tag{12.7}$$

The distinguishing feature of Eqs. (12.7) is that the n ODEs are coupled through the derivatives in t. In other words, *more than one derivative in t appears in each ODE* and so it is not possible to solve explicitly for individual t derivatives using each equation one at a time. Rather we must *solve the entire ODE system simultaneously* for the three t derivatives in each ODE – a total of n derivatives in t.

Since the t derivatives in Eqs. (12.7) appear implicitly, ODEs of the form of Eqs. (12.7) are termed *implicit* ODEs (in contrast to ODE systems discussed in earlier chapters in which the ODE t derivatives appeared explicitly, i.e., individually, or only one derivative in each ODE, and which are therefore termed *explicit* ODEs). This classification of ODEs should not be confused with explicit and implicit integration algorithms, which relates to a different issue, namely nonstiff and stiff ODEs.

Another important issue is that we have so far used the Matlab integrators for explicit ODE systems. Although some of the Matlab integrators will accept implicit ODEs, we will not address this feature directly, but rather we will still call the Matlab integrators, in the present case `ode45`, as we did previously, but in the ODE routine, for example, pde_1, we will convert the implicit ODEs to the explicit form. The details are given in the discussion of pde_1 to follow, and involve the coefficient matrix cm defined in pde_1_main.

Some additional terminology can be explained at this point. The implicit ODEs of Eqs. (12.7) are coupled in the sense that any given t derivative cannot be evaluated without solving for all n t derivatives. Thus, the ODEs are simultaneous or *coupled*. Further, in the case of Eqs. (12.7), the derivatives are to the first power, so Eqs. (12.7) are termed *linearly implicit* ODEs (which are relatively easy to solve in contrast to ODEs in which the t derivatives appear nonlinearly – in that case, a nonlinear algebraic solver, usually based on a variant of Newton's method, must also be used). The coefficient matrix that couples the ODEs of Eqs. (12.7), cm, is therefore also called a *coupling matrix* or possibly a *mass matrix* reflecting the origin of implicit ODEs from application to mechanical systems with discrete masses.

The uncoupling of the simultaneous ODEs of Eqs. (12.7) can be accomplished by the use of some standard matrix methods of linear algebra. First, Eqs. (12.7) can be written as

$$\begin{aligned}&(1/\Delta x^2)u_t((i-1)\Delta x,t)+(1-2/\Delta x^2)u_t(i\Delta x,t)+(1/\Delta x^2)u_t((i+1)\Delta x,t)\\&\quad=u(i\Delta x,t)\end{aligned}\tag{12.8}$$

or in matrix form,

$$\begin{bmatrix}(1-\frac{2}{\Delta x^2}) & \frac{1}{\Delta x^2} & & & \\ \frac{1}{\Delta x^2} & (1-\frac{2}{\Delta x^2}) & \frac{1}{\Delta x^2} & & \\ & \ddots & \ddots & \ddots & \\ & & \frac{1}{\Delta x^2} & (1-\frac{2}{\Delta x^2}) & \frac{1}{\Delta x^2} \\ & & & \frac{1}{\Delta x^2} & (1-\frac{2}{\Delta x^2})\end{bmatrix}\begin{bmatrix}u_t(\Delta x,t)\\u_t(2\Delta x,t)\\ \vdots \\ u_t((n-1)\Delta x,t)\\u_t(n\Delta x,t)\end{bmatrix}$$
$$=\begin{bmatrix}u(\Delta x,t)\\u(2\Delta x,t)\\ \vdots \\ u((n-1)\Delta x,t)\\u(n\Delta x,t)\end{bmatrix}\tag{12.9}$$

We can then solve Eq. (12.9) for the vector of derivatives in t

$$\begin{bmatrix}u_t(\Delta x,t)\\u_t(2\Delta x,t)\\ \vdots \\ u_t((n-1)\Delta x,t)\\u_t(n\Delta x,t)\end{bmatrix}$$

$$
= \begin{bmatrix} (1-\frac{2}{\Delta x^2}) & \frac{1}{\Delta x^2} & & & \\ \frac{1}{\Delta x^2} & (1-\frac{2}{\Delta x^2}) & \frac{1}{\Delta x^2} & & \\ & \ddots & \ddots & \ddots & \\ & & \frac{1}{\Delta x^2} & (1-\frac{2}{\Delta x^2}) & \frac{1}{\Delta x^2} \\ & & & \frac{1}{\Delta x^2} & (1-\frac{2}{\Delta x^2}) \end{bmatrix}^{-1} \begin{bmatrix} u(\Delta x, t) \\ u(2\Delta x, t) \\ \vdots \\ u((n-1)\Delta x, t) \\ u(n\Delta x, t) \end{bmatrix} \tag{12.10}
$$

Equation (12.10) is programmed in the ODE routine pde_1 through the use of the Matlab matrix inverse operator (as applied to the RHS of Eq. (12.10)). Once the vector of (explicit) derivatives $u_t(1\Delta x, t), u_t(2\Delta x, t), \ldots, u_t(n\Delta x, t)$ is computed (from Eq. (12.10)), it can be sent to a Matlab integrator such as ode45 in the usual fashion.

To illustrate how this is done, we now consider pde_1, as given in Listing 12.2.

```
  function ut=pde_1(t,u)
%
% Function yt computes the t derivative vector of the PDE with
% a mixed partial derivative
%
  global  a     x     xl     xu     dx     cm     n  ncall
%
% PDE
  ut=cm\u;
%
% Increment calls to pde_1
  ncall=ncall+1;
```

Listing 12.2. MOL ODE routine pde_1 called by main program pde_1_main

After the function is defined and some parameters and variables are declared global, Eq. (12.10) is programmed as ut=cm\u; (which perhaps qualifies as some of the most compact coding of a PDE to be found anywhere!). Note that the Matlab inverse operator \ is used in conjunction with Eq. (12.10) (you might check that the coefficient matrix of Eq. (12.10) is in fact cm programmed in pde_1_main of Listing 12.1). Even the usual transpose is not required since \ returns a column vector, as required by ode45. cm was not evaluated in pde_1 since it is a constant matrix and is therefore evaluated just once in the main program and passed as a global variable to pde_1. If the elements of this coupling matrix were functions of the dependent variable (so that the ODE system would be nonlinear), cm would then be evaluated in pde_1.

The initialization routine, inital_1, that implements Eq. (12.2) is given in Listing 12.3.

```
  function u0=inital_1(t0)
%
% Function inital_1 sets the initial condition for the PDE
% with a mixed partial derivative
%
  global  a     x    xl    xu    dx    cm    n
%
% Grid spacing, constant a
  dx=(xu-xl)/(n+1);
  a=1.0/(1.0-(pi/(xu-xl))^2);
%
% IC over the spatial grid
  for i=1:n
%
%   Uniform grid
    x(i)=xl+i*dx;
%
%   Initial condition
    u0(i)=sin(pi*x(i)/(xu-xl));
  end
```

Listing 12.3. Initial condition routine inital_1

We can note the following points about inital_1:

1. After the function is defined and certain parameters and variables are declared global, the grid spacing dx is computed based on the n interior points (and thus the use of $n+1$ in computing the grid spacing dx). Also, the constant a of Eq. (12.5) is computed.


```
  function u0=inital_1(t0)
%
% Function inital_1 sets the initial condition for the PDE
% with a mixed partial derivative
%
  global  a     x    xl    xu    dx    cm    n
%
% Grid spacing, constant a
  dx=(xu-xl)/(n+1);
  a=1.0/(1.0-(pi/(xu-xl))^2);
```

2. IC (2) is then implemented in a `for` loop.

```
%
% IC over the spatial grid
  for i=1:n
%
%   Uniform grid
    x(i)=xl+i*dx;
%
%   Initial condition
    u0(i)=sin(pi*x(i)/(xu-xl));
  end
```

Finally, `ua` is a straightforward coding of the analytical solution of Eq. (12.5).

```
  function uanal=ua(t,x)
%
% Function ua computes the analytical solution to the PDE with
% a mixed partial derivative
%
  global  a    xl    xu
%
% Analytical solution
  uanal=sin(pi*x/(xu-xl))*exp(a*t);
```

This completes the Matlab coding of Eqs. (12.1)–(12.4). We now consider the output from `pde_1_main`. Selected numerical output is given in Table 12.1. Agreement between the numerical and analytical solution is to three figures (with only 67 calls to `pde_1`) so that the three-point FDs of Eq. (12.6) provide satisfactory accuracy. Clearly, the 49 ODEs are not stiff (i.e., `ncall=67`).

The plotted output from `pde_1` is shown in Figure 12.1. The maximum peak corresponds to $t = 0$ and decays according to Eq. (12.5). The long-term solution is $u(x, t = \infty) = 0$ (provided $a < 0$).

We conclude this chapter with the following observations:

1. Mixed partial derivatives like that of Eq. (12.1) (a combination of an initial-value independent variable such as t and a boundary-value independent variable such as x) appear in PDEs that model a spectrum of physical systems.
2. When there is an initial-value independent variable in the mixed partial, the methods presented in this chapter can generally be applied, including the use of an initial-value ODE integrator such as `ode45`.

Table 12.1. Partial output from `pde_1_main` and `pde_1`

```
    t        x        u(x,t)      u(x,t)        err
                       num         anal
  0.00    0.020    0.062791    0.062791    0.000000
  0.00    0.120    0.368125    0.368125    0.000000
  0.00    0.220    0.637424    0.637424    0.000000
  0.00    0.320    0.844328    0.844328    0.000000
  0.00    0.420    0.968583    0.968583    0.000000
  0.00    0.520    0.998027    0.998027    0.000000
  0.00    0.620    0.929776    0.929776    0.000000
  0.00    0.720    0.770513    0.770513    0.000000
  0.00    0.820    0.535827    0.535827    0.000000
  0.00    0.920    0.248690    0.248690    0.000000


    t        x        u(x,t)      u(x,t)        err
                       num         anal
  4.00    0.020    0.039991    0.039998   -0.000007
  4.00    0.120    0.234458    0.234497   -0.000039
  4.00    0.220    0.405975    0.406042   -0.000067
  4.00    0.320    0.537752    0.537841   -0.000089
  4.00    0.420    0.616890    0.616992   -0.000102
  4.00    0.520    0.635643    0.635748   -0.000105
  4.00    0.620    0.592174    0.592272   -0.000098
  4.00    0.720    0.490739    0.490820   -0.000081
  4.00    0.820    0.341268    0.341324   -0.000056
  4.00    0.920    0.158390    0.158417   -0.000026
                .                       .
                .                       .
                .                       .

   Output for t = 8, 12, 16 has been removed
                .                       .
                .                       .
                .                       .

    t        x        u(x,t)      u(x,t)        err
                       num         anal
 20.00    0.020    0.006580    0.006586   -0.000005
 20.00    0.120    0.038579    0.038611   -0.000032
 20.00    0.220    0.066801    0.066856   -0.000055
 20.00    0.320    0.088484    0.088557   -0.000073
 20.00    0.420    0.101506    0.101590   -0.000084
 20.00    0.520    0.104592    0.104678   -0.000086
 20.00    0.620    0.097439    0.097519   -0.000080
 20.00    0.720    0.080749    0.080815   -0.000067
 20.00    0.820    0.056154    0.056200   -0.000046
 20.00    0.920    0.026062    0.026084   -0.000022

 ncall =    67
```

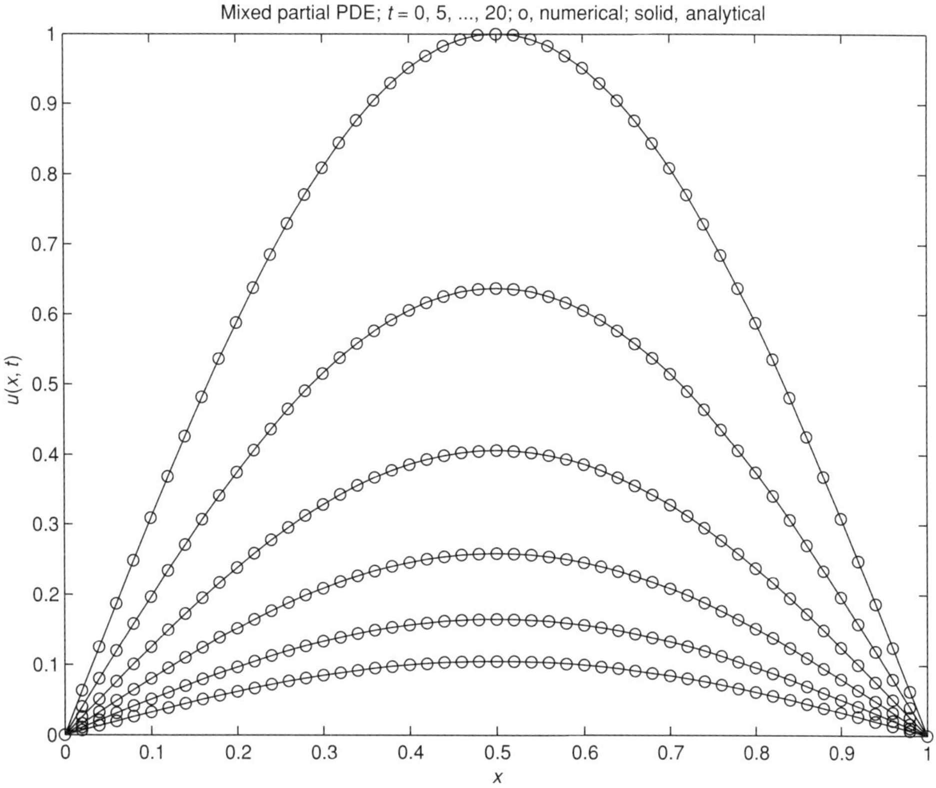

Figure 12.1. Output of main program `pde_1_main`

3. A more difficult problem results from a mixed partial that involves two or more boundary-value independent variables. For example, a PDE in two dimensions, with spatial variables x and y, could have a mixed partial $\partial^2 u/\partial x\, \partial y$. This mixed partial could be approximated by FDs. One approach would be to use *stagewise differentiation*; that is, differentiate u with respect to x to produce $\partial u/\partial x$ and then differentiate this derivative with respect to y to produce $\partial^2 u/\partial x\, \partial y$. The difficulty with this type of mixed partial originates with the stability of the PDE solution, but we will not pursue this matter further.
4. The FD approximation of the PDE with a mixed partial derivative such as Eq. (12.1) generally leads to implicitly coupled ODEs such as Eqs. (12.8). However, coupled ODEs can arise in a variety of applications, particularly in finite element analysis. Again, we will not go into this matter here other than to point out that implicitly coupled ODEs can be handled with the matrix methods illustrated previously or by using ODE integrators particularly designed for this type of ODE system; in fact, some of the Matlab integrators have this feature.
5. The matrix approach considered here in which the ODEs were uncoupled in `pde_1` through the use of the linear algebraic operator (solver) \ is based on full matrix methods. Clearly, however, the coupling matrix `cm` defined numerically in `pde_1_main` is banded (it is tridiagonal from Eq. (12.6)), which is typical in physical applications. Thus, the use of full-matrix linear algebra

can be inefficient when applied to banded matrix systems, particularly as the number of ODEs increases. ($n = 49$ is a very modest problem, so the computational efficiency was quite acceptable.)

6. Another type of differential equation formulation was introduced through BCs (12.3) and (12.4). Specifically, the algebraic equation $u(x = 0, t) = u(x = L, t) = 0$ was added to the ODE system of Eqs. (12.7) (you might check how these BCs were included in Eqs. (12.7)). In other words, algebraic equations were included at the external points $x = 0, x = L$ of the grid, which is a common occurrence in MOL analysis to produce a DAE system.

   However, algebraic equations can also occur at the interior points of a spatial grid. For example, for a reaction–diffusion system, the reactions may be more rapid than the diffusion. The reactions can therefore be considered at equilibrium, so that the equations describing the reactions are algebraic, while the equations describing the diffusion are ODEs (through the MOL analysis of the PDEs). We will not consider DAE systems further here other than to mention that algorithms and associated computer codes for DAE systems are available. Also, depending on a measure of difficulty of a DAE system, the so-called *index*, an algebraic equation solver can possibly be added at the beginning of the MOL ODE routine to solve the algebraic equations first, followed by the coding of the derivative vector of the ODEs.
7. a in Eq. (12.5) can be positive (if $L > \pi$), in which case, the solution of Eq. (12.1) would increase exponentially in t (according to Eq. (12.5)). Thus, this problem can be stable or unstable in its long-time behavior (and this is a property of the PDE problem, not the numerical method used to solve the problem).

   Finally, to further elucidate the numerical solution, we also include a 3D plot produced with the following code:

```
%
% 3D Plot
  figure(2)
  surfl(x,t,uplot, 'light'); shading interp
  axis tight
  title('Mixed partials equation');
  set(get(gca,'XLabel'),'String','space, x')
  set(get(gca,'YLabel'),'String','time, t')
  set(get(gca,'ZLabel'),'String','u(x,t)')
  set(gca,'XLim',[xl xu],'YLim',[t0 tf],'ZLim', [0 1]);
  view(60,40);
  colormap('cool');
  print -dpng -r300 fig2.png;
```

The resulting 3D plot is shown in Figure 12.2.

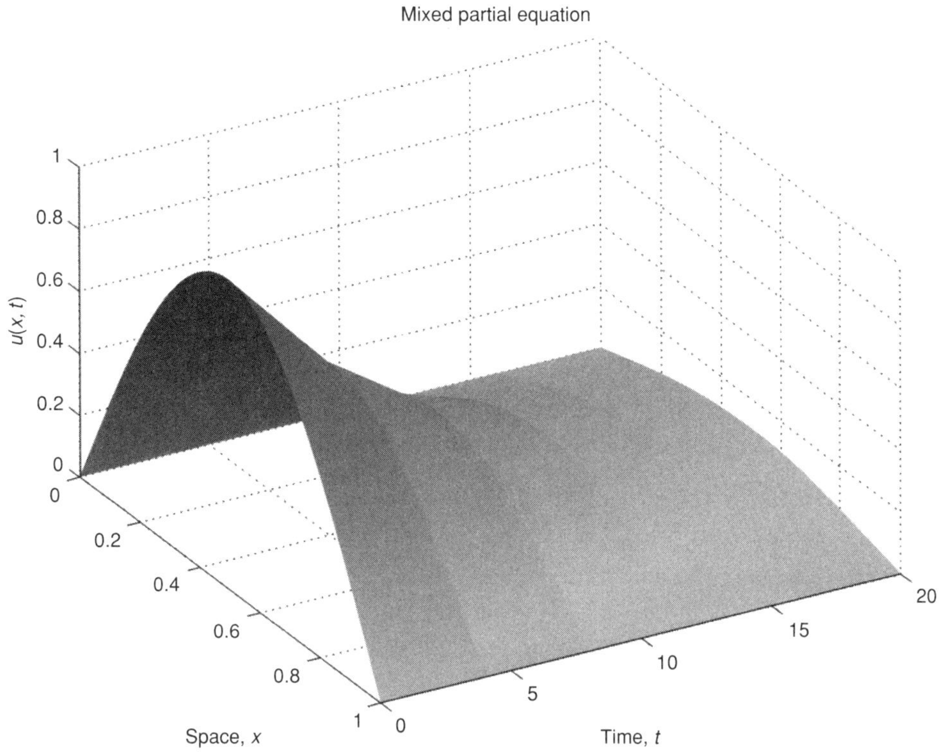

Figure 12.2. Three-dimensional output from main program `pde_1_main`

REFERENCE

[1] Polyanin, A. D. and V. F. Zaitsev (2004), *Handbook of Nonlinear Partial Differential Equations*, Chapman and Hall/CRC, Boca Raton, FL

13

Simultaneous, Nonlinear, Two-Dimensional Partial Differential Equations in Cylindrical Coordinates

This partial differential equation (PDE) application introduces the following mathematical concepts and computational methods:

1. Method of lines (MOL) solution of two simultaneous, nonlinear, two-dimensional (2D) PDEs in cylindrical coordinates.
2. The solution of PDEs with variable coefficients.
3. Analysis to regularize a singular point ($r = 0$ in cylindrical coordinates).
4. Detailed derivation of the PDEs based on conservation principles.
5. The origin of convection–diffusion–reaction PDEs.
6. Analysis of physical units to ensure consistency of a mathematical model.
7. Implementation of Dirichlet, Neumann, and third-type (mixed, Robin) boundary conditions.
8. Inclusion of nonlinear source terms in the PDE model.
9. Examination of the numerical and graphical output from the computer code to gain insight into the performance of the physical system described by the PDE model, which then suggests other conditions (e.g., model parameters) that might be investigated using the model and associated code.

We start with the derivation of the PDEs. The physical system is a convection–diffusion–reaction system, for example, a tubular reactor, which is modeled in cylindrical coordinates, (r, z) (we assume angular symmetry so that the third cylindrical coordinate, θ, is not required). The PDE system has two dependent variables, c_a and T_k, which physically are the reacting fluid concentration, c_a, and the fluid temperature, T_k. Thus we require two PDEs.

These two dependent variables are functions of three independent variables: r, the radial coordinate; z, the axial coordinate; and t, time, as illustrated by Figure 13.1. The solution will be $c_a(r, z, t)$ and $T_k(r, z, t)$ in numerical form as a function of r, z, t.

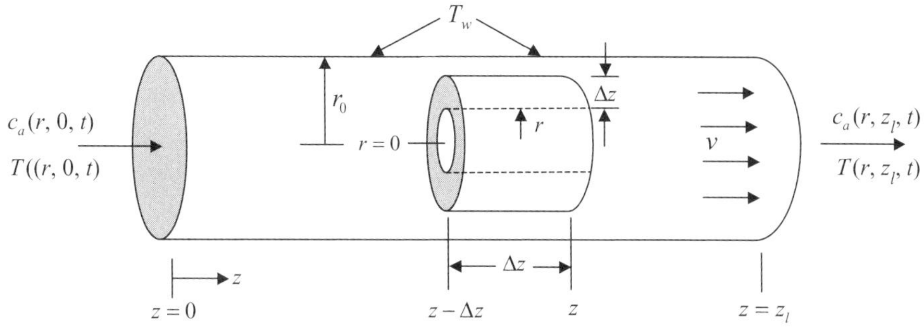

Figure 13.1. Representation of a convection–diffusion–reaction system in cylindrical coordinates

The mass balance for an incremental element of length Δz is (see Figure 13.1)

$$2\pi r\Delta r\Delta z\frac{\partial c_a}{\partial t} = 2\pi r\Delta z q_m|_r - 2\pi(r+\Delta r)\Delta z q_m|_{r+\Delta r}$$

$$+ 2\pi r\Delta r v c_a|_{z-\Delta z} - 2\pi r\Delta r v c_a|_z - 2\pi r\Delta r\Delta z k_r c_a^2$$

Division by $2\pi r\Delta r\Delta z$ and minor rearrangement gives

$$\frac{\partial c_a}{\partial t} = -\frac{(r+\Delta r)q_m|_{r+\Delta r} - rq_m|_r}{r\Delta r} - v\left(\frac{c_a|_z - c_a|_{z-\Delta z}}{\Delta z}\right) - k_r c_a^2$$

or in the limit $r \to 0$, $\Delta z \to 0$,

$$\frac{\partial c_a}{\partial t} = -\frac{1}{r}\frac{\partial(rq_m)}{\partial r} - v\frac{\partial c_a}{\partial z} - k_r c_a^2$$

We discuss in an appendix to this chapter the physical meaning of each of the terms in this PDE. Briefly, here is a summary. For the differential volume in r and z,

$\frac{\partial c_a}{\partial t}$	rate of mass accumulation per unit volume
$-\frac{1}{r}\frac{\partial(rq_m)}{\partial r}$	rate of radial mass diffusion per unit volume
$-v\frac{\partial c_a}{\partial z}$	rate of axial mass convection per unit volume
$-k_r c_a^2$	rate of reaction mass generation per unit volume

If we now assume *Fick's first law* for the flux with a mass diffusivity, D_c,

$$q_m = -D_c\frac{\partial c_a}{\partial r} \tag{13.1}$$

we obtain

$$\frac{\partial c_a}{\partial t} = \frac{D_c}{r}\frac{\partial\left(\frac{r\partial c_a}{\partial r}\right)}{\partial r} - v\frac{\partial c_a}{\partial z} - k_r c_a^2$$

or expanding the radial group,

$$\frac{\partial c_a}{\partial t} = D_c\left(\frac{\partial^2 c_a}{\partial r^2} + \frac{1}{r}\frac{\partial c_a}{\partial r}\right) - v\frac{\partial c_a}{\partial z} - k_r c_a^2 \tag{13.2}$$

Equation (13.2) is the required material balance for $c_a(r, z, t)$ and would be sufficient for the calculation of $c_a(r, z, t)$ except that we will consider the case of an *exothermic reaction* that liberates heat so that system is *not isothermal*. That is, the reacting fluid temperature varies as a function of r, z, and t so that a second PDE is required, an energy balance, for the computation of the temperature $T_k(r, z, t)$. Further, Eq. (13.2) is coupled to the temperature through the reaction rate coefficient k_r as explained in the derivation of the energy balance.

Note also that the volumetric rate of reaction in Eq. (13.2), $k_r c_a^2$, is *second order*; that is, the rate of reaction is proportional to the square of the concentration c_a. This *kinetic term is therefore a source of nonlinearity* in the model.

The thermal energy balance for the incremental section is

$$\begin{aligned} 2\pi r\Delta r\Delta z\rho C_p\frac{\partial T_k}{\partial t} &= 2\pi r\Delta z q_h|_r - 2\pi(r+\Delta r)\Delta z q_h|_{r+\Delta r} \\ &\quad + 2\pi r\Delta r v\rho C_p T_k|_{z-\Delta z} - 2\pi r\Delta r v\rho C_p T_k|_z \\ &\quad - \Delta H 2\pi r\Delta r\Delta z k_r c_a^2 \end{aligned}$$

Division by $2\pi r\Delta r\Delta z\rho$ gives

$$C_p\frac{\partial T_k}{\partial t} = -\frac{(r+\Delta r)q_h|_{r+\Delta r} - rq_h|_r}{r\Delta r\rho} - vC_p\frac{(T_k|_z - T_k|_{z-\Delta z})}{\Delta z} - \frac{\Delta H k_r}{\rho}c_a^2$$

or in the limit $\Delta r \to 0$, $\Delta z \to 0$,

$$C_p\frac{\partial T_k}{\partial t} = -\frac{1}{\rho r}\frac{\partial(rq_h)}{\partial r} - vC_p\frac{\partial T_k}{\partial z} - \frac{\Delta H k_r}{\rho}c_a^2$$

We discuss in an appendix to this chapter the physical meaning of each of the terms in this PDE. Briefly, here is a summary. For the differential volume in r and z,

$C_p\frac{\partial T_k}{\partial t}$	rate of thermal energy accumulation per unit mass
$-\frac{1}{\rho r}\frac{\partial(rq_h)}{\partial r}$	rate of radial heat conduction per unit mass
$-vC_p\frac{\partial T_k}{\partial z}$	rate of axial thermal energy convection per unit mass
$-\frac{\Delta H k_r}{\rho}c_a^2$	rate of reaction thermal energy generation per unit mass

Also, there is one technical detail about this energy balance we should mention. The derivative in t is based on the specific heat C_p reflecting the fluid enthalpy. More generally, an energy balance is based on the specific heat C_v reflecting the fluid internal energy. For a liquid, for which pressure effects are negligible, the two are essentially the same, and therefore we use C_p.

If we now assume *Fourier's first law* for the flux with a thermal conductivity, k,

$$q_h = -k\frac{\partial T_k}{\partial r} \tag{13.3}$$

we obtain (after division by C_p)

$$\frac{\partial T_k}{\partial t} = \frac{D_t}{r}\frac{\partial\left(r\frac{\partial T_k}{\partial r}\right)}{\partial r} - v\frac{\partial T_k}{\partial z} - \frac{\Delta H k_r}{\rho C_p}c_a^2$$

or expanding the radial group,

$$\frac{\partial T_k}{\partial t} = D_t\left(\frac{\partial^2 T_k}{\partial r^2} + \frac{1}{r}\frac{\partial T_k}{\partial r}\right) - v\frac{\partial T_k}{\partial z} - \frac{\Delta H k_r}{\rho C_p}c_a^2 \tag{13.4}$$

where $D_t = k/\rho C_p$. Equation (13.4) is the required energy balance for $T(r, z, t)$.

The reaction rate constant, k_r, is given by

$$k_r = k_0\, e^{-E/(RT_k)} \tag{13.5}$$

Note from Eq. (13.5) that k_r is also nonlinearly related to the temperature T_k as e^{-E/RT_k}. This *Arrhenius temperature dependency* is a strong nonlinearity, and explains why chemical reaction rates are generally strongly dependent on temperature; of course, this depends, to some extent, on the magnitude of the *activation energy*, E. The variables and parameters of Eqs. (13.1)–(13.5) and the subsequent auxiliary conditions are summarized in Table 13.1 (in cgs units).

Equation (13.2) requires one initial condition (IC) in t (since it is first order in t), two boundary conditions (BCs) in r (since it is second order in r), and one BC in z (since it is first order in z).

$$c_a(r, z, t = 0) = c_{a0} \tag{13.6}$$

$$\frac{\partial c_a(r = 0, z, t)}{\partial r} = 0, \tag{13.7}$$

$$\frac{\partial c_a(r = r_0, z, t)}{\partial r} = 0 \tag{13.8}$$

$$c_a(r, z = 0, t) = c_{ae} \tag{13.9}$$

Equation (13.4) also requires one IC in t, two BCs in r, and one BC in z.

$$T_k(r, z, t = 0) = T_{k0} \tag{13.10}$$

$$\frac{\partial T_k(r = 0, z, t)}{\partial r} = 0 \tag{13.11}$$

$$k\frac{\partial T_k(r = r_0, z, t)}{\partial r} = h(T_w - T_k(r = r_0, z, t)) \tag{13.12}$$

$$T_k(r, z = 0, t) = T_{ke} \tag{13.13}$$

The solution to Eqs. (13.1)–(13.13) gives $c_a(r, z, t)$, $T_k(r, z, t)$ as a function of the independent variables r, z, t.

Table 13.1. Variables and parameters of the 2D PDE model

Variable/parameter	Symbol	Units/value
Reactant concentration	$c_a(r, z, t)$	gmol/cm^3
Temperature	$T_k(r, z, t)$	K
Reaction rate constant	k_r	$\text{cm}^3/(\text{gmol}\cdot\text{s})$
Time	t	s
Radial position	r	cm
Axial position	z	cm
Mass diffusion flux	q_m	$\text{gmol}/(\text{cm}^2\cdot\text{s})$
Energy conduction flux	q_h	$\text{cal}/(\text{cm}^2\cdot\text{s})$
Entering concentration	c_{ae}	$0.01\,\text{gmol/cm}^3$
Entering temperature	T_{ke}	305 K
Initial concentration	c_{a0}	$0\,\text{gmol/cm}^3$
Initial temperature	T_{k0}	305 K
Wall temperature	T_w	355 K
Reactor radius	r_0	2 cm
Reactor length	z_l	100 cm
Linear fluid velocity	v	1.0 cm/s
Mass diffusivity	D_c	$0.1\,\text{cm}^2/\text{s}$
Thermal diffusivity	$D_t = k/(\rho C_p)$	$0.1\,\text{cm}^2/\text{s}$
Fluid density	ρ	$1.0\,\text{g/cm}^3$
Fluid specific heat	C_p	$0.5\,\text{cal}/(\text{g}\cdot\text{K})$
Heat of reaction	ΔH	−10,000 cal/gmol
Specific rate constant	k_0	$1.5\times 10^9\,\text{cm}^3/(\text{gmol}\cdot\text{s})$
Activation energy	E	$15{,}000\,\text{cal}/(\text{gmol}\cdot\text{K})$
Gas constant	R	$1.987\,\text{cal/gmol}\cdot\text{K}$
Thermal conductivity	k	$0.01\,(\text{cal}\cdot\text{cm})/(\text{s}\cdot\text{cm}^2\cdot\text{K})$
Heat transfer coefficient	h	$0.01\,\text{cal}/(\text{s}\cdot\text{cm}^2\cdot\text{K})$

For physical problems, an important check of a PDE is *a check on the units of each term in the PDE*. Specifically, the units should be *consistent throughout each equation and make sense physically*. A units check of Eqs. (13.1)–(13.13) is detailed in an appendix to this chapter. For further detailed discussion related to chemical reactors, the reader is referred to [1].

A main program for the MOL of the preceding equations is given in Listing 13.1.

```
%
% Clear previous files
  clear all
  clc
%
% Global area
  global      nr     nz     dr     dz    drs    dzs...
               r      z     Dc     Dt     ca     Tk...
             cae    Tke      h      k      E      R...
             rk0      v    rho     Cp     Tw     dH...
            ncall
```

```
%
% Model parameters
  ca0=0.0;
  cae=0.01;
  Tk0=305.0;
  Tke=305.0;
  Tw=355.0;
  r0=2.0;
  zl=100.0;
  v=1.0;
  Dc=0.1;
  Dt=0.1;
  k=0.01;
  h=0.01;
  rho=1.0;
  Cp=0.5;
  rk0=1.5e+09;
  dH=-10000.0;
  E=15000.0;
  R=1.987;
%
% Grid in axial direction
  nz=20;
  dz=zl/nz;
  for i=1:nz
    z(i)=i*dz;
  end
%
% Grid in radial direction
  nr=7;
  dr=r0/(nr-1);
  for j=1:nr
    r(j)=(j-1)*dr;
  end
  drs=dr^2;
%
% Independent variable for ODE integration
  tf=200.0;
  tout=[0.0:50.0:tf]';
  nout=5;
  ncall=0;
%
% Initial condition
  for i=1:nz
  for j=1:nr
    ca(i,j)=ca0;
    Tk(i,j)=Tk0;
```

```
    y0((i-1)*nr+j)=ca(i,j);
    y0((i-1)*nr+j+nz*nr)=Tk(i,j);
  end
  end
%
% ODE integration
  reltol=1.0e-04; abstol=1.0e-04;
  options=odeset('RelTol',reltol,'AbsTol',abstol);
  mf=2;
  if(mf==1)[t,y]=ode15s(@pde_1,tout,y0,options);end
  if(mf==2)[t,y]=ode15s(@pde_2,tout,y0,options);end
%
% 1D to 2D matrices
  for it=1:nout
  for i=1:nz
  for j=1:nr
    ca(it,i,j)=y(it,(i-1)*nr+j);
    Tk(it,i,j)=y(it,(i-1)*nr+j+nz*nr);
  end
  end
  end
%
% Display a heading and centerline output
  fprintf('\n  nr = %2d   nz = %2d\n',nr,nz);
  for it=1:nout
    fprintf('\n t = %4.1f\n',t(it));
  for i=1:nz
    fprintf(' z = %5.1f   ca(r=0,z,t) = %8.5f
              Tk(r=0,z,t) = %8.2f\n',z(i),ca(it,i,1), ...
              Tk(it,i,1));
  end
  end
  fprintf('\n ncall = %5d\n',ncall);
%
% Parametric plots
%
% Axial profiles
  figure(1);
  subplot(2,2,1)
  plot(z,ca(2,:,1)); axis([0 zl 0 0.01]);
  title('ca(r=0,z,t=50)'); xlabel('z');
         ylabel('ca(r=0,z,t=50)')
  subplot(2,2,2)
  plot(z,Tk(2,:,1)); axis([0 zl 305 450]);
  title('Tk(r=0,z,t=50)'); xlabel('z');
         ylabel('Tk(r=0,z,t=50)')
  subplot(2,2,3)
```

```
  plot(z,ca(3,:,1)); axis([0 zl 0 0.01]);
  title('ca(r=0,z,t=100)'); xlabel('z');
         ylabel('ca(r=0,z,t=100)')
  subplot(2,2,4)
  plot(z,Tk(3,:,1)); axis([0 zl 305 450]);
  title('Tk(r=0,z,t=100)'); xlabel('z');
         ylabel('Tk(r=0,z,t=100)')
  figure(2);
  subplot(2,2,1)
  plot(z,ca(4,:,1)); axis([0 zl 0 0.01]);
  title('ca(r=0,z,t=150)'); xlabel('z');
         ylabel('ca(r=0,z,t=150)')
  subplot(2,2,2)
  plot(z,Tk(4,:,1)); axis([0 zl 305 450]);
  title('Tk(r=0,z,t=150)'); xlabel('z');
         ylabel('Tk(r=0,z,t=150)')
  subplot(2,2,3)
  plot(z,ca(5,:,1)); axis([0 zl 0 0.01]);
  title('ca(r=0,z,t=200)'); xlabel('z');
         ylabel('ca(r=0,z,t=200)')
  subplot(2,2,4)
  plot(z,Tk(5,:,1)); axis([0 zl 305 450]);
  title('Tk(r=0,z,t=200)'); xlabel('z');
         ylabel('Tk(r=0,z,t=200)')
%
% Radial profiles (at t = tf)
  for i=1:nz,
  for j=1:nr
    ca_rad(i,j)=ca(5,i,j);
    Tk_rad(i,j)=Tk(5,i,j);
  end
  end
  figure(3);
  subplot(2,2,1)
  plot(r,ca_rad(1,:)); axis([0 r0 0.0 0.01]);
  title('ca(r,z=5,t=200)'); xlabel('r');
         ylabel('ca(r,z=5,t=200)')
  subplot(2,2,2)
  plot(r,ca_rad(7,:)); axis([0 r0 0.0 0.01]);
  title('ca(r,z=35,t=200)'); xlabel('r');
         ylabel('ca(r,z=35,t=200)')
  subplot(2,2,3)
  plot(r,ca_rad(14,:)); axis([0 r0 0.0 0.01]);
  title('ca(r,z=70,t=200)'); xlabel('r');
         ylabel('ca(r,z=70,t=200)')
  subplot(2,2,4)
  plot(r,ca_rad(20,:)); axis([0 r0 0.0 0.01]);
```

```
  title('ca(r,z=100,t=200)'); xlabel('r');
         ylabel('ca(r,z=100,t=200)')
  figure(4);
  subplot(2,2,1)
  plot(r,Tk_rad(1,:)); axis([0 r0 305 450]);
  title('Tk(r,z=5,t=200)'); xlabel('r');
         ylabel('Tk(r,z=5,t=200)')
  subplot(2,2,2)
  plot(r,Tk_rad(7,:)); axis([0 r0 305 450]);
  title('Tk(r,z=35,t=200)'); xlabel('r');
         ylabel('Tk(r,z=35,t=200)')
  subplot(2,2,3)
  plot(r,Tk_rad(14,:)); axis([0 r0 305 450]);
  title('Tk(r,z=70,t=200)'); xlabel('r');
         ylabel('Tk(r,z=70,t=200)')
  subplot(2,2,4)
  plot(r,Tk_rad(20,:)); axis([0 r0 305 450]);
  title('Tk(r,z=100,t=200)'); xlabel('r');
         ylabel('Tk(r,z=100,t=200)')
%
% 3D plotting
  figure()
  surf(r,z,Tk_rad)
  axis([0 r0 0 zl 0 500]);
  xlabel('r - radial distance');
  ylabel('z - axial distance');
  zlabel('Tk - reactant temperature');
  title([{'Surface & contour plot of reactant temperature,
        Tk(r,z),'},{ 'at time t=200.0'}]);
  view(-130,62);
  figure()
  surf(r,z,ca_rad)
  axis([0 r0 0 zl 0 0.01]);
  xlabel('r - radial distance');
  ylabel('z - axial distance');
  zlabel('ca - reactant concentration');
  title([{'Surface & contour plot of reactant concentration,
        ca(r,z),'},{ 'at time t=200.0'}]);
  view(-130,62);
  rotate3d on
```

Listing 13.1. Main program pde_1_main.m for the solution of Eqs. (13.1)–(13.13)

We can note the following points about this main program:

1. After specification of a *global* area, the model parameters are defined numerically in accordance with the values listed in Table 13.1. These numerical values are then passed as global variables to the MOL ODE routine, as well as the dependent variables CA and TK of Eqs. (13.2) and (13.4), and parameters relating to the spatial grid in r and z (discussed next).

```
%
% Clear previous files
  clear all
  clc
%
% Global area
  global     nr    nz    dr    dz    drs    dzs...
              r     z    Dc    Dt     ca     Tk...
            cae   Tke     h     k      E      R...
            rk0     v   rho    Cp     Tw     dH...
          ncall
%
% Model parameters
  ca0=0.0;
  cae=0.01;
  Tk0=305.0;
  Tke=305.0;
  Tw=355.0;
  r0=2.0;
  zl=100.0;
  v=1.0;
  Dc=0.1;
  Dt=0.1;
  k=0.01;
  h=0.01;
  rho=1.0;
  Cp=0.5;
  rk0=1.5e+09;
  dH=-10000.0;
  E=15000.0;
  R=1.987;
```

2. The spatial grids in z and r are defined. Note that in the case of z, the grid starts at z(1)=dz, while for r, the grid starts at r(1)=0. The reason for this difference (the grid in z does not start at z=0) is explained subsequently. The total number of grid points is (7)(20) = 140, and since there are two PDEs, Eqs. (13.2) and (13.4), there will be (2)(140) = 280 ordinary differential equations (ODEs) programmed in the ODE routine.

```
%
% Grid in axial direction
  nz=20;
  dz=zl/nz;
  for i=1:nz
    z(i)=i*dz;
  end
%
% Grid in radial direction
  nr=7;
  dr=r0/(nr-1);
  for j=1:nr
    r(j)=(j-1)*dr;
  end
  drs=dr^2;
```

3. The interval in t is $0 \le t \le 200$ with an output interval of 50 so that the solution will be displayed five times.

```
%
% Independent variable for ODE integration
  tf=200.0;
  tout=[0.0:50.0:tf]';
  nout=5;
  ncall=0;
```

4. ICs (13.6) and (13.10) are programmed as

```
%
% Initial condition
  for i=1:nz
  for j=1:nr
    ca(i,j)=ca0;
    Tk(i,j)=Tk0;
    y0((i-1)*nr+j)=ca(i,j);
    y0((i-1)*nr+j+nz*nr)=Tk(i,j);
  end
  end
```

The two 2D ICs are then converted to a single 1D IC vector, y0 (with 280 elements). The logic of this conversion of the two 2D matrices, ca(i,j) and

Tk(i,j), to a 1D vector (matrix) y0 in the two nested for loops should be studied to ensure an understanding of this process. Also, refer to Chapter 10 where this operation is discussed more fully and the Matlab reshape function has been used for a similar conversion.

5. The system of 280 ODEs is then integrated by a call to Matlab integrator ode15s. The ODE routine is pde_1.

```
%
% ODE integration
  reltol=1.0e-04; abstol=1.0e-04;
  options=odeset('RelTol',reltol,'AbsTol',abstol);
  mf=2;
  if(mf==1) [t,y]=ode15s(@pde_1,tout,y0,options);end
  if(mf==2) [t,y]=ode15s(@pde_2,tout,y0,options);end
%
% 1D to 2D matrices
  for it=1:nout
  for i=1:nz
  for j=1:nr
    ca(it,i,j)=y(it,(i-1)*nr+j);
    Tk(it,i,j)=y(it,(i-1)*nr+j+nz*nr);
  end
  end
  end
```

The solution matrix y returned by ode15s is then converted to matrices ca and Tk that are now 3D to include the variation in t (as the third subscript it).

6. The centerline c_a and T_k (at $r = 0$ corresponding to j=1) are then displayed as a function of z with t as a parameter (by two nested for loops).

```
%
% Display a heading and centerline output
  fprintf('\n  nr = %2d   nz = %2d\n',nr,nz);
  for it=1:nout
    fprintf('\n t = %4.1f\n',t(it));
  for i=1:nz
    fprintf(' z = %5.1f   ca(r=0,z,t) = %8.5f   Tk(r=0,z,t)
           = %8.2f\n',z(i),ca(it,i,1),Tk(it,i,1));
  end
  end
  fprintf('\n ncall = %5d\n',ncall);
```

7. The centerline c_a and T_k are plotted as profiles in z with t as a parameter for the plots (at $t = 50, 100, 150, 200$).

```
%
% Parametric plots
%
% Axial profiles
  figure(1);
  subplot(2,2,1)
  plot(z,ca(2,:,1)); axis([0 zl 0 0.01]);
  title('ca(r=0,z,t=50)'); xlabel('z');
        ylabel('ca(r=0,z,t=50)')
  subplot(2,2,2)
  plot(z,Tk(2,:,1)); axis([0 zl 305 450]);
  title('Tk(r=0,z,t=50)'); xlabel('z');
        ylabel('Tk(r=0,z,t=50)')
  subplot(2,2,3)
  plot(z,ca(3,:,1)); axis([0 zl 0 0.01]);
  title('ca(r=0,z,t=100)'); xlabel('z');
        ylabel('ca(r=0,z,t=100)')
  subplot(2,2,4)
  plot(z,Tk(3,:,1)); axis([0 zl 305 450]);
  title('Tk(r=0,z,t=100)'); xlabel('z');
        ylabel('Tk(r=0,z,t=100)')
  figure(2);
  subplot(2,2,1)
  plot(z,ca(4,:,1)); axis([0 zl 0 0.01]);
  title('ca(r=0,z,t=150)'); xlabel('z');
        ylabel('ca(r=0,z,t=150)')
  subplot(2,2,2)
  plot(z,Tk(4,:,1)); axis([0 zl 305 450]);
  title('Tk(r=0,z,t=150)'); xlabel('z');
        ylabel('Tk(r=0,z,t=150)')
  subplot(2,2,3)
  plot(z,ca(5,:,1)); axis([0 zl 0 0.01]);
  title('ca(r=0,z,t=200)'); xlabel('z');
        ylabel('ca(r=0,z,t=200)')
  subplot(2,2,4)
  plot(z,Tk(5,:,1)); axis([0 zl 305 450]);
  title('Tk(r=0,z,t=200)'); xlabel('z');
        ylabel('Tk(r=0,z,t=200)')
```

8. The radial profiles in c_a and T_k are plotted at $t = 200$ and $z = 5, 35, 70, 100$.

```
%
% Radial profiles (at t = tf)
  for i=1:nz,
  for j=1:nr
    ca_rad(i,j)=ca(5,i,j);
    Tk_rad(i,j)=Tk(5,i,j);
  end
  end
  figure(3);
  subplot(2,2,1)
  plot(r,ca_rad(1,:)); axis([0 r0 0.0 0.01]);
  title('ca(r,z=5,t=200)'); xlabel('r');
        ylabel('ca(r,z=5,t=200)')
  subplot(2,2,2)
  plot(r,ca_rad(7,:)); axis([0 r0 0.0 0.01]);
  title('ca(r,z=35,t=200)'); xlabel('r');
        ylabel('ca(r,z=35,t=200)')
  subplot(2,2,3)
  plot(r,ca_rad(14,:)); axis([0 r0 0.0 0.01]);
  title('ca(r,z=70,t=200)'); xlabel('r');
        ylabel('ca(r,z=70,t=200)')
  subplot(2,2,4)
  plot(r,ca_rad(20,:)); axis([0 r0 0.0 0.01]);
  title('ca(r,z=100,t=200)'); xlabel('r');
        ylabel('ca(r,z=100,t=200)')
  figure(4);
  subplot(2,2,1)
  plot(r,Tk_rad(1,:)); axis([0 r0 305 450]);
  title('Tk(r,z=5,t=200)'); xlabel('r');
        ylabel('Tk(r,z=5,t=200)')
  subplot(2,2,2)
  plot(r,Tk_rad(7,:)); axis([0 r0 305 450]);
  title('Tk(r,z=35,t=200)'); xlabel('r');
        ylabel('Tk(r,z=35,t=200)')
  subplot(2,2,3)
  plot(r,Tk_rad(14,:)); axis([0 r0 305 450]);
  title('Tk(r,z=70,t=200)'); xlabel('r');
        ylabel('Tk(r,z=70,t=200)')
  subplot(2,2,4)
  plot(r,Tk_rad(20,:)); axis([0 r0 305 450]);
  title('Tk(r,z=100,t=200)'); xlabel('r');
        ylabel('Tk(r,z=100,t=200)')
```

The intention of this plotting is to indicate the detailed picture of the PDE solution that is available.

Part of the numerical output is listed in Table 13.2.

Table 13.2. Selected numerical output from `pde_1_main` of Listing 13.1

```
nr =   7   nz = 20

t =   0.0
z =    5.0   ca(r=0,z,t) =  0.00000   Tk(r=0,z,t) =   305.00
z =   10.0   ca(r=0,z,t) =  0.00000   Tk(r=0,z,t) =   305.00
z =   15.0   ca(r=0,z,t) =  0.00000   Tk(r=0,z,t) =   305.00
z =   20.0   ca(r=0,z,t) =  0.00000   Tk(r=0,z,t) =   305.00
z =   25.0   ca(r=0,z,t) =  0.00000   Tk(r=0,z,t) =   305.00
z =   30.0   ca(r=0,z,t) =  0.00000   Tk(r=0,z,t) =   305.00
z =   35.0   ca(r=0,z,t) =  0.00000   Tk(r=0,z,t) =   305.00
z =   40.0   ca(r=0,z,t) =  0.00000   Tk(r=0,z,t) =   305.00
z =   45.0   ca(r=0,z,t) =  0.00000   Tk(r=0,z,t) =   305.00
z =   50.0   ca(r=0,z,t) =  0.00000   Tk(r=0,z,t) =   305.00
z =   55.0   ca(r=0,z,t) =  0.00000   Tk(r=0,z,t) =   305.00
z =   60.0   ca(r=0,z,t) =  0.00000   Tk(r=0,z,t) =   305.00
z =   65.0   ca(r=0,z,t) =  0.00000   Tk(r=0,z,t) =   305.00
z =   70.0   ca(r=0,z,t) =  0.00000   Tk(r=0,z,t) =   305.00
z =   75.0   ca(r=0,z,t) =  0.00000   Tk(r=0,z,t) =   305.00
z =   80.0   ca(r=0,z,t) =  0.00000   Tk(r=0,z,t) =   305.00
z =   85.0   ca(r=0,z,t) =  0.00000   Tk(r=0,z,t) =   305.00
z =   90.0   ca(r=0,z,t) =  0.00000   Tk(r=0,z,t) =   305.00
z =   95.0   ca(r=0,z,t) =  0.00000   Tk(r=0,z,t) =   305.00
z =  100.0   ca(r=0,z,t) =  0.00000   Tk(r=0,z,t) =   305.00

          .                                                .
          .                                                .
          .                                                .

              Output from t = 50, 100, 150 removed

          .                                                .
          .                                                .
          .                                                .

t = 200.0
z =    5.0   ca(r=0,z,t) =  0.00997   Tk(r=0,z,t) =   311.14
z =   10.0   ca(r=0,z,t) =  0.00990   Tk(r=0,z,t) =   320.31
z =   15.0   ca(r=0,z,t) =  0.00977   Tk(r=0,z,t) =   329.97
z =   20.0   ca(r=0,z,t) =  0.00956   Tk(r=0,z,t) =   339.31
z =   25.0   ca(r=0,z,t) =  0.00925   Tk(r=0,z,t) =   348.46
z =   30.0   ca(r=0,z,t) =  0.00882   Tk(r=0,z,t) =   357.94
z =   35.0   ca(r=0,z,t) =  0.00821   Tk(r=0,z,t) =   368.60
z =   40.0   ca(r=0,z,t) =  0.00736   Tk(r=0,z,t) =   381.88
z =   45.0   ca(r=0,z,t) =  0.00611   Tk(r=0,z,t) =   400.19
z =   50.0   ca(r=0,z,t) =  0.00454   Tk(r=0,z,t) =   421.72
z =   55.0   ca(r=0,z,t) =  0.00340   Tk(r=0,z,t) =   432.13
z =   60.0   ca(r=0,z,t) =  0.00283   Tk(r=0,z,t) =   429.28
z =   65.0   ca(r=0,z,t) =  0.00259   Tk(r=0,z,t) =   420.06
z =   70.0   ca(r=0,z,t) =  0.00247   Tk(r=0,z,t) =   409.35
```

```
z =  75.0   ca(r=0,z,t) =  0.00239   Tk(r=0,z,t) =   399.31
z =  80.0   ca(r=0,z,t) =  0.00233   Tk(r=0,z,t) =   390.68
z =  85.0   ca(r=0,z,t) =  0.00228   Tk(r=0,z,t) =   383.55
z =  90.0   ca(r=0,z,t) =  0.00224   Tk(r=0,z,t) =   377.79
z =  95.0   ca(r=0,z,t) =  0.00220   Tk(r=0,z,t) =   373.20
z = 100.0   ca(r=0,z,t) =  0.00217   Tk(r=0,z,t) =   369.56

ncall =   414
```

We can note the following points about this output:

1. ICs (13.6) and (13.10) are verified for t=0 (with the initial values from Table 13.1 of $c_{a0} = 0$, $T_{k0} = 305$).
2. The solution at t=200 indicates (a) a reduction in c_a from 0.00997 at $z = \Delta z = 5$ to 0.00217 at $z_l = 100$ due to consumption of the reactant, and (b) an increase in the temperature from 311.14 at $z = \Delta z = 5$ to 369.56 at $z_l = 100$ due to the exothermic reaction; also, the temperature goes through a maximum of 432.13 at $z = 55$ with subsequent cooling (beyond $z = 55$) due to the wall temperature $T_w = 355$). Note also that the first value of z for the solution is $z = \Delta z = 5$ since $z = 0$ is not part of the grid in z (where the entering conditions are $c_{ae} = 0.01$, $T_{ke} = 305$ from Table 13.1).

The properties of the solutions are also displayed in Figures 13.2–13.5

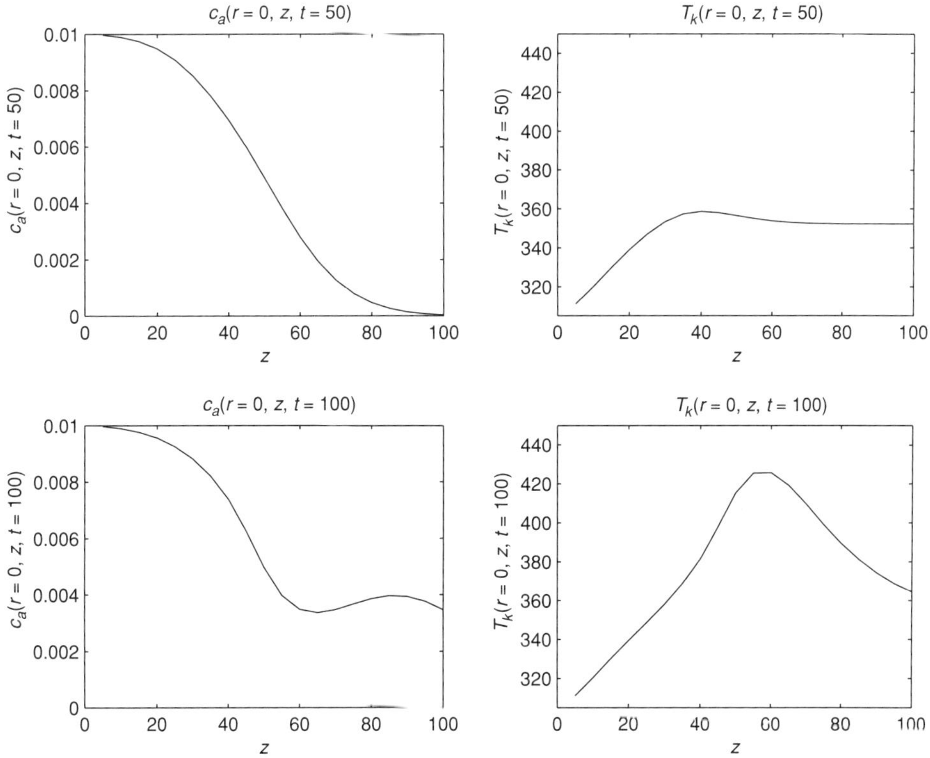

Figure 13.2. Plot of the solution of Eqs. (13.1)–(13.13) for $r = 0$, $0 \leq z \leq z_l$, $t = 50, 100$

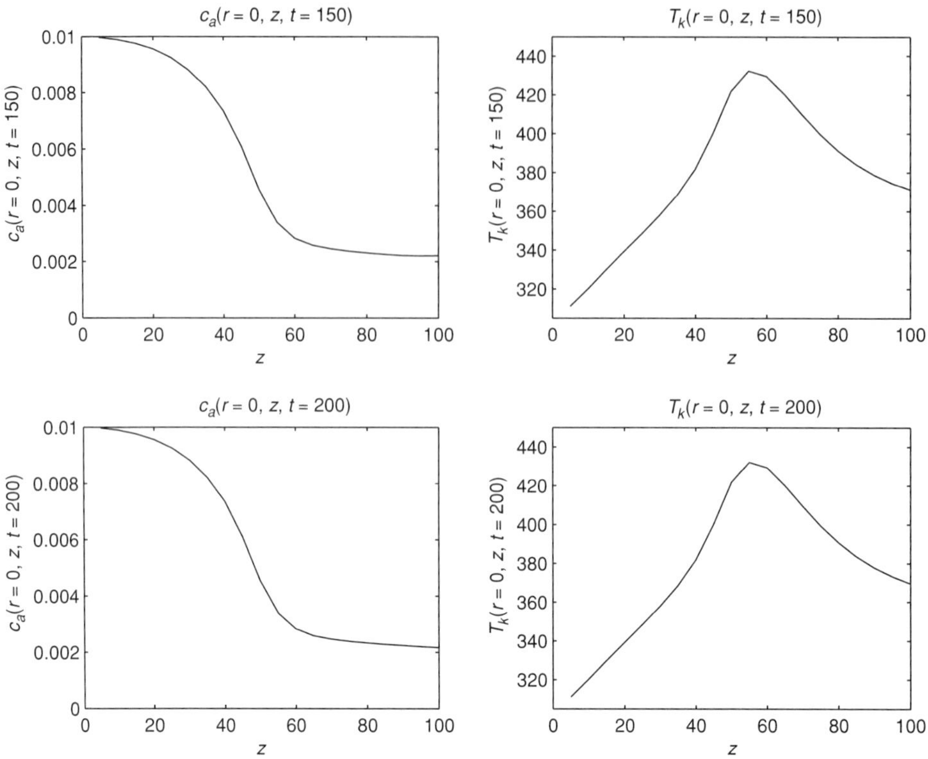

Figure 13.3. Plot of the solution of Eqs. (13.1)–(13.13) for $r = 0$, $0 \le z \le z_l$, $t = 150, 200$

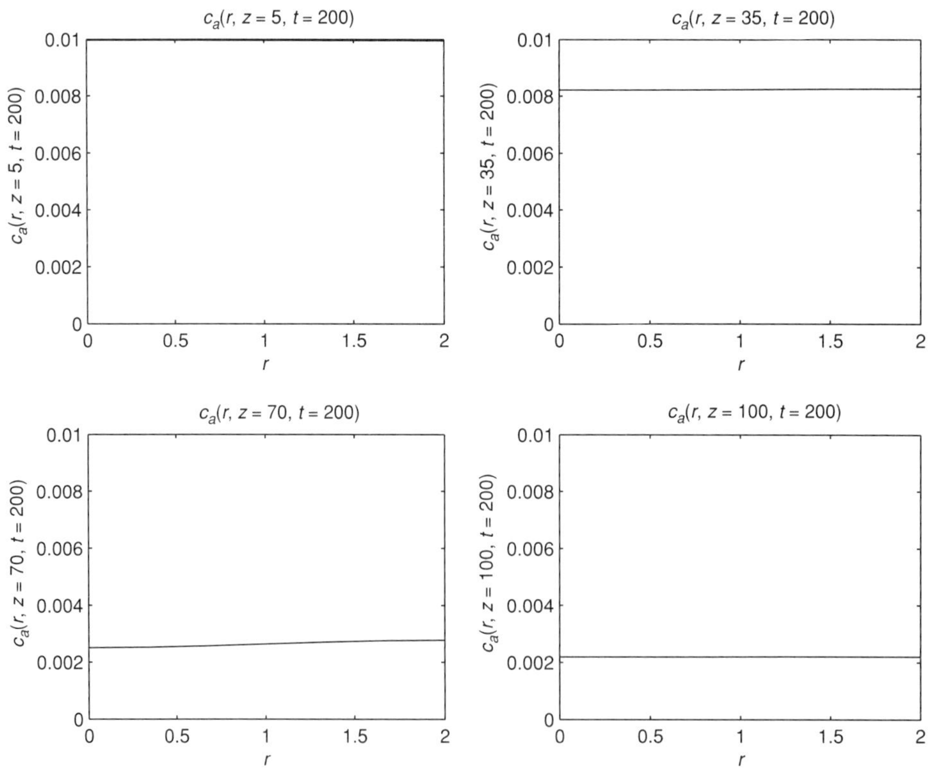

Figure 13.4. Plot of the solution c_a from Eqs. (13.1)–(13.13) for $0 \le r \le r_0$, $z = 5, 35, 70, 100$, $t = 200$

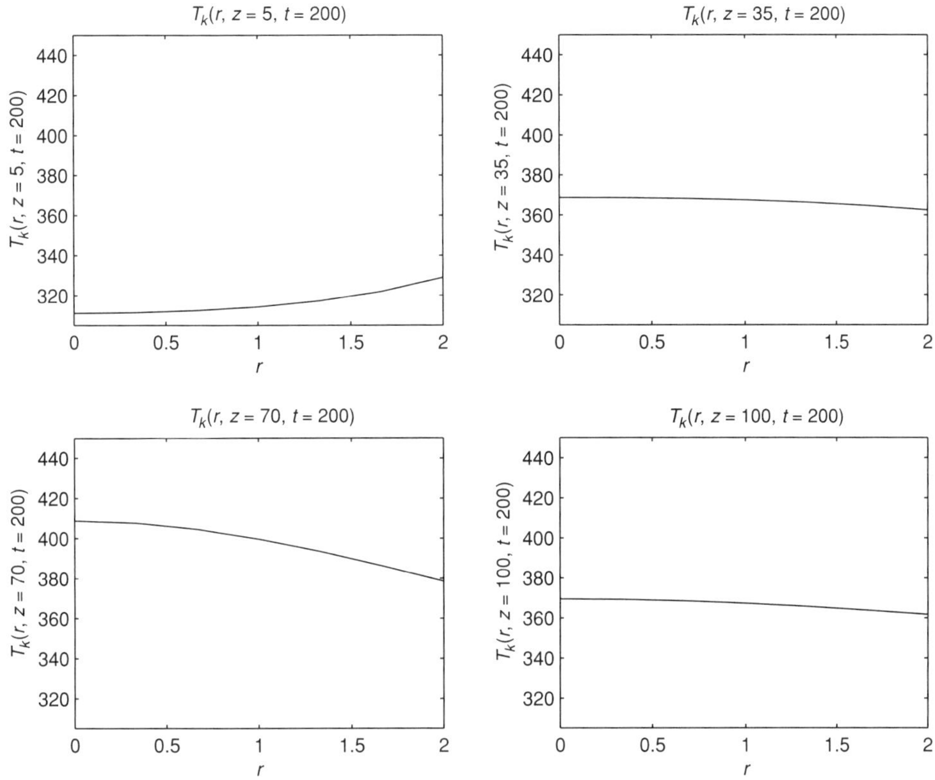

Figure 13.5. Plot of the solution T_k from Eqs. (13.1)–(13.13) for $0 \le r \le r_0$, $z = 5, 35, 70, 100$, $t = 200$

We can note the following points about these plots:

1. In Figure 13.2, the c_a and T_k profiles in z change appreciably with t, but in Figure 13.3, they have reached essentially a steady state (no further change with t). This result is also confirmed by the numerical output (although the output for $t = 150$ does not appear in Table 13.2 to save some space).
2. In Figure 13.4, the profiles in r indicate the consumption of the reactant with increasing z as expected; that is the reactor is accomplishing significant conversion of the reactant.
3. In Figure 13.5, the increase in temperature with z is evident; that is, as the reactant flows through the reactor, heat is liberated by the exothermic reaction causing a temperature increase.
4. In Figures 13.4 and 13.5, the radial profiles in c_a and T_k are relatively flat (relative to the case where the resistance to radial heat and mass transfer as reflected in D_c and D_t is greater, i.e., when these coefficients are smaller).
5. Figures 13.6 and 13.7 show clearly the radial variation of concentration and temperature along the axial length of the reactor at final time $t = 200$, when a steady-state profile has been reached. The concentration c_a (Figure 13.6) falls from the inlet value of 0.01 as the reactants flow along the length of the

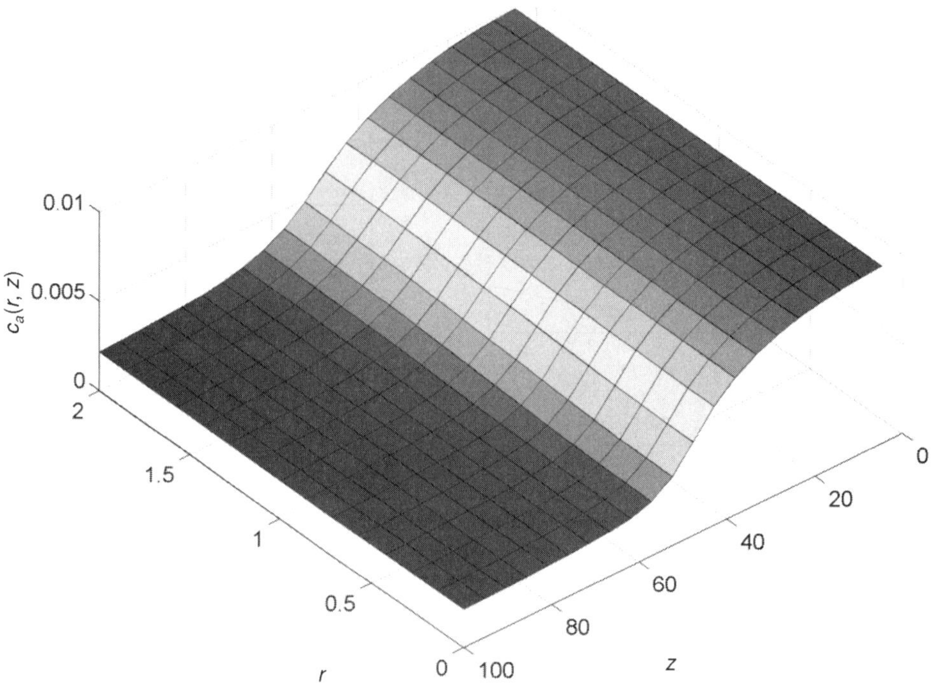

Figure 13.6. Surface plot of reactant concentration $c_a(z, t)$ at $t = 200$

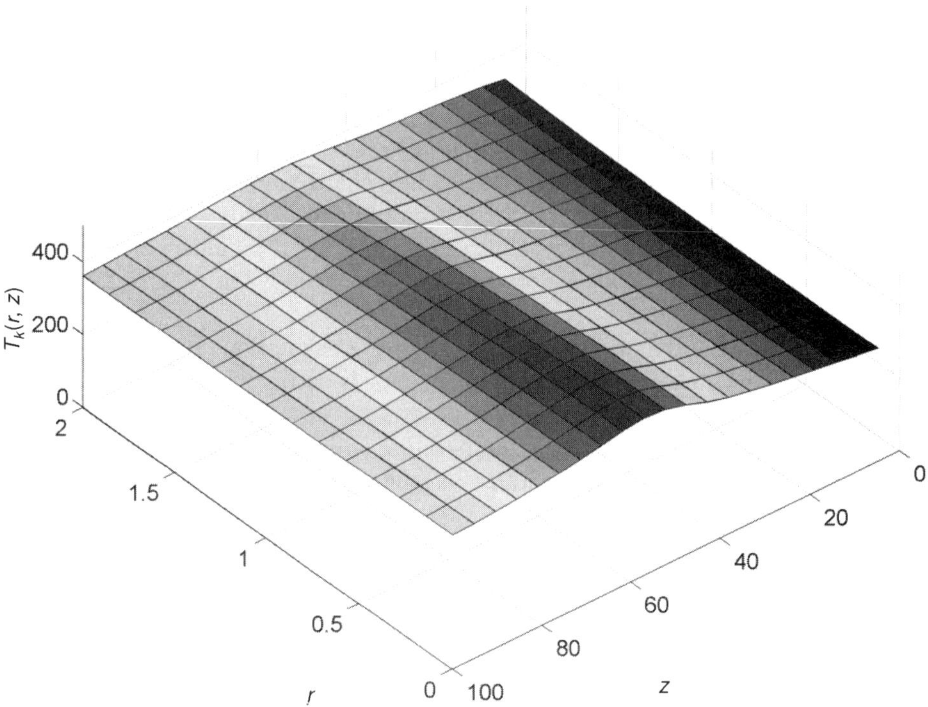

Figure 13.7. Surface plot of reactant temperature $T_k(r, z)$ at $t = 200$

reactor and steadies out to an equilibrium value of 0.00217 at the exit; the radial variation is relatively small. However, the temperature T_k (Figure 13.7) of the reactant rises initially as the reaction proceeds until it reaches a maximum half way along the reactor and then falls back due to the reactant being consumed as it continues its journey toward the reactor exit. We note that the radial temperature variation is significant around the axial midpoint of the reactor, being ≈40 K. This is because the rate at which heat is generated by the exothermic reaction is greater than the rate at which heat is carried away by radial mass and thermal diffusion – see also notes to Figures 13.4 and 13.5 given previously.

We now consider the programming of the ODEs in pde_1 (see Listing 13.2).

```
  function yt=pde_1(t,y)
%
% Global area
  global      nr     nz     dr     dz    drs     dzs...
               r      z     Dc     Dt     ca      Tk...
             cae    Tke      h      k      E       R...
             rk0      v    rho     Cp     Tw     dH...
            ncall
%
% 1D to 2D matrices
  for i=1:nz
  for j=1:nr
    ij=(i-1)*nr+j;
    ca(i,j)=y(ij);
    Tk(i,j)=y(ij+nr*nz);
  end
  end
%
% Step through the grid points in r and z
  for i=1:nz
  for j=1:nr
%
%   (1/r)*car, (1/r)*Tkr
    if(j==1)
      car(i,j)=2.0*(ca(i,j+1)-ca(i,j))/drs;
      Tkr(i,j)=2.0*(Tk(i,j+1)-Tk(i,j))/drs;
    elseif(j==nr)
      car(i,j)=0.0;
      Tkr(i,j)=(1.0/r(j))*(h/k)*(Tw-Tk(i,j));
    else
      car(i,j)=(1.0/r(j))*(ca(i,j+1)-ca(i,j-1))/(2.0*dr);
      Tkr(i,j)=(1.0/r(j))*(Tk(i,j+1)-Tk(i,j-1))/(2.0*dr);
    end
```

```
%
% carr, Tkrr
    if(j==1)
      carr(i,j)=2.0*(ca(i,j+1)-ca(i,j))/drs;
      Tkrr(i,j)=2.0*(Tk(i,j+1)-Tk(i,j))/drs;
    elseif(j==nr)
      carr(i,j)=2.0*(ca(i,j-1)-ca(i,j))/drs;
      Tkf=Tk(i,j-1)+2.0*dr*h/k*(Tw-Tk(i,j));
      Tkrr(i,j)=(Tkf-2.0*Tk(i,j)+Tk(i,j-1))/drs;
    else
      carr(i,j)=(ca(i,j+1)-2.0*ca(i,j)+ca(i,j-1))/drs;
      Tkrr(i,j)=(Tk(i,j+1)-2.0*Tk(i,j)+Tk(i,j-1))/drs;
    end
%
% caz, Tkz
    if(i==1)
      caz(i,j)=(ca(i,j)-cae)/dz;
      Tkz(i,j)=(Tk(i,j)-Tke)/dz;
    else
      caz(i,j)=(ca(i,j)-ca(i-1,j))/dz;
      Tkz(i,j)=(Tk(i,j)-Tk(i-1,j))/dz;
    end
%
% PDEs
    rk=rk0*exp(-E/(R*Tk(i,j)))*ca(i,j)^2;
    cat(i,j)=Dc*(carr(i,j)+car(i,j))-v*caz(i,j)-rk;
    Tkt(i,j)=Dt*(Tkrr(i,j)+Tkr(i,j))-v*Tkz(i,j)...
            -dH/(rho*Cp)*rk;
  end
  end
%
% 2D to 1D matrices
  for i=1:nz
  for j=1:nr
    ij=(i-1)*nr+j;
    yt(ij)=cat(i,j);
    yt(ij+nr*nz)=Tkt(i,j);
  end
  end
%
% Transpose and count
  yt=yt';
  ncall=ncall+1;
```

Listing 13.2. Function pde_1 called by main program pde_1_main

We can note the following details about pde_1:

1. After the definition of the function and a *global* area for parameters and variables shared between routines, a 1D to 2D transformation permits programming in terms of the problem-oriented variables ca and Tk.

```
  function yt=pde_1(t,y)
%
% Global area
  global      nr      nz      dr      dz     drs      dzs...
               r       z      Dc      Dt      ca       Tk...
             cae     Tke       h       k       E        R...
             rk0       v     rho      Cp      Tw       dH...
           ncall
%
% 1D to 2D matrices
  for i=1:nz
  for j=1:nr
    ij=(i-1)*nr+j;
    ca(i,j)=y(ij);
    Tk(i,j)=y(ij+nr*nz);
  end
  end
```

2. Computation of the nr*nz = 280 MOL ODEs is accomplished with a pair of nested for loops.

```
%
% Step through the grid points in r and z
  for i=1:nz
  for j=1:nr
%
%   (1/r)*car, (1/r)*Tkr
    if(j==1)
      car(i,j)=2.0*(ca(i,j+1)-ca(i,j))/drs;
      Tkr(i,j)=2.0*(Tk(i,j+1)-Tk(i,j))/drs;
    elseif(j==nr)
      car(i,j)=0.0;
      Tkr(i,j)=(1.0/r(j))*(h/k)*(Tw-Tk(i,j));
   else
      car(i,j)=(1.0/r(j))*(ca(i,j+1)-ca(i,j-1))/(2.0*dr);
      Tkr(i,j)=(1.0/r(j))*(Tk(i,j+1)-Tk(i,j-1))/(2.0*dr);
    end
```

The programming of the derivatives (in t) begins with the first-derivative radial groups in Eqs. (13.2) and (13.4). This programming requires some explanation.

(a) At $r = 0$ (`j=1`), the first-derivative radial group in eq. (13.2) is indeterminant, that is,

$$\frac{1}{r}\frac{\partial c_a}{\partial r} = 0/0$$

as a consequence of *homogeneous Neumann BC* (13.7). We therefore apply *l'Hospital's rule to resolve (regularize)* this indeterminant form.

$$\lim_{r\to 0}\frac{1}{r}\frac{\partial c_a}{\partial r} = \frac{\partial^2 c_a}{\partial r^2}$$

This second derivative can then be approximated as

$$\begin{aligned}\frac{\partial^2 c_a}{\partial r^2} &\approx \frac{c_a(r=\Delta r, z, t) - 2c_a(r=0, z, t) + c_a(r=-\Delta r, z, t)}{\Delta r^2}\\ &= 2\frac{c_a(r=\Delta r, z, t) - c_a(r=0, z, t)}{\Delta r^2}\end{aligned}$$

where we have used BC (13.7) to eliminate the fictitious value $c_a(r = -\Delta r, z, t)$. The corresponding programming is (and also for T_k using BC (13.11))

```
if(j==1)
  car(i,j)=2.0*(ca(i,j+1)-ca(i,j))/drs;
  Tkr(i,j)=2.0*(Tk(i,j+1)-Tk(i,j))/drs;
```

The use of BC (13.7) is as follows. The finite-difference (FD) approximation of BC (13.7) is

$$\frac{\partial c_a(r=0, z, t)}{\partial r} \approx \frac{c_a(r=\Delta r, z, t) - c_a(r=-\Delta r, z, t)}{2\Delta r} = 0$$

Thus, the fictitious value $c_a(r = -\Delta r, z, t)$ is

$$c_a(r = -\Delta r, z, t) = c_a(r = \Delta r, z, t)$$

which is then used in the FD approximation of $\partial^2 c_a/\partial r^2$ (as earlier).

(b) At $r = r_0$ (`j=nr`), the first-derivative radial group in Eq. (13.2) is programmed as

```
elseif(j==nr)
  car(i,j)=0.0;
```

which reflects homogeneous Neumann BC (13.8).

(c) For T_k at $r = r_0$ (`j=nr`), the first-derivative radial group in Eq. (13.4) is first approximated through the use of *third-type BC (13.12)*. The approximation of this BC is

$$k\frac{\partial T_k(r = r_0, z, t)}{\partial r} \approx k\frac{T_k(r_0 + \Delta r, z, t) - T_k(r_o - \Delta r, z, t)}{2\Delta r}$$
$$= h(T_w - T_k(r = r_0, z, t))$$

which can be rearranged to give the fictitious value

$$T_k(r_0 + \Delta r, z, t) = T_k(r_0 - \Delta r, z, t) + (2\Delta rh/k)(T_w - T_k(r = r_0, z, t)) \tag{13.14}$$

This result can then be substituted in the approximation of the radial group in T_k in Eq. (13.4).

$$\frac{1}{r_0}\frac{\partial T_k}{\partial r}$$
$$\approx \frac{1}{r_0}\frac{T_k(r_0 + \Delta r, z, t) - T_k(r_0 - \Delta r, z, t)}{2\Delta r}$$
$$= \frac{1}{r_0}\frac{T_k(r_0 - \Delta r, z, t) + (2\Delta rh/k)(T_w - T_k(r = r_0, z, t)) - T_k(r_0 - \Delta r, z, t)}{2\Delta r}$$
$$= \frac{1}{r_0}(h/k)(T_w - T_k(r = r_0, z, t))$$

The corresponding programming is (with `j=nr`)

```
    Tkr(i,j)=(1.0/r(j))*(h/k)*(Tw-Tk(i,j));
```

For $r \neq 0, r_0$, the approximation of the first-derivative radial group in Eq. (13.2) is

$$\frac{1}{r}\frac{\partial T_k}{\partial r} \approx \frac{1}{r}\frac{\partial T_k(r + \Delta r, z, t) - T_k(r - \Delta r, z, t)}{2\Delta r}$$

This approximation is programmed as (and also, for c_a)

```
else
    car(i,j)=(1.0/r(j))*(ca(i,j+1)-ca(i,j-1))/(2.0*dr);
    Tkr(i,j)=(1.0/r(j))*(Tk(i,j+1)-Tk(i,j-1))/(2.0*dr);
```

The programming of the second derivative in Eqs. (13.2) and (13.4)

```
%
%    carr, Tkrr
     if(j==1)
       carr(i,j)=2.0*(ca(i,j+1)-ca(i,j))/drs;
       Tkrr(i,j)=2.0*(Tk(i,j+1)-Tk(i,j))/drs;
     elseif(j==nr)
       carr(i,j)=2.0*(ca(i,j-1)-ca(i,j))/drs;
       Tkf=Tk(i,j-1)+2.0*dr*h/k*(Tw-Tk(i,j));
       Tkrr(i,j)=(Tkf-2.0*Tk(i,j)+Tk(i,j-1))/drs;
     else
       carr(i,j)=(ca(i,j+1)-2.0*ca(i,j)+ca(i,j-1))/drs;
       Tkrr(i,j)=(Tk(i,j+1)-2.0*Tk(i,j)+Tk(i,j-1))/drs;
     end
```

also requires some elaboration.

(a) At $r = 0$, BCs (13.7) and (13.11) apply. Thus the approximation of the second derivative for c_a becomes (with Eq. (13.7))

$$\frac{\partial^2 c_a(r=0,z,t)}{\partial r^2} = \frac{c_a(r=\Delta r,z,t) - 2c_a(r=0,z,t) + c_a(r=-\Delta r,z,t)}{\Delta r^2}$$
$$= 2\frac{c_a(r=\Delta r,z,t) - c_a(r=0,z,t)}{\Delta r^2}$$

This approximation and a similar one for T_k are coded as

```
     if(j==1)
       carr(i,j)=2.0*(ca(i,j+1)-ca(i,j))/drs;
       Tkrr(i,j)=2.0*(Tk(i,j+1)-Tk(i,j))/drs;
```

(b) For $r = r_0$, a similar argument for the second derivative of c_a in r gives

$$\frac{\partial^2 c_a(r=r_0,z,t)}{\partial r^2} = 2\frac{c_a(r=r_0-\Delta r,z,t) - c_a(r=r_0,z,t)}{\Delta r^2}$$

which is coded as

```
     elseif(j==nr)
       carr(i,j)=2.0*(ca(i,j-1)-ca(i,j))/drs;
```

(c) For the second derivative of T_k at $r = r_0$, we can use the fictitious value of $T_k(r_0 + \Delta r, z, t)$ from Eq. (13.14) that includes BC (13.12). This value can then be substituted in the usual approximation of a second derivative

$$\frac{\partial^2 T_k}{\partial r^2} \approx \frac{\partial T_k(r_0+\Delta r,z,t) - 2T_k(r_0,z,t) + T_k(r_0-\Delta r,z,t)}{2\Delta r}$$

The coding for these equations is (with j=nr)

```
Tkf=Tk(i,j-1)+2.0*dr*h/k*(Tw-Tk(i,j));
Tkrr(i,j)=(Tkf-2.0*Tk(i,j)+Tk(i,j-1))/drs;
```

(d) For $r \neq 0, r_0$, the usual approximation of a second derivative can be used without concern for BCs. The coding is

```
else
  carr(i,j)=(ca(i,j+1)-2.0*ca(i,j)+ca(i,j-1))/drs;
  Tkrr(i,j)=(Tk(i,j+1)-2.0*Tk(i,j)+Tk(i,j-1))/drs;
```

3. The coding for the derivatives in z in Eqs. (13.2) and (13.4), subject to BCs (13.9) and (13.13), is straightforward.

```
%
%    caz, Tkz
     if(i==1)
       caz(i,j)=(ca(i,j)-cae)/dz;
       Tkz(i,j)=(Tk(i,j)-Tke)/dz;
     else
       caz(i,j)=(ca(i,j)-ca(i-1,j))/dz;
       Tkz(i,j)=(Tk(i,j)-Tk(i-1,j))/dz;
     end
```

The only special requirement is to use the entering values of c_a and T_k at the first grid point (i=1), which, again, is for $z = \Delta z$. These entering values are cae and Tke; they do not require ODEs (since they are specified as parameters), and therefore the gridding does not have to start at $z = 0$. In other words, including $z = 0$ would require trivial ODEs for the condition $(\partial c_a(r, z = 0, t))/\partial t = (\partial T_k(r, z = 0, t))/\partial t = 0$; there would be $(2)(7) = 14$ such ODEs for the 2 PDEs (Eqs. (13.2) and (13.4)) each approximated over 7 points in r.

Note that the first-order derivatives in z in Eqs. (13.2) and (13.4) are approximated with *two-point upwind FDs*. This name comes from the fact that these derivatives represent convection, and we therefore use values of the dependent variables that are "upwind" (point i-1) of the point of the approximation (point i). We will not discuss the reasons for this choice of an FD other than to indicate that *some form of upwinding is generally required for convective terms to account for the influence of the flow*.

Also, these two-point upwind approximations have the significant limitation of being only first-order correct (the principal truncation error term is $O(\Delta z)$). Thus, a higher-order FD approximation or nonlinear approximations for the convective derivatives in PDEs could lead to a significant increase in the accuracy of the numerical solution. However, a discussion of the numerical integration of convective (hyperbolic) PDEs that propagate moving fronts is not included in this discussion. Numerical methods for the resolution of discontinuities and moving fronts are discussed extensively in the literature, for example, in references [2–4] and are implemented in available computer codes.

This completes the coding for all of the spatial derivatives in Eqs. (13.2) and (13.4). All that remains to program these equations is to put the derivatives together along with the reaction term. The coding for this is

```
%
%   PDEs
    rk=rk0*exp(-E/(R*Tk(i,j)))*ca(i,j)^2;
    cat(i,j)=Dc*(carr(i,j)+car(i,j))-v*caz(i,j)-rk;
    Tkt(i,j)=Dt*(Tkrr(i,j)+Tkr(i,j))-v*Tkz(i,j)...
            -dH/(rho*Cp)*rk;
  end
  end
```

The double end concludes the pair of nested for loops at the beginning that step the calculations through all 280 points (i.e., 280 ODEs).

4. Finally, there is a 2D to 1D matrix conversion to put all of the derivatives in t (cat(i,j), Tkt(i,j)) in a single vector (yt), which is then transposed as required by ode15s.

```
%
% 2D to 1D matrices
  for i=1:nz
  for j=1:nr
    ij=(i-1)*nr+j;
    yt(ij)=cat(i,j);
    yt(ij+nr*nz)=Tkt(i,j);
  end
  end
%
% Transpose and count
  yt=yt';
  ncall=ncall+1;
```

The counter ncall displayed at the end of the solution has the value 414 (see Table 13.2), indicating a rather modest computational effort in computing the

MOL solution; in other words, integrator ode15s handled this problem with quite acceptable efficiency.

pde_1 in Listing 13.2 is based on the explicit programming of the FDs for the MOL ODEs. We could, however, also use library routines for this purpose. To illustrate this idea, we consider an alternative ODE routine as given in Listing 13.3.

```
  function yt=pde_2(t,y)
%
% Global area
  global      nr      nz      dr      dz     drs      dzs...
               r      r0       z      zl      Dc       Dt...
              ca      Tk     cae     Tke       h        k...
               E       R     rk0       v     rho       Cp...
              Tw      dH   ncall
%
% 1D to 2D matrices
  for i=1:nz
  for j=1:nr
    ij=(i-1)*nr+j;
    ca(i,j)=y(ij);
    Tk(i,j)=y(ij+nr*nz);
  end
  end
%
% Step through the grid points in r and z
  for i=1:nz
    ca1d=ca(i,:);
    Tk1d=Tk(i,:);
%
%   car, Tkr
    car1d=dss004(0.0,r0,nr,ca1d);
    car(i,:)=car1d;
    car(i,1)= 0.0;
    car(i,nr)=0.0;
    Tkr1d=dss004(0.0,r0,nr,Tk1d);
    Tkr(i,:)=Tkr1d;
    Tkr(i,1)= 0.0;
    Tkr(i,nr)=(h/k)*(Tw-Tk(i,nr));
%
%   carr, Tkrr
    car1d( 1)=0.0;
    car1d(nr)=0.0;
    nl=2;
    nu=2;
    carr1d=dss044(0.0,r0,nr,ca1d,car1d,nl,nu);
    carr(i,:)=carr1d;
    Tkr1d( 1)=0.0;
```

```
      Tkr1d(nr)=(h/k)*(Tw-Tk1d(nr));
      nl=2;
      nu=2;
      Tkrr1d=dss044(0.0,r0,nr,Tk1d,Tkr1d,nl,nu);
      Tkrr(i,:)=Tkrr1d;
    for j=1:nr
%
%     (1/r)*car, (1/r)*Tkr
      if(j~=1)
        car(i,j)=(1.0/r(j))*car(i,j);
        Tkr(i,j)=(1.0/r(j))*Tkr(i,j);
      end
      if(j==1)carr(i,j)=2.0*carr(i,j);end
      if(j==1)Tkrr(i,j)=2.0*Tkrr(i,j);end
%
%     caz, Tkz
      if(i==1)
        caz(i,j)=(ca(i,j)-cae)/dz;
        Tkz(i,j)=(Tk(i,j)-Tke)/dz;
      else
        caz(i,j)=(ca(i,j)-ca(i-1,j))/dz;
        Tkz(i,j)=(Tk(i,j)-Tk(i-1,j))/dz;
      end
%
%     PDEs
      rk=rk0*exp(-E/(R*Tk(i,j)))*ca(i,j)^2;
      cat(i,j)=Dc*(carr(i,j)+car(i,j))-v*caz(i,j)-rk;
      Tkt(i,j)=Dt*(Tkrr(i,j)+Tkr(i,j))-v*Tkz(i,j)...
              -dH/(rho*Cp)*rk;
    end
    end
%
% 2D to 1D matrices
  for i=1:nz
  for j=1:nr
    ij=(i-1)*nr+j;
    yt(ij)=cat(i,j);
    yt(ij+nr*nz)=Tkt(i,j);
  end
  end
%
% Transpose and count
  yt=yt';
  ncall=ncall+1;
```

Listing 13.3. Function pde_2 called by main program pde_1_main

We can note the following details about pde_2:

1. As in pde_1 in Listing 13.2, the function and a global area are defined, followed by a 1D to 2D matrix transformation.

```
  function yt=pde_1(t,y)
%
% Global area
  global      nr      nz      dr      dz     drs     dzs...
               r      r0       z      zl      Dc      Dt...
              ca      Tk     cae     Tke       h       k...
               E       R     rk0       v     rho      Cp...
              Tw      dH   ncall
%
% 1D to 2D matrices
  for i=1:nz
  for j=1:nr
    ij=(i-1)*nr+j;
    ca(i,j)=y(ij);
    Tk(i,j)=y(ij+nr*nz);
  end
  end
```

2. A for loop is used to step through z. Since the spatial differentiation dss routines compute numerical derivatives of 1D array (vector), a conversion from 2D to 1D is first required at each value of z.

```
%
% Step through the grid points in r and z
  for i=1:nz
    ca1d=ca(i,:);
    Tk1d=Tk(i,:);
```

3. Then the first-order radial derivatives can be computed by a call to dss004 ($\partial c_a/\partial r$ = car1d, $\partial T_k/\partial r$ = Tkr1d).

```
%
%   car, Tkr
    car1d=dss004(0.0,r0,nr,ca1d);
    car(i,:)=car1d;
    car(i,1)= 0.0;
    car(i,nr)=0.0;
```

```
Tkr1d=dss004(0.0,r0,nr,Tk1d);
Tkr(i,:)=Tkr1d;
Tkr(i,1)= 0.0;
Tkr(i,nr)=(h/k)*(Tw-Tk(i,nr));
```

Note the conversion back to 2D arrays and the use of BCs (13.7) and (13.8) (for c_a) and (13.11) and (13.12) (for T_k). Thus, through the use of the compact Matlab indexing (subscripting), the calculation of 2D spatial derivatives in r through the use of 1D routines (dss004) is facilitated.

4. Similarly, the second derivatives in r can be computed by calls to dss044. Note again the use of BCs (13.7), (13.8), (13.11), and (13.12) that are all declared Neumann (through nl = nu = 2) and the conversion from 2D to 1D before the calls to dss044 and the 1D to 2D conversions after the calls to dss044.

```
%
%    carr, Tkrr
     car1d( 1)=0.0;
     car1d(nr)=0.0;
     nl=2;
     nu=2;
     carr1d=dss044(0.0,r0,nr,ca1d,car1d,nl,nu);
     carr(i,:)=carr1d;
     Tkr1d( 1)=0.0;
     Tkr1d(nr)=(h/k)*(Tw-Tk1d(nr));
     nl=2;
     nu=2;
     Tkrr1d=dss044(0.0,r0,nr,Tk1d,Tkr1d,nl,nu);
     Tkrr(i,:)=Tkrr1d;
```

5. To this point, the variable coefficient $1/r$ in the terms $(1/r)\ (\partial c_a/\partial r)$ and $(1/r)\ (\partial T_k/\partial r)$ has not been included. The following code accomplishes this (the complication is due to the singular point $r = 0$):

```
%
%    (1/r)*car, (1/r)*Tkr
     if(j~=1)
       car(i,j)=(1.0/r(j))*car(i,j);
       Tkr(i,j)=(1.0/r(j))*Tkr(i,j);
     end
     if(j==1)carr(i,j)=2.0*carr(i,j);end
     if(j==1)Tkrr(i,j)=2.0*Tkrr(i,j);end
```

For $r = 0$ (j=1), the indeterminate forms reduce to the second derivative as discussed previously; that is,

$$\lim_{r \to 0} \frac{1}{r}\frac{\partial c_a}{\partial r} = \frac{\partial^2 c_a}{\partial r^2}$$

$$\lim_{r \to 0} \frac{1}{r}\frac{\partial T_k}{\partial r} = \frac{\partial^2 T_k}{\partial r^2}$$

which are then combined with the preexisting second derivatives to give twice the second derivative (only at $r = 0$). Otherwise ($r \neq 0$), the variable coefficient $1/r$ can be programmed directly.

6. The first-order convective derivatives in z are approximated by the two-point upwind FDs discussed previously.

```
%
%    caz, Tkz
     if(i==1)
       caz(i,j)=(ca(i,j)-cae)/dz;
       Tkz(i,j)=(Tk(i,j)-Tke)/dz;
     else
       caz(i,j)=(ca(i,j)-ca(i-1,j))/dz;
       Tkz(i,j)=(Tk(i,j)-Tk(i-1,j))/dz;
     end
```

7. All of the derivatives in Eqs. (13.2) and (13.4) are now computed, so the PDEs can be programmed.

```
%
%    PDEs
     rk=rk0*exp(-E/(R*Tk(i,j)))*ca(i,j)^2;
     cat(i,j)=Dc*(carr(i,j)+car(i,j))-v*caz(i,j)-rk;
     Tkt(i,j)=Dt*(Tkrr(i,j)+Tkr(i,j))-v*Tkz(i,j)...
             -dH/(rho*Cp)*rk;
```

The resemblance of this coding to the PDEs, Eqs. (13.2) and (13.4), including Eq. (13.5), is one of the salient features of MOL analysis.

The call to pde_2 through ode15s is accomplished with the following code (mf = 2) in pde_1_main:

```
%
% ODE integration
  reltol=1.0e-04; abstol=1.0e-04;
  options=odeset('RelTol',reltol,'AbsTol',abstol);
```

```
mf=2;
if(mf==1)[t,y]=ode15s(@pde_1,tout,y0,options);end
if(mf==2)[t,y]=ode15s(@pde_2,tout,y0,options);end
```

The output from `pde_1_main` and `pde_2` is very similar to the output from `pde_1_main` and `pde_1`, so we will not review it here. The concluding points we make are (1) library routines (e.g., `dss004, dss044`) can be used to calculate the spatial derivatives in PDEs, and (2) since calls to these routines can easily be changed (in the case of the `dss` routines, only the number changes, while the arguments remain the same), the effect of variation in the FD order is easily investigated (e.g., switching from `dss004, dss044` to `dss006, dss046`); in other words, we can easily investigate p-refinement as well as h-refinement (in which the number of grid points is changed).

We conclude this chapter with a few additional points (but it is not our intention to discuss here the design and control of a chemical reactor):

- Since in the analysis of a chemical reactor we would be particularly interested in the exiting conditions (at $z = z_l$), we could focus on $c_a(r, z = z_l, t)$ and $T(r, z = z_l, t)$ as the output to be listed and plotted (this output is included in Figures 13.4 and 13.5). Also, there may be significant radial variation at the exit $z = z_l$, so we could also compute the integrals

$$c_{a,avg}(t) = \int_0^{r_0} 2\pi r c_a(r, z_l, t) dr / (\pi r_0^2)$$

$$T_{k,avg}(t) = \int_0^{r_0} 2\pi r T_k(r, z_l, t) dr / (\pi r_0^2)$$

  to give the average exiting concentration and temperature $c_{a,avg}(t)$ and $T_{k,avg}(t)$. These integrals could be evaluated numerically using a well-established *quadrature method* such as *Simpson's rule*. In the case of Figures 13.4 and 13.5, the radial variation in $c_a(r, z = z_l, t)$ and $T_k(r, z = z_l, t)$ is not very great, but this could change substantially for a more highly exothermic reaction or an internal structure that increases the resistance to radial heat and mass transfer. These conditions could in turn lead to adverse higher peak temperatures.
- An important safety concern in the operation of an exothermic reactor may be the maximum internal temperature. For example, an excessive temperature might damage the reactor or its contents, or possibly even cause an explosion. This analysis could be used to predict the maximum temperature, which would occur along the centerline $r = 0$. A controller could then be added to the model that would reduce the wall temperature T_w as a function of the maximum temperature to cool the reactor as needed to limit the maximum temperature.
- Interesting parametric studies can be performed with the preceding code. For example, $T_w = 355$ was selected in order to start the chemical reaction. If $T_w = 305$ is used instead (corresponding to the initial and entering temperatures), the reaction does not take place so that the concentration for $0 \le z \le z_l$ eventually

reaches the entering value $c_{ae} = 0.01$ (starting at $c_{a0} = 0$), and the temperature for $0 \leq z \leq z_l$ essentially remains at the initial and entering values of 305.

- This model is rather basic in the sense that only one reactant is considered. More realistically, multiple reactants and products would be of interest and a mass balance would be required for each such component. Thus, additional PDEs would be required, but the procedure of the preceding analysis could be directly extended for a more complex model (with more PDEs).
- The name *convection–diffusion–reaction* system could also be given a more mathematical designation, for example, *hyperbolic–parabolic* system where the convective terms (involving the velocity v) are hyperbolic and the diffusion terms (involving the diffusivities D_c, D_t) are parabolic.
- Actually, the term *diffusivity* may not always be appropriate (particularly for a high-Reynolds-number situation) in the sense that the radial dispersion in a reactor or similar physical system is probably not due solely to molecular diffusion (which is generally a small effect) as much as stream splitting and flow around internal obstructions; this effect is also known as *eddy diffusion*. Thus the term *dispersion coefficient* might be more appropriate. Also, dispersion coefficients are frequently measured experimentally since they typically reflect the effects of complex internal flow patterns.
- The accuracy of the numerical solution can be investigated by changing the number of grid points in r and z and observing the effect on the output. For example, doubling `nr` has little effect on the solution, implying that `nr=7` is an adequate number of grid points in r. This *h-refinement* in which the grid spacing is changed is an important method for assessing the solution accuracy (*spatial convergence in r and z*) when an analytical solution is not available (usually the case in applications). But keep in mind that the total number of ODEs is `2*nr*nz` that increases rapidly with increasing *nr* and *nz*, so this practical limitation must be considered when changing the number of grid points.
- The accuracy of the integration in t could also be assessed by changing the error tolerances `abstol` and `reltol` before the call to `ode15s` and observing the effect on the solution. This check on the *temporal convergence* is also an important method for assessing the solution accuracy.
- In summary, an error analysis in both time and space is an important part of the numerical solution of a PDE model.

APPENDIX A

A.1. Units Check for Eqs. (13.1)–(13.13)

We detail here a check on the units of each term in the formulation of the PDE model (Eqs. (13.1)–(13.13)). This is an essential step in the formulation of a mathematical model for at least two reasons: (a) consistency of units throughout an equation is necessary for the equation to be correct, and (b) the units give an insight into the physical (also, chemical, and biological) meaning of each term of the equation and therefore an overall perspective of the applicability and usefulness of the equation.

Here, we tabulate the terms and associated units that are the starting point for Eqs. (13.1)–(13.13). Starting with the terms in the mass balance (leading to Eq. (13.2)), we have

Term	Units
$2\pi r\Delta r\Delta z\frac{\partial c_a}{\partial t}$	(cm)(cm)(cm)(gmol/cm^3)(1/s) = gmol/s
$2\pi r\Delta z q_m\|_r - 2\pi(r+\Delta r)\Delta z q_m\|_{r+\Delta r}$	(cm)(cm)(gmol/(cm^2 · s)) = gmol/s
$2\pi r\Delta r v c_a\|_{z-\Delta z} - 2\pi r\Delta r v c_a\|_z$	(cm)(cm)(cm/s)(gmol/cm^3) = gmol/s
$-2\pi r\Delta r\Delta z k_r c_a^2$	(cm)(cm)(cm)(cm^3/(gmol·s)(gmol/cm^3)^2 = gmol/s

We see that each term has the net units of gmol/s. The physical interpretation of these units is (a) net accumulation or depletion (depending on whether the derivative in t is positive or negative) of the reactant in the incremental volume with dimensions Δr and Δz, (b) net diffusion rate into or out of the incremental volume, (c) net convection rate into or out of the incremental volume, and (d) rate of consumption of the reactant in the incremental volume (consumption because of the minus sign in $-2\pi\ldots$). Thus, we have ensured that the units are consistent throughout the equation and in the process, gained an understanding of the physical significance of each term. Also, we have used *cgs units*. Any other choice would be fine, for example, *mks or SI* (as long as they are metric!).

We now proceed with the units analysis of the remaining equations (Eqs. (13.1)–(13.13)).

Term	Units
$q_m = -D_c\frac{\partial c_a}{\partial r}$	(cm^2/s)(gmol/cm^3)(1/cm) = gmol/(cm^2 · s)

This unit analysis of Eq. (13.1) is for *Fick's first law* that provides a mass diffusion flux, q_m; "flux" means a transfer rate *per unit area*.

Term	Units
$2\pi r\Delta r\Delta z\rho C_p\frac{\partial T_k}{\partial t}$	(cm)(cm)(cm)(g/cm^3)(cal/(g·K))(K/s) = cal/s
$2\pi r\Delta z q_h\|_r - 2\pi(r+\Delta r)\Delta z q_h\|_{r+\Delta r}$	(cm)(cm)(cal/(cm^2 · s) = cal/s
$2\pi r\Delta r v\rho C_p T_k\|_{z-\Delta z} - 2\pi r\Delta r v\rho C_p T_k\|_z$	(cm)(cm)(cm/s)(g/cm^3)(cal/(g·K)(K) = cal/s
$-\Delta H 2\pi r\Delta r\Delta z k_r c_a^2$	(cal/gmol)(cm)(cm)(cm)(cm^3/(gmol·s)(gmol/cm^3)^2 = cal/s

The net units are the cal/s accumulating, diffusing, flowing, and being generated (by the exothermic reaction) in the incremental value with dimensions Δr and Δz. The heat of reaction, ΔH, by convention, is negative for an exothermic reaction so that the reaction term is positive (due to the minus in $-\Delta H$); see the numerical value of ΔH in Table 13.1.

Term	Units
$q_h = -k\frac{\partial T_k}{\partial r}$	((cal·cm)/(s·cm^2·K))(K/cm) = cal/(s·cm^2)

This unit analysis of Eq. (13.3) is for *Fourier's first law* that provides a thermal conduction flux, q_h.

Term	Units
$k_r = k_0\, e^{-E/(RT_k)}$	$(\mathrm{cm^3/(gmol{\cdot}s)})\left[\exp \dfrac{\mathrm{cal/gmol}}{\mathrm{(cal/(gmol{\cdot}K))(K)}}\right] = \mathrm{cm^3/(gmol{\cdot}s)}$

Note that the argument of the exponential function (from Eq. (13.5)) is dimensionless. Equations (13.6)–(13.11) are essentially self-explanatory with respect to units.

Equation (13.12) has the following units:

Term	Units
$k\dfrac{\partial T_k(r = r_0, z, t)}{\partial r} = h(T_w - T_k(r = r_0, z, t))$	(LHS) $\mathrm{((cal{\cdot}cm)/s \cdot cm^2{\cdot}K))(K/cm) = cal/(s \cdot cm^2)}$
	(RHS) $\mathrm{(cal/(s{\cdot}cm^2{\cdot}K))(K) = cal/(s{\cdot}cm^2)}$

Also, Eq. (13.13) is self-explanatory.

REFERENCES

[1] Rosner, D. E. (1986), *Transport Processes in Chemically Reacting Flow Systems*, Butterworth-Heinemann, Berlin

[2] Leveque, R. J. (2002), *Finite Volume Methods for Hyperbolic Problems*, Cambridge University Press, Cambridge

[3] Shu, C.-W. (1998), Essentially Non-Oscillatory and Weighted Essential Non-Oscillatory Schemes for Hyperbolic Conservation Laws, In: B. Cockburn, C. Johnson, C.-W. Shu, and E. Tadmor (Eds.), *Advanced Numerical Approximation of Nonlinear Hyperbolic Equations*, Lecture Notes in Mathematics, vol. 1697, Springer, Berlin, pp. 325–432

[4] Wesseling, P. (2001), *Principles of Computational Fluid Dynamics*, Springer, Berlin

14

Diffusion Equation in Spherical Coordinates[1]

An important goal of current cellular biophysics is the characterization of signal-dependent redistribution of macromolecules in living cells. In this chapter we present a simplified system of equations governing diffusional redistribution in three dimensions (3D) of green fluorescent protein (GFP). However, the biophysical details are not discussed here, and the reader is referred to the original papers for more details [1, 2].

This partial differential equation (PDE) application introduces the following mathematical concepts and computational methods:

1. To demonstrate the origin of a PDE starting with the relevant differential operators in spherical coordinates.
2. These operators are then substituted in the coordinate-free PDE to arrive at the PDE in spherical coordinates.
3. Consideration is given to the variable coefficients in the PDE, specifically how they lead to various singularities.
4. Methods for resolving or regularizing the singularities are reviewed and included in the Matlab routines for the method of lines (MOL) solution of the PDE.
5. Library routines for the calculation of the PDE spatial derivatives are discussed and called from the PDE routines.

Since we are interested in computing a numerical solution to the diffusion equation in spherical coordinates, we start with some differential operators in spherical coordinates (r, θ, ϕ).

[1] This chapter was written in collaboration with Dr. Peter D. Calvert, Department of Ophthalmology, Upstate New York Medical University, Syracuse, NY, and Dr. E. N. Pugh, Jr., Department of Ophthalmology, University of Pennsylvania, Philadelphia, PA.

$\nabla\cdot$ (*gradient of a vector*):

$$[\nabla]_r = \frac{1}{r^2}\frac{\partial}{\partial r}(r^2)$$

$$[\nabla]_\theta = \frac{1}{r\,\sin\theta}\frac{\partial}{\partial\theta}(\sin\theta) \tag{14.1}$$

$$[\nabla]_\phi = \frac{1}{r\,\sin\theta}\frac{\partial}{\partial\phi}$$

∇ (*gradient of a scalar*):

$$[\nabla]_r = \frac{\partial}{\partial r}$$

$$[\nabla]_\theta = \frac{1}{r}\frac{\partial}{\partial\theta} \tag{14.2}$$

$$[\nabla]_\phi = \frac{1}{r\,\sin\theta}\frac{\partial}{\partial\phi}$$

$\nabla\cdot\nabla$ (*divergence of the gradient of a scalar*):

$$\frac{1}{r^2}\frac{\partial}{\partial r}\left(r^2\frac{\partial}{\partial r}\right) + \frac{1}{r^2\sin\theta}\frac{\partial}{\partial\theta}\left(\sin\theta\frac{\partial}{\partial\theta}\right) + \frac{1}{r^2\sin^2\theta}\frac{\partial^2}{\partial\phi^2} \tag{14.3}$$

$\nabla\cdot\nabla$ ($=\nabla^2$ the *Laplacian*) follows directly from the preceding components of $\nabla\cdot$ (divergence of a vector) and ∇ (gradient of a scalar).

$$\begin{aligned}\nabla\cdot\nabla &= \left(\mathbf{i}_r\frac{1}{r^2}\frac{\partial}{\partial r}(r^2) + \mathbf{j}_\theta\frac{1}{r\sin\theta}\frac{\partial}{\partial\theta}(\sin\theta) + \mathbf{k}_\phi\frac{1}{r\sin\theta}\frac{\partial}{\partial\phi}\right)\cdot\left(\mathbf{i}_r\frac{\partial}{\partial r} + \mathbf{j}_\theta\frac{1}{r}\frac{\partial}{\partial\theta} + \mathbf{k}_\phi\frac{1}{r\sin\theta}\frac{\partial}{\partial\phi}\right)\\ &= \frac{1}{r^2}\frac{\partial}{\partial r}\left(r^2\frac{\partial}{\partial r}\right) + \frac{1}{r\sin\theta}\frac{\partial}{\partial\theta}\left(\sin\theta\frac{1}{r}\frac{\partial}{\partial\theta}\right) + \frac{1}{r\sin\theta}\frac{\partial}{\partial\phi}\left(\frac{1}{r\sin\theta}\frac{\partial}{\partial\phi}\right)\\ &= \frac{1}{r^2}\frac{\partial}{\partial r}\left(r^2\frac{\partial}{\partial r}\right) + \frac{1}{r^2\sin\theta}\frac{\partial}{\partial\theta}\left(\sin\theta\frac{\partial}{\partial\theta}\right) + \frac{1}{r^2\sin^2\theta}\frac{\partial^2}{\partial\phi^2}\end{aligned}$$

The *diffusion equation in coordinate-free form* is

$$\frac{\partial u}{\partial t} = D\nabla^2 u \tag{14.4}$$

Substitution of ∇^2 from Eq. (14.3) gives the *diffusion equation in spherical coordinates:*

$$\frac{\partial u}{\partial t} = D\left\{\frac{1}{r^2}\frac{\partial}{\partial r}\left(r^2\frac{\partial u}{\partial r}\right) + \frac{1}{r^2\sin\theta}\frac{\partial}{\partial\theta}\left(\sin\theta\frac{\partial u}{\partial\theta}\right) + \frac{1}{r^2\sin^2\theta}\frac{\partial^2 u}{\partial\phi^2}\right\} \tag{14.5}$$

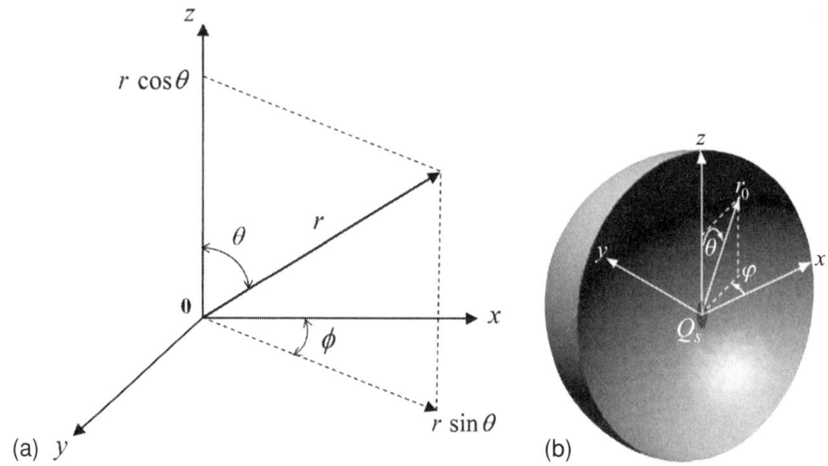

Figure 14.1. (a) Relationship between Cartesian (x, y, z) and spherical (r, θ, ϕ) coordinates. (b) The spherical coordinate system used in the analysis of diffusion in 3D, where a source or sink of protein mass Q_s is created at the cell center. Simplified protein redistribution by diffusion is illustrated in the text by two examples: the first by solving at spherical grid points defined on the radial vector r, assuming angular symmetry, that is, no variations in polar angle θ or azimuthal angle ϕ; and the second by solving at spherical grid points on r and polar angle grid points on θ, but assuming azimuthal angular symmetry, that is, no variations in angle ϕ (image reproduced from [2] with permission of the authors)

To define the angles θ and ϕ, the relationship between *Cartesian coordinates* (x, y, z) and spherical coordinates (r, θ, ϕ) is

$$\begin{aligned} x &= r \sin\theta \cos\phi \\ y &= r \sin\theta \sin\phi \\ z &= r \cos\theta \\ r &= \sqrt{x^2 + y^2 + z^2} \end{aligned} \tag{14.6}$$

These relationships are illustrated in Figure 14.1a. Thus, ϕ is the angle of r in the x–y plane, and θ is the angle of r relative to the z axis.

Equation (14.5) can be written in expanded form (by differentiating the radial group in r and the angular group in θ as a product) as

$$\frac{\partial u}{\partial t} = D\left\{\frac{\partial^2 u}{\partial r^2} + \frac{2}{r}\frac{\partial u}{\partial r} + \frac{1}{r^2}\left(\frac{\partial^2 u}{\partial \theta^2} + \frac{\cos\theta}{\sin\theta}\frac{\partial u}{\partial \theta}\right) + \frac{1}{r^2 \sin^2\theta}\frac{\partial^2 u}{\partial \phi^2}\right\} \tag{14.7}$$

Equation (14.7) is first order in t, and second order in r, θ, and ϕ. However, we will select initial conditions (ICs) and boundary conditions (BCs) so that the 3D PDE of Eq. (14.7) has no variation in ϕ and therefore the third RHS term in ϕ is zero. Additionally, we will add an *inhomogeneous term* to Eq. (14.7), $f(r, \theta, t)$. Thus, Eq. (14.7) becomes

$$\frac{\partial u}{\partial t} = D\left\{\frac{\partial^2 u}{\partial r^2} + \frac{2}{r}\frac{\partial u}{\partial r} + \frac{1}{r^2}\left(\frac{\partial^2 u}{\partial \theta^2} + \frac{\cos\theta}{\sin\theta}\frac{\partial u}{\partial \theta}\right)\right\} + f(r, \theta, t) \tag{14.8}$$

Equation (14.8) is the starting point for our MOL analysis. It is first order in t, and second order in r and θ. We take as the IC (for t)

$$u(r, \theta, t = 0) = 0 \tag{14.9}$$

The BCs in r are

$$\frac{\partial u(r = 0, \theta, t)}{\partial r} = 0 \tag{14.10a}$$

$$\frac{\partial u(r = r_o, \theta, t)}{\partial r} = 0 \tag{14.10b}$$

where r_o is the outer radius of the spherical system.

The BCs in θ are

$$\frac{\partial u(r, \theta = 0, t)}{\partial \theta} = 0 \tag{14.11a}$$

$$\frac{\partial u(r, \theta = \pi/2, t)}{\partial \theta} = 0 \tag{14.11b}$$

As the starting point for the MOL solution of Eqs. (14.8)–(14.11), we start with a simpler problem that does not include the angular variable θ. Thus, Eq. (14.8) reduces to

$$\frac{\partial u}{\partial t} = D\left\{\frac{\partial^2 u}{\partial r^2} + \frac{2}{r}\frac{\partial u}{\partial r}\right\} + f(r, t) \tag{14.12}$$

which has only the radial variation, and the solution is therefore $u(r, t)$. u corresponds to *protein concentration* and f to a *source or sink term representing the volumetric rate of local photoconversion*. Figure 14.1b is a diagrammatic view of the spherical coordinate system used in the diffusion analysis.

We will proceed in two steps: (1) only the singularity $2/r$ in Eq. (14.12) is considered and (2) the singularity in r and θ in Eq. (14.8), $1/r^2((\partial^2 u/\partial\theta^2) + (\cos\theta/\sin\theta)(\partial u/\partial\theta))$, is also considered. In other words, after the solution of Eq. (14.12) is concluded, we will return to Eq. (14.8) that has this considerably more complicated singularity (in r and θ).

A main program for the MOL solution of Eqs. (14.9), (14.10), and (14.12) is given in Listing 14.1.

```
%
% Clear previous files
  clear all
  clc
%
% Global area
  global r r0 nr D f tau ncall
%
% Model parameters
```

```
  D=0.1;
  r0=1.0;
  std=1.0;
  tau=1.0;
%
% Radial grid and inhomogeneous term
  nr=21;
  dr=r0/(nr-1);
  for i=1:nr
    r(i)=(i-1)*dr;
    f(i)=exp(-r(i)^2/std^2);
  end
%
% Independent variable for ODE integration
  tf=2.5;
  tout=[0.0:0.25:tf]';
  nout=11;
  ncall=0;
%
% Initial condition
  u0=zeros(nr,1);
%
% ODE integration
  reltol=1.0e-04; abstol=1.0e-04;
  options=odeset('RelTol',reltol,'AbsTol',abstol);
  [t,u]=ode15s(@pde_1,tout,u0,options);
%
% Display a heading and selected output
  fprintf('\n nr = %2d  r0 = %4.2f  std = %4.2f\n',nr,r0,std);
  for it=1:nout
    fprintf('\n t = %4.2f\n',t(it));
    for i=1:2:nr
      fprintf(' r = %4.1f   u(r,t) = %8.5f\n',r(i),u(it,i));
    end
    for i=1:nr
      u1d(i)=4.0*pi*r(i)^2*u(it,i);
    end
    I1=simp(0.0,r0,nr,u1d)/(4.0/3.0*pi*r0^3);
    fprintf(' \n integral = %7.5f\n',I1);
  end
  fprintf('\n ncall = %5d\n',ncall);
%
% Parametric plot
  figure(1);
  plot(r,u); axis([0 r0 0 1]);
  title('u(r,t), t = 0, 0.25, ..., 2.5');
        xlabel('r'); ylabel('u(r,t)')
```

```
%
% Extend integration for closer approach to steady state
  fprintf('\n Extended solution\n');
  u0=u(nout,:);
  t0=t(nout);
  tf=t0+5.0;
  tout=[t0:2.5:tf]';
  nout=3;
  [t,u]=ode15s(@pde_1,tout,u0,options);end
  for it=1:nout
    fprintf('\n t = %4.1f\n',t(it));
    for i=1:2:nr
      fprintf(' r = %4.1f   u(r,t) = %8.5f\n',r(i),u(it,i));
    end
    for i=1:nr
      u1d(i)=4.0*pi*r(i)^2*u(it,i);
    end
    I1=simp(0.0,r0,nr,u1d)/(4.0/3.0*pi*r0^3);
    fprintf(' \n integral = %7.5f\n',I1);
  end
  fprintf('\n ncall = %5d\n',ncall);
```

Listing 14.1. Main program pde_1_main.m for the solution of Eqs. (14.9), (14.10), and (14.12)

We can note the following points about this main program:

1. A *global* area with parameters and variables that can be shared with other routines is defined. Then some model parameters are defined numerically.

```
%
% Clear previous files
  clear all
  clc
%
% Global area
  global r r0 nr D f tau ncall
%
% Model parameters
  D=0.1;
  r0=1.0;
  std=1.0;
  tau=1.0;
```

2. A 21-point grid in r is then defined.

```
%
% Radial grid and inhomogeneous term
  nr=21;
  dr=r0/(nr-1);
  for i=1:nr
    r(i)=(i-1)*dr;
    f(i)=exp(-r(i)^2/std^2);
  end
```

At the same time, the inhomogeneous term $f(r,t)$ in Eq. (14.12) is programmed. This function is

$$f(r,t) = e^{[-r^2/\sigma^2]}[h(t) - h(t-\tau)] \tag{14.13}$$

where $h(t)$ is the *Heaviside step function*

$$h(t) = \begin{Bmatrix} 0, t < 0 \\ 1, t > 0 \end{Bmatrix} \tag{14.14}$$

The exponential in Eq. (14.13) is a *Gaussian function* with a *standard deviation* σ. τ is a delay in t and therefore $h(t) - h(t-\tau)$ is a pulse with a value of 1 over the interval $0 \le t \le \tau$ and zero otherwise. In other words, $h(t) - h(t-\tau)$ "turns on" the Gaussian function for the interval $0 \le t \le \tau$.

3. A timescale is defined over the interval $0 \le t \le 2.5$ with an output interval of 0.25 so that there are 11 output points in t (including the IC $t = 0$).

```
%
% Independent variable for ODE integration
  tf=2.5;
  tout=[0.0:0.25:tf]';
  nout=11;
  ncall=0;
%
% Initial condition
  u0=zeros(nr,1);
```

Also, homogeneous IC (14.9) is defined over the `nr` points in r.

4. The $nr = 21$ ordinary differential equations (ODEs) are integrated by `ode15s` through the ODE routine `pde_1` (discussed subsequently).

```
%
% ODE integration
  reltol=1.0e-04; abstol=1.0e-04;
  options=odeset('RelTol',reltol,'AbsTol',abstol);
  [t,u]=ode15s(@pde_1,tout,u0,options);
```

5. A heading and selected output are displayed.

```
%
% Display a heading and selected output
  fprintf('\n nr = %2d  r0 = %4.2f  std = %4.2f\n',nr,r0,std);
  for it=1:nout
    fprintf('\n t = %4.2f\n',t(it));
    for i=1:2:nr
      fprintf(' r = %4.1f    u(r,t) = %8.5f\n',r(i),u(it,i));
    end
    for i=1:nr
      u1d(i)=4.0*pi*r(i)^2*u(it,i);
    end
    I1=simp(0.0,r0,nr,u1d)/(4.0/3.0*pi*r0^3);
    fprintf(' \n integral = %7.5f\n',I1);
  end
  fprintf('\n ncall = %5d\n',ncall);
```

Also, the normalized integral

$$I_1 = \int_0^{r_o} 4\pi r^2 u(r,t)dr \Big/ \left(\frac{4}{3}\pi r_o^3\right) \tag{14.15}$$

is evaluated numerically by a call to function `simp` (discussed subsequently) that implements *Simpson's rule* for *numerical quadrature* (integration). The integral of Eq. (14.15) is an average value of $u(r,t)$ over $0 \le r \le r_o$. As the solution to Eq. (14.12) approaches a final state ($t \to \infty$), it approaches a uniform value given by Eq. (14.15). This property of the solution of Eq. (14.12) is demonstrated in the numerical output discussed subsequently.

The *conservation principle* or *invariant* of Eq. (14.15) is therefore a check on the numerical solution. That is, the numerical solution should approach the constant value of I_1; if it does not, the numerical solution does not *conserve mass* as stated in physical terms. The reason mass is conserved in the present case (Eq. (14.12)) is

(a) For $t > \tau$, mass is no longer added to the system through the inhomogeneous term $f(r,t)$ in Eq. (14.12).

(b) Mass cannot leave the system because of BC (14.10b), that is, this is a *no-flux* BC according to *Fick's first law*, which states that the mass flux q_m at $r = r_o$ is

$$q_m = -D\frac{\partial u(r = r_o, t)}{\partial r} \tag{14.16}$$

Since $\partial u(r = r_o, t)/\partial r = 0$ according to BC (14.10b), $q_m = 0$.

(c) Thus, since mass is not entering (for $t > \tau$) or leaving the system, the mass within the system must remain constant, which is expressed by I_1 in Eq. (14.15) remaining constant.

(d) This is an important test of the preceding code. If I_1 does not remain constant, "mass" must be accumulating in or leaking from the system due to an incorrect approximation of Eqs. (14.9), (14.10), and (14.12), that is, the numerical analysis and/or associated coding has an error.

The number of calls to the ODE routine, `pde_1`, is also displayed to give an indication of the computational effort required to compute the solution. `ncall` is incremented in `pde_1` each time this routine is called.

6. A plot of $u(r, t)$ versus r with t as a parameter ($t = 0, 0.25, \ldots, 2.5$) indicates the evolution of the solution toward the steady state of Eq. (14.15).

```
%
% Parametric plots
  figure(1);
  plot(r,u); axis([0 r0 0 1]);
  title('u(r,t), t = 0, 0.25, ..., 2.5'); xlabel('r');
        ylabel('u(r,t)')
```

The plot is discussed subsequently.

7. When the solution reaches $t = 2.5$, it has not quite reached the steady-state value of Eq. (14.15). Therefore, it is extended to $t = 7.5$ through a second call to `ode15s`.

```
%
% Extend integration for closer approach to steady state
  fprintf('\n Extended solution\n');
  u0=u(nout,:);
  t0=t(nout);
  tf=t0+5.0;
  tout=[t0:2.5:tf]';
  nout=3;
  [t,u]=ode15s(@pde_1,tout,u0,options);end
  for it=1:nout
    fprintf('\n t = %4.1f\n',t(it));
    for i=1:2:nr
      fprintf(' r = %4.1f   u(r,t) = %8.5f\n',r(i),u(it,i));
```

```
    end
    for i=1:nr
      u1d(i)=4.0*pi*r(i)^2*u(it,i);
    end
    I1=simp(0.0,r0,nr,u1d)/(4.0/3.0*pi*r0^3);
    fprintf(' \n integral = %7.5f\n',I1);
  end
  fprintf('\n ncall = %5d\n',ncall);
```

This extension illustrates an important detail. The numerical solution to an ODE system can start from a previous solution, in this case, at $t = 2.5$. Thus, a new IC is used, which is the solution at the final point of the preceding solution (i.e., $u(r, t = 2.5)$). The solution then continues on to $t = 2.5 + 5.0 = 7.5$; at this final value, the solution is close to the value from Eq. (14.15), as demonstrated in the output discussed subsequently.

During this continuation from $t = 2.5$ to $t = 7.5$, the integral of Eq. (14.15) is computed (at $t = 2.5, 5.0, 7.5$), and the integral along with the continuing `ncall` is displayed.

We briefly review the output before going on to discussions of the routines `pde_1`, `simp`. A portion of the output is given in Table 14.1.

We can note the following details about the output given in Table 14.1:

1. The integral of Eq. (14.15) increases from $t = 0$ (where it is zero from IC (14.9)) to an essentially constant value for $t > 1$ (the inhomogeneous term $f(r, t)$ of Eq. (14.12) is zero for $t > 1$ so that it does not contribute to $\partial u/\partial t$ in Eq. (14.12)).

```
 t = 0.00, integral = 0.00000

 t = 0.25, integral = 0.14210

 t = 0.50, integral = 0.28421

 t = 0.75, integral = 0.42631

 t = 1.00, integral = 0.56835

 t = 1.25, integral = 0.56860

       .            .
       .            .
       .            .

 t = 2.50, integral = 0.56860
```

Table 14.1. Selected output from pde_1_main

```
nr = 21   r0 = 1.00   std = 1.00

 t = 0.00
 r =  0.0   u(r,t) =  0.00000
 r =  0.1   u(r,t) =  0.00000
 r =  0.2   u(r,t) =  0.00000
 r =  0.3   u(r,t) =  0.00000
 r =  0.4   u(r,t) =  0.00000
 r =  0.5   u(r,t) =  0.00000
 r =  0.6   u(r,t) =  0.00000
 r =  0.7   u(r,t) =  0.00000
 r =  0.8   u(r,t) =  0.00000
 r =  0.9   u(r,t) =  0.00000
 r =  1.0   u(r,t) =  0.00000

 integral = 0.00000

 t = 0.25
 r =  0.0   u(r,t) =  0.23268
 r =  0.1   u(r,t) =  0.23047
 r =  0.2   u(r,t) =  0.22397
 r =  0.3   u(r,t) =  0.21355
 r =  0.4   u(r,t) =  0.19978
 r =  0.5   u(r,t) =  0.18346
 r =  0.6   u(r,t) =  0.16556
 r =  0.7   u(r,t) =  0.14736
 r =  0.8   u(r,t) =  0.13061
 r =  0.9   u(r,t) =  0.11788
 r =  1.0   u(r,t) =  0.11268

 integral = 0.14210

 t = 0.50
 r =  0.0   u(r,t) =  0.43582
 r =  0.1   u(r,t) =  0.43189
 r =  0.2   u(r,t) =  0.42036
 r =  0.3   u(r,t) =  0.40194
 r =  0.4   u(r,t) =  0.37783
 r =  0.5   u(r,t) =  0.34970
 r =  0.6   u(r,t) =  0.31967
 r =  0.7   u(r,t) =  0.29033
 r =  0.8   u(r,t) =  0.26475
 r =  0.9   u(r,t) =  0.24647
 r =  1.0   u(r,t) =  0.23945

 integral = 0.28421
```

```
t = 0.75
r =  0.0   u(r,t) =  0.61680
r =  0.1   u(r,t) =  0.61166
r =  0.2   u(r,t) =  0.59660
r =  0.3   u(r,t) =  0.57270
r =  0.4   u(r,t) =  0.54171
r =  0.5   u(r,t) =  0.50601
r =  0.6   u(r,t) =  0.46851
r =  0.7   u(r,t) =  0.43259
r =  0.8   u(r,t) =  0.40198
r =  0.9   u(r,t) =  0.38064
r =  1.0   u(r,t) =  0.37264

integral = 0.42631

t = 1.00
r =  0.0   u(r,t) =  0.78277
r =  0.1   u(r,t) =  0.77685
r =  0.2   u(r,t) =  0.75955
r =  0.3   u(r,t) =  0.73219
r =  0.4   u(r,t) =  0.69690
r =  0.5   u(r,t) =  0.65653
r =  0.6   u(r,t) =  0.61450
r =  0.7   u(r,t) =  0.57464
r =  0.8   u(r,t) =  0.54107
r =  0.9   u(r,t) =  0.51795
r =  1.0   u(r,t) =  0.50937

integral = 0.56835

t = 1.25
r =  0.0   u(r,t) =  0.70719
r =  0.1   u(r,t) =  0.70299
r =  0.2   u(r,t) =  0.69079
r =  0.3   u(r,t) =  0.67170
r =  0.4   u(r,t) =  0.64751
r =  0.5   u(r,t) =  0.62059
r =  0.6   u(r,t) =  0.59368
r =  0.7   u(r,t) =  0.56963
r =  0.8   u(r,t) =  0.55100
r =  0.9   u(r,t) =  0.53953
r =  1.0   u(r,t) =  0.53581

integral = 0.56860
```

(continued)

Table 14.1 (*continued*)

```
     .                 .
     .                 .
     .                 .

Output for t = 1.5, 1.75, 2, 2.25
             removed
     .                 .
     .                 .
     .                 .

t = 2.50
r =  0.0   u(r,t) =  0.58037
r =  0.1   u(r,t) =  0.57997
r =  0.2   u(r,t) =  0.57884
r =  0.3   u(r,t) =  0.57711
r =  0.4   u(r,t) =  0.57498
r =  0.5   u(r,t) =  0.57269
r =  0.6   u(r,t) =  0.57048
r =  0.7   u(r,t) =  0.56859
r =  0.8   u(r,t) =  0.56717
r =  0.9   u(r,t) =  0.56632
r =  1.0   u(r,t) =  0.56605

integral = 0.56860

ncall =    98

Extended solution

t =  2.5
r =  0.0   u(r,t) =  0.58037
r =  0.1   u(r,t) =  0.57997
r =  0.2   u(r,t) =  0.57884
r =  0.3   u(r,t) =  0.57711
r =  0.4   u(r,t) =  0.57498
r =  0.5   u(r,t) =  0.57269
r =  0.6   u(r,t) =  0.57048
r =  0.7   u(r,t) =  0.56859
r =  0.8   u(r,t) =  0.56717
r =  0.9   u(r,t) =  0.56632
r =  1.0   u(r,t) =  0.56605

integral = 0.56860
```

```
t =  5.0
r =  0.0   u(r,t) =  0.56868
r =  0.1   u(r,t) =  0.56867
r =  0.2   u(r,t) =  0.56866
r =  0.3   u(r,t) =  0.56865
r =  0.4   u(r,t) =  0.56864
r =  0.5   u(r,t) =  0.56863
r =  0.6   u(r,t) =  0.56861
r =  0.7   u(r,t) =  0.56860
r =  0.8   u(r,t) =  0.56859
r =  0.9   u(r,t) =  0.56859
r =  1.0   u(r,t) =  0.56859

integral = 0.56860

t =  7.5
r =  0.0   u(r,t) =  0.56861
r =  0.1   u(r,t) =  0.56860
r =  0.2   u(r,t) =  0.56860
r =  0.3   u(r,t) =  0.56860
r =  0.4   u(r,t) =  0.56860
r =  0.5   u(r,t) =  0.56860
r =  0.6   u(r,t) =  0.56860
r =  0.7   u(r,t) =  0.56860
r =  0.8   u(r,t) =  0.56860
r =  0.9   u(r,t) =  0.56860
r =  1.0   u(r,t) =  0.56861

integral = 0.56860

ncall =   145
```

2. The extended output indicates a convergence of the solution to the constant value of Eq. (14.15). Thus, for $t = 7.5$,

```
t =  7.5
r =  0.0   u(r,t) =  0.56861
r =  0.1   u(r,t) =  0.56860
r =  0.2   u(r,t) =  0.56860
r =  0.3   u(r,t) =  0.56860
r =  0.4   u(r,t) =  0.56860
r =  0.5   u(r,t) =  0.56860
r =  0.6   u(r,t) =  0.56860
```

```
r =  0.7   u(r,t) =  0.56860
r =  0.8   u(r,t) =  0.56860
r =  0.9   u(r,t) =  0.56860
r =  1.0   u(r,t) =  0.56861

integral = 0.56860

ncall =   145
```

Even with this extension to $t = 7.5$, the number of calls to `pde_1` is quite modest (`ncall = 145`). The plotted output is indicated in Figure 14.2. The effect of BCs (14.10a) and (14.10b) is clear from this plot (zero derivatives in r at the boundaries $r = 0, r_o$). The convergence of the solution to $u(r, t = \infty) = 0.56860$ is clear.

We now consider the remaining routines to produce the solution of Table 14.1 and Figure 14.1. The ODE routine, `pde_1`, is given in Listing 14.2.

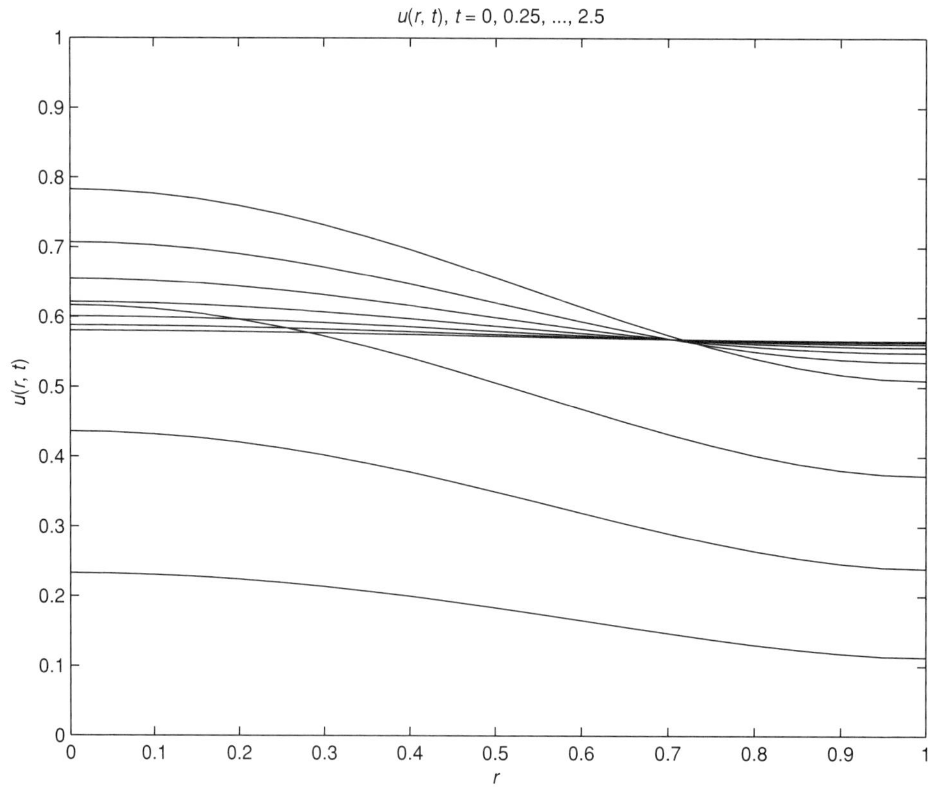

Figure 14.2. Plot of the numerical solution to Eq. (14.12) from pde_1_main and pde_1

```
  function ut=pde_1(t,u)
%
% Global area
  global r r0 nr D f tau ncall
%
% Spatial grid in r
  for i=1:nr
%
%   ur
    ur=dss004(0.0,r0,nr,u);
    ur(nr)=0.0;
    ur( 1)=0.0;
%
%   urr
    urr=dss004(0.0,r0,nr,ur);
  end
%
% PDE
  for i=1:nr
    if(t<=tau)
      fs=f(i);
    else
      fs=0.0;
    end
    if(i==1)
      ut(i)=D*3.0*urr(i)+fs;
    else
      ut(i)=D*(urr(i)+2.0/r(i)*ur(i))+fs;
    end
  end
  ut=ut';
%
% Increment number of calls to pde_1
  ncall=ncall+1;
```

Listing 14.2. pde_1 for Eq. (14.12)

We can note the following points about pde_1:

1. The function is first defined and a global area is then defined to share the problem parameters and variables with main program pde_1_main in Listing 14.1.

```
  function ut=pde_1(t,u)
%
% Global area
  global r r0 nr D f tau ncall
```

Note that the dependent variable u and its derivative ut are given the names used in Eq. (14.12) since there is only one such variable (i.e., the programming in pde_1 can be directly in terms of u).

2. A for loop is then used to step through the nr radial grid points (to define the nr ODEs in the MOL solution of Eq. (14.12)).

```
%
% Spatial grid in r
  for i=1:nr
%
%   ur
    ur=dss004(0.0,r0,nr,u);
    ur(nr)=0.0;
    ur( 1)=0.0;
```

The first derivative $\partial u/\partial r$ = ur is then computed at each of the radial points by a call to dss004. After this call, BCs (14.10a) and (14.10b) are imposed to ensure the correct (zero) values of ur at $r = 0, r_o$.

3. The second derivative in r in Eq. (14.12), (urr), is computed by differentiating the first derivative (using *stagewise differentiation*).

```
%
%   urr
    urr=dss004(0.0,r0,nr,ur);
  end
```

The for loop is terminated with the end statement.

4. A second for loop is used for the programming of the nr MOL ODEs. This loop starts with the addition of the inhomogeneous term $f(r, t)$ in Eq. (14.12) (which is available through the global area from the main program and is nonzero for $0 \leq t \leq \tau$).

```
%
% PDE
  for i=1:nr
    if(t<=tau)
```

```
      fs=f(i);
    else
      fs=0.0;
    end
```

5. At $r = 0$ (`i=1`), the first-derivative radial group in Eq. (14.12) is indeterminate; that is,

$$\frac{2}{r}\frac{\partial u}{\partial r} = 0/0$$

as a consequence of *homogeneous Neumann BC* (14.10a). We therefore apply *l'Hospital's rule to resolve (regularize)* this indeterminant form.

$$\lim_{r \to 0} \frac{2}{r}\frac{\partial u}{\partial r} = 2\frac{\partial u^2}{\partial r^2}$$

Thus, the programming of the complete radial group in Eq. (14.12) is

$$\frac{\partial^2 u}{\partial r^2} + 2\frac{\partial^2 u}{\partial r^2} = 3\frac{\partial^2 u}{\partial r^2}$$

and the ODE at $r = 0$ is (with $f(r, t)$)

```
  if(i==1)
    ut(i)=D*3.0*urr(i)+fs;
```

Again, BC (14.10a) is included in this result (from when the indeterminate form was evaluated).

6. The programming of the ODEs at the other grid points ($r \neq 0$ or $i \neq 1$) follows directly from Eq. (14.12).

```
    else
      ut(i)=D*(urr(i)+2.0/r(i)*ur(i))+fs;
    end
  end
  ut=ut';
%
% Increment number of calls to pde_1
  ncall=ncall+1;
```

BC (14.10b) is included in this computation of `ut(i)` from `ur(nr)=0` in the preceding `for` loop. The routine then concludes with a transpose of `ut` as required by `ode15s` and incrementing of `ncall` (to record `pde_1` has been called).

Numerical integration routine `simp` is given in Listing 14.3.

```
  function uint=simp(xl,xu,n,u)
%
% Function simp computes an integral numerically by Simpson's
% rule
%
    h=(xu-xl)/(n-1);
    uint(1)=u(1)-u(n);
      for i=3:2:n
        uint(1)=uint(1)+4.0*u(i-1)+2.0*u(i);
      end
    uint=h/3.0*uint;
```

Listing 14.3. `simp` for the Evaluation of the Integral I_1 of Eq. (14.15)

`simp` is a straightforward implementation of Simpson's rule applied to Eq. (14.15):

$$I_1 = \int_0^{r_o} 4\pi r^2 u(x,t)dr \approx \frac{h}{3}\left[u(1) + \sum_{i=2}^{n-2} 4u(i) + 2u(i+1) + u(n)\right] \tag{14.17}$$

where the factor $4\pi r^2$ has been included in u (in `pde_1_main`).

This concludes the MOL solution of Eqs. (14.9), (14.10), and (14.12). We now go on to the 2D problem of Eqs. (14.8)–(14.11) for the calculation of $u(r,\theta,t)$. The reason this was deferred until the simpler problem was completed is because the second RHS term in θ of Eq. (14.8) requires special attention.

In particular, this term is singular for both $r = 0$ (due to the factor $1/r^2$) and $\theta = 0, \pi/2$ (due to the factor $1/\sin\theta$). The complexity of this r–θ group makes the regularization by a l'Hospital's rule analysis considerably more complex than for the $2/r$ singularity of Eq. (14.12), so we proceed with an alternative approach [3]. Returning to the general form of the PDE, Eq. (14.4), with the inhomogeneous term included

$$\frac{\partial u}{\partial t} = D\nabla^2 u + f(r,\theta,t) \tag{14.18}$$

all that is required to compute a solution of Eq. (14.18) is the accurate evaluation of the RHS. In other words, *it is not necessary to use a particular coordinate system* for the RHS (again, ∇^2 applies to any coordinate system). Thus, at the origin $r = 0$ where a singularity of the θ group in Eq. (14.8) occurs, we can *evaluate the θ group* *in Cartesian coordinates*. The reason for this choice is that *this group in Cartesian coordinates does not have a singularity at the origin*.

Thus, at the origin, Eq. (14.18) becomes (partly in Cartesian coordinates)

$$\frac{\partial u}{\partial t} = D\left[\frac{\partial^2 u}{\partial x^2} + \frac{\partial^2 u}{\partial y^2} + \frac{\partial^2 u}{\partial z^2}\right] + f\,(r = 0, \theta, t) \tag{14.19}$$

The RHS of Eq. (14.19) appears to be a mixture of Cartesian and spherical coordinates. However, the two are related through Eqs. (14.6). In other words, we can express the derivative in x on a grid in r and θ by realizing that the x axis corresponds to $\theta = \pi/2$ (again, we have formulated the problem based on Eq. (14.8) with no variation in ϕ so that $\theta = \pi/2$ also applies to the y axis – see Figure 14.1). Also, the z axis corresponds to $\theta = 0$. With the value of θ specified, the r vector in spherical coordinates corresponds to the x, y, or z axis in Cartesian coordinates.

With these relations in mind, we can easily link the solution (of Eq. (14.19)) in r, θ to the solution in x, y, z at the origin (at $x = y = z = r = 0$). All of this might seem complicated, but we will now illustrate how this can easily be done in the ODE routine pde_2. We start first with a main program (from which pde_2 is called by ode15s) (see Listing 14.4).

```
%
% Clear previous files
  clear all
  clc
%
% Global area
  global r r0 nr th th0 nth D std_0 std_pi2 f tau ncall
%
% Model parameters
  D=0.1;
  r0=1.0;
  th0=pi/2.0;
  std_0  =1.0;
  std_pi2=1.0;
  tau=1.0;
%
% Radial grid
  nr=11;
  dr=r0/(nr-1);
  for i=1:nr
    r(i)=(i-1)*dr;
  end
%
% Angular grid
  nth=11;
  dth=th0/(nth-1);
```

```
    for j=1:nth
      th(j)=(j-1)*dth;
    end
%
% Inhomogeneous term
    for i=1:nr
    for j=1:nth
      f(i,j)=exp(-(r(i)*sin(th(j))/std_pi2)^2 ...
                 -(r(i)*cos(th(j))/std_0  )^2);
    end
    end
%
% Independent variable for ODE integration
    tf=1.0;
    tout=[0.0:0.25:tf]';
    nout=5;
    ncall=0;
%
% Initial condition
    for i=1:nr
    for j=1:nth
      y0((i-1)*nth+j)=0.0;
    end
    end
%
% ODE integration
    reltol=1.0e-04; abstol=1.0e-04;
    options=odeset('RelTol',reltol,'AbsTol',abstol);
    [t,y]=ode15s(@pde_2,tout,y0,options);
%
% 1D to 2D (3D with t)
    for it=1:nout
    for  i=1:nr
    for  j=1:nth
       u(it,i,j)=y(it,(i-1)*nth+j);
    end
    end
    end
%
% Display a heading and selected output
    fprintf('\n  nr = %2d    r0 = %4.2f', nr, r0);
    fprintf('\n nth = %2d   th0 = %4.2f',nth,th0);
    fprintf('\n   std_0 = %4.2f   ',std_0);
    fprintf('\n std_pi2 = %4.2f\n',std_pi2);
    for it=1:nout
      fprintf('\n t = %4.2f',t(it));
      for i=1:2:nr
```

```
      fprintf('\n');
     for j=1:2:nth
      fprintf(' r = %4.1f  th = %4.2f  u(r,th,t) = %8.5f\n',...
               r(i),th(j),u(it,i,j));
     end
     end
   end
   fprintf('\n ncall = %5d\n',ncall);
%
% Parametric plots
   figure(1);
   subplot(2,2,1)
   uplot(1:nr,1:5:nth)=u(2,1:nr,1:5:nth);
   plot(r,uplot); axis([0 r0 0 1]);
   title('u(r,th,t), th = 0, \pi/4, \pi/2; t = 0.25');
   xlabel('r'); ylabel('u(r,th,t)');
   subplot(2,2,2)
   uplot(1:nr,1:5:nth)=u(3,1:nr,1:5:nth);
   plot(r,uplot); axis([0 r0 0 1]);
   title('u(r,th,t), th = 0, \pi/4, \pi/2; t = 0.5');
   xlabel('r'); ylabel('u(r,th,t)');
   subplot(2,2,3)
   uplot(1:nr,1:5:nth)=u(4,1:nr,1:5:nth);
   plot(r,uplot); axis([0 r0 0 1]);
   title('u(r,th,t), th = 0, \pi/4, \pi/2; t = 0.75');
   xlabel('r'); ylabel('u(r,th,t)');
   subplot(2,2,4)
   uplot(1:nr,1:5:nth)=u(5,1:nr,1:5:nth);
   plot(r,uplot); axis([0 r0 0 1]);
   title('u(r,th,t), th = 0, \pi/4, \pi/2; t = 1');
   xlabel('r'); ylabel('u(r,th,t)');
```

Listing 14.4. Main program `pde_2_main.m` for the Solution of Eqs. (14.8)–(14.11)

`pde_2_main` in Listing 14.4 has several common features with `pde_1_main` in Listing 14.1. The most significant difference is that the inclusion of θ (Eq. (14.12) has only the spatial variable r, while Eq. (14.8) has the spatial variables r, θ). Thus, we have gone from a 1D to a 2D problem.

We can note the following points about this main program:

1. Again, a global area is defined followed by the specification of some problem parameters.

```
%
% Clear previous files
  clear all
  clc
%
% Global area
  global r r0 nr th th0 nth D std_0 std_pi2 f tau ncall
%
% Model parameters
  D=0.1;
  r0=1.0;
  th0=pi/2.0;
  std_0  =1.0;
  std_pi2=1.0;
  tau=1.0;
```

The interval in r is $0 \leq r \leq r_o$ and in θ is $0 \leq \theta \leq \pi/2$. Note that for θ, from symmetry, we can consider just one quadrant (θ could extend to 2π). The two parameters `std_0, std_pi2` are discussed next.

2. Equation (14.13) is now extended to 2D as

$$f\,(r,\theta,t) = e^{\{-[(r\,\cos\theta)^2/\sigma_0{}^2+(r\,\sin\theta)^2/\sigma_{\pi/2}{}^2]\}}[h(t)-h(t-\tau)] \tag{14.20}$$

where σ_0 = `std_0`, $\sigma_{\pi/2}$ = `std_pi2`. The exponential in Eq. (14.20) is programmed as

```
%
% Inhomogeneous term
  for i=1:nr
  for j=1:nth
    f(i,j)=exp(-(r(i)*sin(th(j))/std_pi2)^2 ...
               -(r(i)*cos(th(j))/std_0  )^2);
  end
  end
```

The indices for r and θ, `i` and `j` respectively, will be used throughout the programming.

3. The radial and angular grids, with `nr=11` and `nth=11` points, are defined (in general, θ will be represented by the characters `th` in the programming).

```
%
% Radial grid
  nr=11;
```

```
  dr=r0/(nr-1);
  for i=1:nr
    r(i)=(i-1)*dr;
  end
%
% Angular grid
  nth=11;
  dth=th0/(nth-1);
  for j=1:nth
    th(j)=(j-1)*dth;
  end
```

4. A timescale is defined over the interval $0 \leq t \leq 1.0$ with an output interval of 0.25 so that there are five output points in t (including the IC $t = 0$).

```
%
% Independent variable for ODE integration
  tf=1.0;
  tout=[0.0:0.25:tf]';
  nout=5;
  ncall=0;
```

5. IC (14.9) follows.

```
%
% Initial condition
  for i=1:nr
  for j=1:nth
    y0((i-1)*nth+j)=0.0;
  end
  end
```

Note in particular the use of the 1D IC array y0.

6. The integration of the $nr \times n\text{th} = 11 \times 11 = 121$ ODEs is by ode15s. Note in particular the use of function pde_2, which is discussed subsequently.

```
%
% ODE integration
  reltol=1.0e-04; abstol=1.0e-04;
  options=odeset('RelTol',reltol,'AbsTol',abstol);
  [t,y]=ode15s(@pde_2,tout,y0,options);
```

```
%
% 1D to 2D (3D with t)
  for it=1:nout
  for  i=1:nr
  for  j=1:nth
     u(it,i,j)=y(it,(i-1)*nth+j);
  end
  end
  end
```

After the solution matrix y is returned by ode15s, it is put into a 3D array (with t index it, r index i and θ index j); an alternative approach would be to use the Matlab reshape utility as illustrated in Chapter 10.

7. Selected output (a subset of the 121 ODE dependent variables) is displayed. This output is discussed subsequently.

```
%
% Display a heading and selected output
  fprintf('\n  nr = %2d    r0 = %4.2f', nr, r0);
  fprintf('\n nth = %2d   th0 = %4.2f',nth,th0);
  fprintf('\n   std_0 = %4.2f   ',std_0);
  fprintf('\n std_pi2 = %4.2f\n',std_pi2);
  for it=1:nout
    fprintf('\n t = %4.2f',t(it));
    for i=1:2:nr
      fprintf('\n');
    for j=1:2:nth
      fprintf(' r = %4.1f th = %4.2f u(r,th,t) = %8.5f\n', ...
              r(i),th(j),u(it,i,j));
    end
    end
  end
  fprintf('\n ncall = %5d\n',ncall);
```

8. A series of four plots for $t = 0.25, 0.5, 0.75, 1.0$ is produced with every fifth value of θ plotted parametrically ($\theta = 0, \pi/4, \pi/2$). The plots are of $u(r, \theta, t)$ versus r.

```
%
% Parametric plots
  figure(1);
  subplot(2,2,1)
  uplot(1:nr,1:5:nth)=u(2,1:nr,1:5:nth);
```

```
plot(r,uplot); axis([0 r0 0 1]);
title('u(r,th,t), th = 0, \pi/4, \pi/2; t = 0.25');
xlabel('r'); ylabel('u(r,th,t)');
subplot(2,2,2)
uplot(1:nr,1:5:nth)=u(3,1:nr,1:5:nth);
plot(r,uplot); axis([0 r0 0 1]);
title('u(r,th,t), th = 0, \pi/4, \pi/2; t = 0.5');
xlabel('r'); ylabel('u(r,th,t)');
subplot(2,2,3)
uplot(1:nr,1:5:nth)=u(4,1:nr,1:5:nth);
plot(r,uplot); axis([0 r0 0 1]);
title('u(r,th,t), th = 0, \pi/4, \pi/2; t = 0.75');
xlabel('r'); ylabel('u(r,th,t)');
subplot(2,2,4)
uplot(1:nr,1:5:nth)=u(5,1:nr,1:5:nth);
plot(r,uplot); axis([0 r0 0 1]);
title('u(r,th,t), th = 0, \pi/4, \pi/2; t = 1');
xlabel('r'); ylabel('u(r,th,t)');
```

This output was selected to emphasize the variation of the solution with θ, which is the essential difference between Eqs. (14.8) and (14.12). (θ is included in Eq. (14.8).) To explore this idea a bit further, for $\sigma_0 = \sigma_{\pi/2} = \sigma$, Eq. (14.20) reduces to Eq. (14.13) (so that there is no variation in θ in the solution of Eq. (14.8) (which is the case programmed in `pde_1_main` of Listing 14.1 to test the code). Selected output for this case is listed in Table 14.2.

We can note the following details about this output:

1. $u(r, \theta, t)$ becomes larger with increasing t; for the final value $t = 1$, the inhomogeneous term of Eq. (14.20) has just switched to zero ($t = \tau = 1$ in `pde_2`), and for larger t, the radial variation will diminish and approach a constant value as in the case of the output in Table 14.1.
2. There is no θ variation as expected (since $\sigma_0 = \sigma_{\pi/2} = \sigma$) in Eq. (14.20)).

The plotted output is shown in Figure 14.3. We note that there is only one curve in each plot since there is no variation in θ. Actually, there are three superimposed plots for $\theta = 0, \pi/4, \pi/2$, which is a good test since all of the *theta* code is executed but it gives the same result.

We can now change the inhomogeneous term of Eq. (14.20) so that $\sigma_0 \neq \sigma_{\pi/2}$. For example, in `pde_2_main`

```
std_0  =2.0;
std_pi2=1.0;
```

(note we are changing just one number, `std_0`, from `1` to `2`). The resulting output is now as given in Table 14.3.

Table 14.2. Selected output from pde_2_main for $\sigma_0 = 1, \sigma_{\pi/2} = 1$

```
 nr = 11    r0 = 1.00
nth = 11   th0 = 1.57
  std_0 = 1.00
std_pi2 = 1.00

t = 0.00
r =  0.0   th = 0.00   u(r,th,t) =  0.00000
r =  0.0   th = 0.31   u(r,th,t) =  0.00000
r =  0.0   th = 0.63   u(r,th,t) =  0.00000
r =  0.0   th = 0.94   u(r,th,t) =  0.00000
r =  0.0   th = 1.26   u(r,th,t) =  0.00000
r =  0.0   th = 1.57   u(r,th,t) =  0.00000
       .                               .
       .                               .
       .                               .

 Output for r = 0.2, 0.4, 0.6, 0.8 removed

       .                               .
       .                               .
       .                               .
r =  1.0   th = 0.00   u(r,th,t) =  0.00000
r =  1.0   th = 0.31   u(r,th,t) =  0.00000
r =  1.0   th = 0.63   u(r,th,t) =  0.00000
r =  1.0   th = 0.94   u(r,th,t) =  0.00000
r =  1.0   th = 1.26   u(r,th,t) =  0.00000
r =  1.0   th = 1.57   u(r,th,t) =  0.00000

t = 0.25
r =  0.0   th = 0.00   u(r,th,t) =  0.23263
r =  0.0   th = 0.31   u(r,th,t) =  0.23263
r =  0.0   th = 0.63   u(r,th,t) =  0.23263
r =  0.0   th = 0.94   u(r,th,t) =  0.23263
r =  0.0   th = 1.26   u(r,th,t) =  0.23263
r =  0.0   th = 1.57   u(r,th,t) =  0.23263
       .                               .
       .                               .
       .                               .

 Output for r = 0.2, 0.4, 0.6, 0.8 removed

       .                               .
       .                               .
       .                               .
```

```
r =  1.0   th = 0.00   u(r,th,t) =  0.11264
r =  1.0   th = 0.31   u(r,th,t) =  0.11264
r =  1.0   th = 0.63   u(r,th,t) =  0.11264
r =  1.0   th = 0.94   u(r,th,t) =  0.11264
r =  1.0   th = 1.26   u(r,th,t) =  0.11264
r =  1.0   th = 1.57   u(r,th,t) =  0.11264
       .                              .
       .                              .
       .                              .

      Output for t = 0.5, 0.75 removed

       .                              .
       .                              .
       .                              .

t = 1.00
r =  0.0   th = 0.00   u(r,th,t) =  0.78288
r =  0.0   th = 0.31   u(r,th,t) =  0.78288
r =  0.0   th = 0.63   u(r,th,t) =  0.78288
r =  0.0   th = 0.94   u(r,th,t) =  0.78288
r =  0.0   th = 1.26   u(r,th,t) =  0.78288
r =  0.0   th = 1.57   u(r,th,t) =  0.78288
       .                              .
       .                              .
       .                              .

  Output for r = 0.2, 0.4, 0.6, 0.8 removed

       .                              .
       .                              .
       .                              .
r =  1.0   th = 0.00   u(r,th,t) =  0.50943
r =  1.0   th = 0.31   u(r,th,t) =  0.50943
r =  1.0   th = 0.63   u(r,th,t) =  0.50943
r =  1.0   th = 0.94   u(r,th,t) =  0.50943
r =  1.0   th = 1.26   u(r,th,t) =  0.50943
r =  1.0   th = 1.57   u(r,th,t) =  0.50943

ncall =   156
```

Table 14.3. Selected output from pde_2_main for $\sigma_0 = 2$, $\sigma_{\pi/2} = 1$

```
 nr = 11     r0 = 1.00
nth = 11    th0 = 1.57
  std_0 = 2.00
std_pi2 = 1.00

t = 0.00
r =  0.0    th = 0.00   u(r,th,t) =  0.00000
r =  0.0    th = 0.31   u(r,th,t) =  0.00000
r =  0.0    th = 0.63   u(r,th,t) =  0.00000
r =  0.0    th = 0.94   u(r,th,t) =  0.00000
r =  0.0    th = 1.26   u(r,th,t) =  0.00000
r =  0.0    th = 1.57   u(r,th,t) =  0.00000
       .                              .
       .                              .

 Output for r = 0.2, 0.4, 0.6, 0.8 removed

       .                              .
       .                              .
       .                              .

r =  1.0    th = 0.00   u(r,th,t) =  0.00000
r =  1.0    th = 0.31   u(r,th,t) =  0.00000
r =  1.0    th = 0.63   u(r,th,t) =  0.00000
r =  1.0    th = 0.94   u(r,th,t) =  0.00000
r =  1.0    th = 1.26   u(r,th,t) =  0.00000
r =  1.0    th = 1.57   u(r,th,t) =  0.00000

t = 0.25
r =  0.0    th = 0.00   u(r,th,t) =  0.24586
r =  0.0    th = 0.31   u(r,th,t) =  0.24586
r =  0.0    th = 0.63   u(r,th,t) =  0.24586
r =  0.0    th = 0.94   u(r,th,t) =  0.24586
r =  0.0    th = 1.26   u(r,th,t) =  0.24586
r =  0.0    th = 1.57   u(r,th,t) =  0.24586
       .                              .
       .                              .
       .                              .

 Output for r = 0.2, 0.4, 0.6, 0.8 removed

       .                              .
       .                              .
       .                              .
```

```
r =  1.0   th = 0.00   u(r,th,t) =  0.21035
r =  1.0   th = 0.31   u(r,th,t) =  0.19360
r =  1.0   th = 0.63   u(r,th,t) =  0.16146
r =  1.0   th = 0.94   u(r,th,t) =  0.13026
r =  1.0   th = 1.26   u(r,th,t) =  0.11164
r =  1.0   th = 1.57   u(r,th,t) =  0.10523
       .                             .
       .                             .
       .                             .

      Output for t = 0.5, 0.75 removed

       .                             .
       .                             .
       .                             .

t = 1.00
r =  0.0   th = 0.00   u(r,th,t) =  0.94674
r =  0.0   th = 0.31   u(r,th,t) =  0.94674
r =  0.0   th = 0.63   u(r,th,t) =  0.94674
r =  0.0   th = 0.94   u(r,th,t) =  0.94674
r =  0.0   th = 1.26   u(r,th,t) =  0.94674
r =  0.0   th = 1.57   u(r,th,t) =  0.94674
       .                             .
       .                             .
       .                             .

  Output for r = 0.2, 0.4, 0.6, 0.8 removed

       .                             .
       .                             .
       .                             .
r =  1.0   th = 0.00   u(r,th,t) =  0.88327
r =  1.0   th = 0.31   u(r,th,t) =  0.77077
r =  1.0   th = 0.63   u(r,th,t) =  0.65852
r =  1.0   th = 0.94   u(r,th,t) =  0.54209
r =  1.0   th = 1.26   u(r,th,t) =  0.47481
r =  1.0   th = 1.57   u(r,th,t) =  0.43531

ncall =   157
```

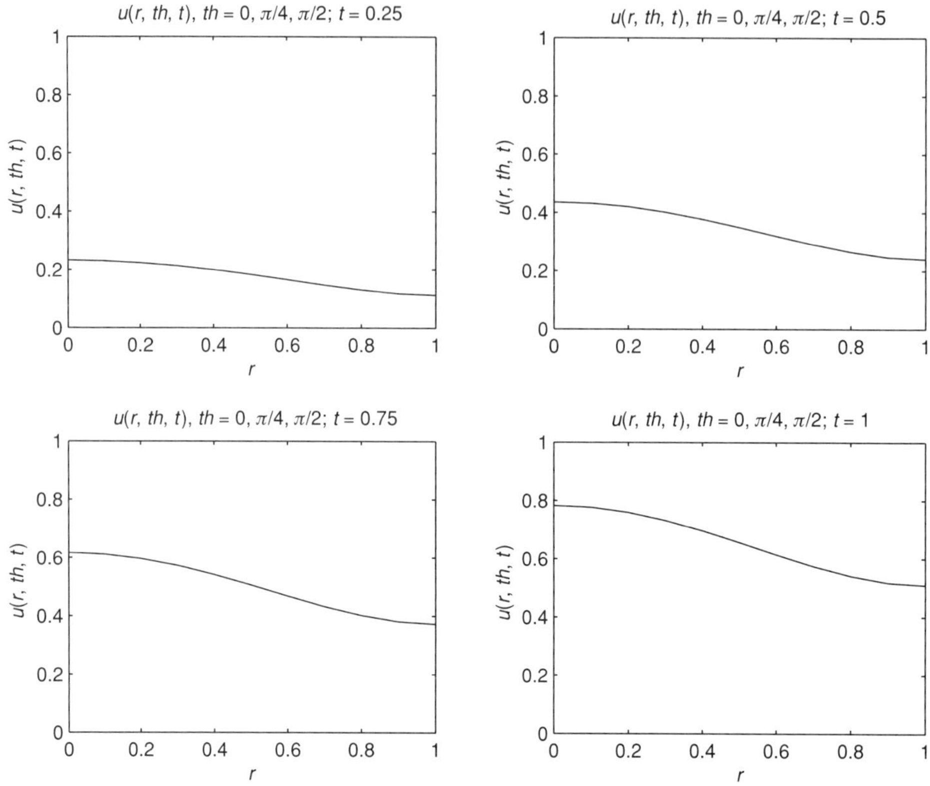

Figure 14.3. Plot of the numerical solution to Eq. (14.8) from pde_2_main and pde_2 for $\sigma_0 = 1, \sigma_{\pi/2} = 1$

We can note the following details about this output:

1. The θ variation is evident (except for $r = 0$, which is just a point value, so the θ variation is not meaningful).
2. The computational effort is quite modest (`ncall = 157`).

The plotted output is shown in Figure 14.4. The three curves in each plot correspond to $\theta = 0, \pi/4, \pi/2$. The overall level of $u(r, \theta, t)$ increases with t as would be expected since the inhomogeneous term in Eq. (14.20) continues to add mass until $t = 1$. Also, the solution is elongated in the $\theta = 0$ (or z) direction (the top curve of the set of three, and also, as demonstrated in Table 14.4) as expected with $\sigma_0 = 2, \sigma_{\pi/2} = 1$ in the Gaussian ellipsoid of Eq. (14.20).

To assist in the interpretation of the numerical solution of Figure 14.4, we have also included a discussion of an animation in Appendix 6.

To conclude this discussion, we now consider pde_2. The challenge here is to approximate the θ group in Eq. (14.8). The r group can be approximated as in the case of Eq. (14.12). The ODE routine for Eq. (14.8) is given in Listing 14.5.

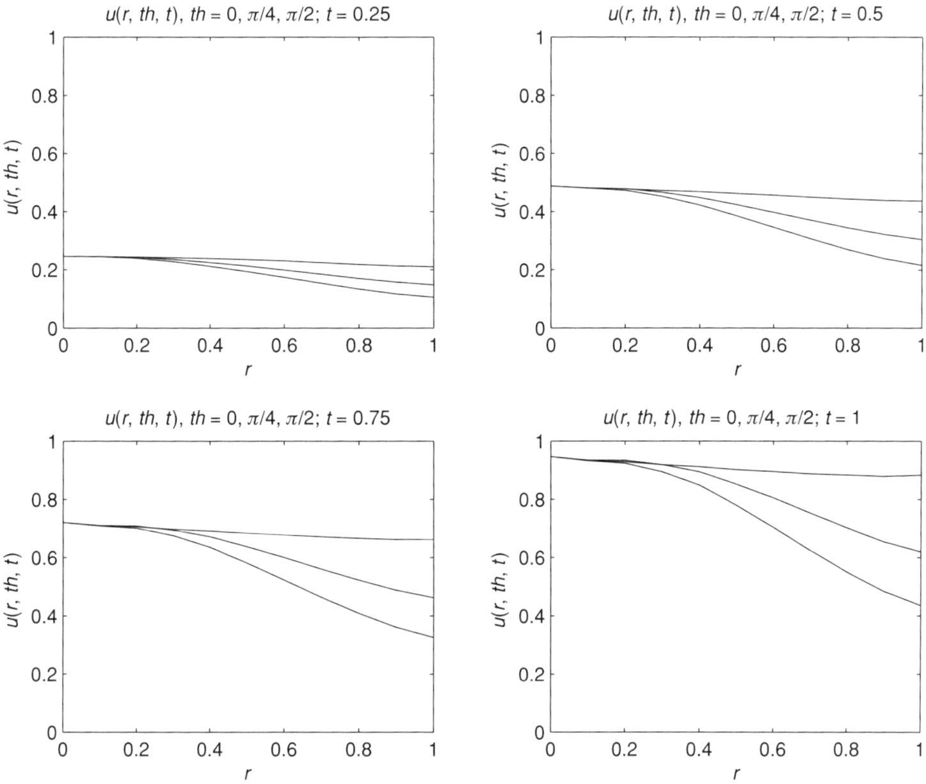

Figure 14.4. Plot of the numerical solution to Eq. (14.8) from pde_2_main and pde_2 for $\sigma_0 = 2, \sigma_{\pi/2} = 1$

```
  function yt=pde_2(t,y)
%
% Global area
  global r r0 nr th th0 nth D std_0 std_pi2 f tau ncall
%
% 1D to 2D
  for i=1:nr
  for j=1:nth
    u(i,j)=y((i-1)*nth+j);
  end
  end
%
% Spatial grids in r and theta
  for i=1:nr
  for j=1:nth
%
%   ur
```

```
        u1d=u(:,j);
        ur1d=dss004(0.0,r0,nr,u1d);
        ur1d(nr)=0.0;
        ur1d( 1)=0.0;
        ur(:,j)=ur1d;
%
%       urr
        urr1d=dss004(0.0,r0,nr,ur1d);
        urr(:,j)=urr1d;
%
%       uth
        u1d=u(i,:);
        uth1d=dss004(0.0,th0,nth,u1d);
        uth1d(nth)=0.0;
        uth1d(  1)=0.0;
        uth(i,:)=uth1d;
%
%       uthth
        uthth1d=dss004(0.0,th0,nth,uth1d);
        uthth(i,:)=uthth1d;
      end
      end
%
% PDE
  for i=1:nr
  for j=1:nth
    if(t<=tau)
      fs=f(i,j);
    else
      fs=0.0;
    end
%
%   r=0
    if(i==1)
%
%     th = 0
      if(j==1)
        u1d(:)=u(:,1);
        ur1d=dss004(0.0,r0,nr,u1d);
        ur1d( 1)=0.0;
        ur1d(nr)=0.0;
        urr1d=dss004(0.0,r0,nr,ur1d);
        urr(:,1)=urr1d;
%
%     th = pi/2
      elseif(j==nth)
        u1d(:)=u(:,nth);
```

```
          ur1d=dss004(0.0,r0,nr,u1d);
          ur1d( 1)=0.0;
          ur1d(nr)=0.0;
          urr1d=dss004(0.0,r0,nr,ur1d);
          urr(:,nth)=urr1d;
        end
%
%       PDE for r = 0
        ut(i,j)=D*(2.0*urr(i,nth)+urr(i,1))+fs;
%
%     r ~= 0
      elseif(i~=1)
%
%       th = 0, pi/2
        if(j==1|j==nth)
          ut(i,j)=D*(urr(i)+2.0/r(i)*ur(i)...
                     +(2.0/r(i)^2)*uthth(i,j))+fs;
        else
%
%       th ~= 0, pi/2
        ut(i,j)=D*(urr(i)+2.0/r(i)*ur(i)+(1.0/r(i)^2)...
                *(uthth(i,j)+cos(j)/sin(j)*uth(i,j)))+fs;
        end
      end
    end
    end
%
% 2D to 1D
    for i=1:nr
    for j=1:nth
      yt((i-1)*nth+j)=ut(i,j);
    end
    end
    yt=yt';
%
% Increment number of calls to pde_2
    ncall=ncall+1;
```

Listing 14.5. pde_2 for eq. (14.8)

We can note the following points about pde_2:

1. The function is defined and a global area is specified for the parameters and variables shared with pde_2_main of Listing 14.4.

```
  function yt=pde_2(t,y)
%
% Global area
  global r r0 nr th th0 nth D std_0 std_pi2 f tau ncall
%
% 1D to 2D
  for i=1:nr
  for j=1:nth
    u(i,j)=y((i-1)*nth+j);
  end
  end
```

The ODE dependent-variable vector `y` is then converted to a 2D array, `u(i,j)`, where the indices for r and θ are `i` and `j`, respectively. Thus, we can program in terms of the problem-oriented variable u of Eq. (14.8).

2. Two nested `for` loops step through the r–θ plane (for a total of $nr \times nth = 121$ ODEs). The derivative $\partial u/\partial r$ is calculated by first storing u in a 1D array, `u1d`, which is then differentiated by a call to `dss004` to give the 1D derivative array `ur1d`. BCs (14.10a) and (14.10b) are imposed and the 1D derivative array is stored in the 2D derivative array `ur` for subsequent use in the programming of Eq. (14.8).

```
%
% Spatial grids in r and theta
  for i=1:nr
  for j=1:nth
%
%   ur
    u1d=u(:,j);
    ur1d=dss004(0.0,r0,nr,u1d);
    ur1d(nr)=0.0;
    ur1d( 1)=0.0;
    ur(:,j)=ur1d;
```

3. The second derivative $\partial^2 u/\partial r^2$ is calculated by a second call to `dss004` applied to the first derivative (using stagewise differentiation). The final result is a 2D array, `urr`, with the second derivative in r.

```
%
%   urr
    urr1d=dss004(0.0,r0,nr,ur1d);
    urr(:,j)=urr1d;
```

4. For the derivatives in θ, we consider the two cases $r = 0$ and $r \neq 0$. For $r = 0$ (i=1), $\theta = 0$ (j=1), the θ group in Eq. (14.8) is approximated with

```
%
%   r=0
    if(i==1)
%
%     th = 0
      if(j==1)
        u1d(:)=u(:,1);
        ur1d=dss004(0.0,r0,nr,u1d);
        ur1d( 1)=0.0;
        ur1d(nr)=0.0;
        urr1d=dss004(0.0,r0,nr,ur1d);
        urr(:,1)=urr1d;
```

We have here used the Laplacian in Cartesian coordinates as explained with Eq. (14.19). The derivative $\partial^2 u/\partial z^2$ is the derivative $\partial^2 u(r, \theta = 0, t)/\partial^2 r$ (refer to Figure 14.1 to see that the radial vector along $\theta = 0$ is the z axis). The array urr(:,1) therefore contains the derivative $\partial^2 u/\partial z^2$. Note also that the preceding code includes BCs (14.11a) and (14.11b).

5. A similar argument equates the $\partial^2 u/\partial x^2$ to $\partial^2 u(r, \theta = \pi/2, t)/\partial^2 r$ (refer to Fig. 14.1 to see that the radial vector along $\theta = \pi/2$ is the x axis; it is also the y axis since we are not considering variations in ϕ). In other words to obtain $\partial^2 u/\partial x^2$ *and* $\partial^2 u/\partial y^2$, we use

```
%
%     th = pi/2
      elseif(j==nth)
        u1d(:)=u(:,nth);
        ur1d=dss004(0.0,r0,nr,u1d);
        ur1d( 1)=0.0;
        ur1d(nr)=0.0;
        urr1d=dss004(0.0,r0,nr,ur1d);
        urr(:,nth)=urr1d;
      end
```

Note the subscript j = nth corresponding to $\theta = \pi/2$. Also, this code includes BCs (14.11a) and (14.11b).

6. Having the three derivatives in Cartesian coordinates (the derivatives in x, y, and z of Eq. (14.19)) we can program Eq. (14.19) (which is a replacement for Eq. (14.8) at $r = 0$ or i=1) as

```
%
%      PDE for r = 0
       ut(i,j)=D*(2.0*urr(i,nth)+urr(i,1))+fs;
```

The factor 2 in the first RHS term reflects the fact that $\partial^2 u/\partial x^2$ and $\partial^2 u/\partial y^2$ in Eq. (14.19) are the same; again, note index `nth` in the first RHS term (for x and y), and `1` in the second RHS term (for z). Thus, we have handled the r and θ groups of Eq. (14.8) at $r = 0$ by using Eq. (14.19) instead.

7. For $r \neq 0$, the situation is somewhat simpler, except that we still need to consider the singularity for $\theta = 0, \pi/2$ due to the $1/\sin\theta$ term. However, this can be handled by l'Hospital's rule.

$$\lim_{\theta \to 0} \frac{\cos\theta}{\sin\theta}\frac{\partial u}{\partial \theta} = \frac{\cos\theta}{\cos\theta}\frac{\partial^2 u}{\partial \theta^2}$$

Thus, the θ group becomes

$$\frac{1}{r^2}\left(\frac{\partial^2 u}{\partial \theta^2} + \frac{\cos\theta}{\sin\theta}\frac{\partial u}{\partial \theta}\right) = \frac{2}{r^2}\frac{\partial^2 u}{\partial \theta^2} \tag{14.21}$$

Equation (14.8) with Eq. (14.21) (for both $\theta = 0, \pi/2$) is programmed as

```
%
%    r ~= 0
     elseif(i~=1)
%
%      th = 0, pi/2
       if(j==1|j==nth)
         ut(i,j)=D*(urr(i)+2.0/r(i)*ur(i)...
                +(2.0/r(i)^2)*uthth(i,j))+fs;
```

8. For $r > 0$ and $\theta \neq 0$ and $\theta \neq \pi/2$, Eq. (14.8) can be programmed directly as

```
       else
%
%      th ~= 0, pi/2
       ut(i,j)=D*(urr(i)+2.0/r(i)*ur(i)+(1.0/r(i)^2)...
               *(uthth(i,j)+cos(j)/sin(j)*uth(i,j)))+fs;
       end
     end
   end
   end
```

Note the similarity of this programming to Eq. (14.8). The double `end` completes the programming of the two nested `for` loops at the beginning (for stepping through r and θ).

9. This completes the programming of the RHS of Eq. (14.8) (or (14.19) where required). The 2D array for $\partial u/\partial t$ (in array `ut(i,j)`) is put into a 1D array `yt` for use by `ode15s`. The usual transpose and incrementing of `ncall` are included.

```
%
% 2D to 1D
  for i=1:nr
  for j=1:nth
    yt((i-1)*nth+j)=ut(i,j);
  end
  end
  yt=yt';
%
% Increment number of calls to pde_2
  ncall=ncall+1;
```

In summary, our intention in presenting this chapter is to illustrate how:

1. A PDE in spherical coordinates can be integrated numerically by the MOL.
2. Singularities in the PDE can be evaluated.
3. To include a no-flux BC (in this case, BC (14.10b)).
4. To use a conservation principle (in this case, Eq. (14.15)) to check the numerical solution.

The complexity of this procedure, particularly the resolution of singularities resulting from variable coefficients in the PDE, is justified if the physical system is spherical, which facilitates the analysis, particularly the use of BCs such as Eqs. (14.10) and (14.11).

REFERENCES

[1] Calvert, P. D., K. J. Strissel, W. E. Schiesser, E. N. Pugh, Jr., and V. Y. Arshavsky (November 2006), Light-Driven Translocation of Signaling Proteins in Vertebrate Photoreceptors, *Trends in Cell Biology*, **16**(11): 560–568

[2] Calvert, P. D., J. A. Peet, A. Bragin, W. E. Schiesser, and E. N. Pugh, Jr. (January 2007), Fluorescence Relaxation in 3D from Diffraction-Limited Sources of PAGFP or Sinks of EGFP Created by Multiphoton Photoconversion, *Journal of Microscopy*, **225**(1): 49–71

[3] Smith, G. D. (1965), *Numerical Solution of Partial Differential Equations*, Oxford University Press, London, pp. 44–45

APPENDIX 1

Partial Differential Equations from Conservation Principles: The Anisotropic Diffusion Equation

In some of the preceding chapters, we discussed the origin of partial differential equations (PDEs) by simplifying general PDE systems. For example, we arrived at the damped wave equation for an electric field as a special case of the Maxwell equations for electromagnetic (EM) fields. Also, we obtained Burgers' equation as a special case of the Euler and Navier Stokes equations of fluid mechanics.

While this approach of starting with a general PDE system is always an important first step to consider in developing a PDE model for a physical application, it also has limitations in the sense that the general system may not encompass all of the physical phenomena we wish to include in a mathematical model. For example, the Navier Stokes equations, as we used them, did not include an energy balance so that thermal effects, for example, a temperature field, could not be included in an analysis of a nonisothermal system. Of course, an energy balance could be (and has been) included with the Navier Stokes equations, but this requires some additional analysis. Also, we might be interested in a physical situation that is not reflected in a general PDE system, and we therefore have to derive the relevant PDEs starting from first principles, usually conservation principles in the case of physical applications.

In this appendix we consider the derivation of the *equations for anisotropic diffusion*, that is, a PDE system for which the diffusivity is a *nine-component tensor*. The final result is a generalization of the usual diffusion equation that includes directional effects that are significant in certain physical systems. The primary intent is to illustrate a method for the derivation of PDEs to include physical effects and phenomena that might not appear in previously available general PDE systems. The analysis leads to PDEs in Cartesian, cylindrical, and spherical coordinates, but the use of other orthogonal coordinate systems follows in the same way as illustrated here. Examples of PDEs in cylindrical and spherical coordinates are given in Chapters 13 and 14.

We start with a mass balance on an incremental cube with sides of length Δx, Δy, Δz.

$$\begin{aligned}
\Delta x \Delta y \Delta z \frac{\partial c}{\partial t} = & -\Delta y \Delta z D_{xx} \frac{\partial c}{\partial x}|_x - \left(-\Delta y \Delta z D_{xx} \frac{\partial c}{\partial x}|_{x+\Delta x}\right) \\
& - \Delta y \Delta z D_{xy} \frac{\partial c}{\partial y}|_x - \left(-\Delta y \Delta z D_{xy} \frac{\partial c}{\partial y}|_{x+\Delta x}\right) \\
& - \Delta y \Delta z D_{xz} \frac{\partial c}{\partial z}|_x - \left(-\Delta y \Delta z D_{xz} \frac{\partial c}{\partial z}|_{x+\Delta x}\right) \\
& - \Delta x \Delta z D_{yx} \frac{\partial c}{\partial x}|_y - \left(-\Delta x \Delta z D_{yx} \frac{\partial c}{\partial x}|_{y+\Delta y}\right) \\
& - \Delta x \Delta z D_{yy} \frac{\partial c}{\partial y}|_y - \left(-\Delta x \Delta z D_{yy} \frac{\partial c}{\partial y}|_{y+\Delta y}\right) \\
& - \Delta x \Delta z D_{yz} \frac{\partial c}{\partial z}|_y - \left(-\Delta x \Delta z D_{yz} \frac{\partial c}{\partial z}|_{y+\Delta y}\right) \\
& - \Delta x \Delta y D_{zx} \frac{\partial c}{\partial x}|_z - \left(-\Delta x \Delta y D_{zx} \frac{\partial c}{\partial x}|_{z+\Delta z}\right) \\
& - \Delta x \Delta y D_{zy} \frac{\partial c}{\partial y}|_z - \left(-\Delta x \Delta y D_{zy} \frac{\partial c}{\partial y}|_{z+\Delta z}\right) \\
& - \Delta x \Delta y D_{zz} \frac{\partial c}{\partial z}|_z - \left(-\Delta x \Delta y D_{zz} \frac{\partial c}{\partial z}|_{z+\Delta z}\right) \qquad \text{(A.1.1)}
\end{aligned}$$

with the interpretation of a component of the diffusivity tensor, D_{ij}, as the diffusive flux in direction i due to a concentration gradient in direction j. Diffusion based on the nine-component diffusivity tensor is termed *anisotropic* because of the directionality imparted through the components with $i \neq j$. For example, D_{xy} produces a diffusion flux in the x direction due to a concentration gradient in the y direction.

A straightforward rearrangement of Eq. (A.1.1) gives

$$\begin{aligned}
\frac{\partial c}{\partial t} = & \frac{D_{xx} \frac{\partial c}{\partial x}|_{x+\Delta x} - D_{xx} \frac{\partial c}{\partial x}|_x}{\Delta x} \\
& + \frac{D_{xy} \frac{\partial c}{\partial y}|_{x+\Delta x} - D_{xy} \frac{\partial c}{\partial y}|_x}{\Delta x} \\
& + \frac{D_{xz} \frac{\partial c}{\partial z}|_{x+\Delta x} - D_{xz} \frac{\partial c}{\partial z}|_x}{\Delta x} \\
& + \frac{D_{yx} \frac{\partial c}{\partial x}|_{y+\Delta y} - D_{yx} \frac{\partial c}{\partial x}|_y}{\Delta y}
\end{aligned}$$

$$+\frac{D_{yy}\frac{\partial c}{\partial y}|_{y+\Delta y}-D_{yy}\frac{\partial c}{\partial y}|_{y}}{\Delta y}$$

$$+\frac{D_{yz}\frac{\partial c}{\partial z}|_{y+\Delta y}-D_{yz}\frac{\partial c}{\partial z}|_{y}}{\Delta y}$$

$$+\frac{D_{zx}\frac{\partial c}{\partial x}|_{z+\Delta z}-D_{zx}\frac{\partial c}{\partial x}|_{z}}{\Delta z}$$

$$+\frac{D_{zy}\frac{\partial c}{\partial y}|_{z+\Delta z}-D_{zy}\frac{\partial c}{\partial y}|_{z}}{\Delta z}$$

$$+\frac{D_{zz}\frac{\partial c}{\partial z}|_{z+\Delta z}-D_{zz}\frac{\partial c}{\partial z}|_{z}}{\Delta z} \tag{A.1.2}$$

In the limit $\Delta x, \Delta y, \Delta z \longrightarrow 0$, Eq. (A.1.2) becomes

$$\frac{\partial c}{\partial t}=\frac{\partial\left(D_{xx}\frac{\partial c}{\partial x}\right)}{\partial x}+\frac{\partial\left(D_{xy}\frac{\partial c}{\partial y}\right)}{\partial x}+\frac{\partial\left(D_{xz}\frac{\partial c}{\partial z}\right)}{\partial x}$$
$$+\frac{\partial\left(D_{yx}\frac{\partial c}{\partial x}\right)}{\partial y}+\frac{\partial\left(D_{yy}\frac{\partial c}{\partial y}\right)}{\partial y}+\frac{\partial\left(D_{yz}\frac{\partial c}{\partial z}\right)}{\partial y}$$
$$+\frac{\partial\left(D_{zx}\frac{\partial c}{\partial x}\right)}{\partial z}+\frac{\partial\left(D_{zy}\frac{\partial c}{\partial y}\right)}{\partial z}+\frac{\partial\left(D_{zz}\frac{\partial c}{\partial z}\right)}{\partial z} \tag{A.1.3}$$

Equation (A.1.3) can be expressed in vector–tensor notation (vectors and tensor expressed in boldface) as

$$\frac{\partial c}{\partial t}=\nabla\cdot(\mathbf{D}\cdot\nabla c) \tag{A.1.4}$$

where

$$\nabla c=\mathbf{i}\frac{\partial c}{\partial x}+\mathbf{j}\frac{\partial c}{\partial y}+\mathbf{k}\frac{\partial c}{\partial z} \tag{A.1.5}$$

and **i**, **j**, and **k** are unit vectors in Cartesian coordinates.

Here we introduce the idea of *dimensions of tensors*, and the *net dimensions of operations on tensors* ([1], p. 808). To start, refer to Table A.1.1 for some terminology.

Table A.1.1. Basic terminology for tensors of various orders

Term	Corresponding tensor order
Scalar	Zero order
Vector	First order
Tensor	Second and higher order

The reason the terminology in Table A.1.1 is useful is that it suggests dimensions that can be used to check expressions involving tensors such as the RHS of Eq. (A.1.4); this idea is explained next.

Various tensor operations, which we call *multiplications*, have associated orders summarized in the Table A.1.2, where we have used the symbol Σ to represent the *sum of the orders of the quantities being multiplied*. For example, using the definitions given in Table A.1.1,

- For the *product* of scalar s with vector $\mathbf{v}$ or the tensor $\mathbf{D}$ of order 2, written as $s\mathbf{v}$ or $s\mathbf{D}$, the orders are given by $\Sigma = 0 + 1 = 1$ or $\Sigma = 0 + 2 = 2$, respectively, and the result is a *vector* or a *tensor*.
- For the *cross product* of two vectors $\mathbf{v}$ and $\mathbf{w}$, written as $\mathbf{v} \times \mathbf{w}$, the order is given by $\Sigma - 1 = 1 + 1 - 1 = 1$, and the result is a *vector*.
- For the *dot product* of two vectors $\mathbf{v}$ and $\mathbf{w}$, written as $\mathbf{v} \cdot \mathbf{w}$, the order is given by $\Sigma - 2 = 1 + 1 - 2 = 0$, and the result is a *scalar*.
- For the *product* of two vectors $\mathbf{v}$ and $\mathbf{w}$ or two tensors $\mathbf{D}$ and $\mathbf{E}$ of order 2 (also known as the *dyadic product*), written as $\mathbf{vw}$ or $\mathbf{DE}$ without a product sign separating the variables, the order is given by $\Sigma = 1 + 1 = 2$ or $\Sigma = 2 + 2 = 4$, respectively, and the result is a *tensor* of order 2 or 4.
- For the *double-dot product* of two tensors $\mathbf{D}$ and $\mathbf{E}$, written as $\mathbf{D} : \mathbf{E}$, the order is given by $\Sigma - 4 = 2 + 2 - 4 = 0$, and the result is a *scalar*.
- For the *dot product* of a tensor $\mathbf{D}$ of order 2 with the vector $\mathbf{v}$, written as $\mathbf{D} \cdot \mathbf{v}$, the order is given by $\Sigma - 2 = 2 + 1 - 2 = 1$, and the result is a *vector*.

We now illustrate the use of Table A.1.2 applied to the LHS of Eq. (A.1.5). ∇c has two components: ∇, a vector with dimension 1, and c, a scalar with dimension 0. Also, it involves a dyadic multiplication, so from Table A.1.2, the net dimension is $\Sigma = 1 + 0 = 1$, a vector.

Table A.1.2. Orders for tensor multiplications

Name	Multiplication sign	Order of result
Product or dyadic product	None	Σ
Vector product or cross product	$\times$	$\Sigma - 1$
Scalar product or dot product	$\cdot$	$\Sigma - 2$
Double-dot product	:	$\Sigma - 4$

Continuing on with the RHS of Eq. (A.1.4),

$$\begin{aligned}
\mathbf{D}\cdot\nabla c &= (\mathbf{ii}D_{xx}+\mathbf{ij}D_{xy}+\mathbf{ik}D_{xz}\\
&\quad+\mathbf{ji}D_{yx}+\mathbf{jj}D_{yy}+\mathbf{jk}D_{yz}\\
&\quad+\mathbf{ki}D_{zx}+\mathbf{kj}D_{zy}+\mathbf{kk}D_{zz})\cdot\left(\mathbf{i}\frac{\partial c}{\partial x}+\mathbf{j}\frac{\partial c}{\partial y}+\mathbf{k}\frac{\partial c}{\partial z}\right)\\
&= \mathbf{i}(\mathbf{i}\cdot\mathbf{i})D_{xx}\frac{\partial c}{\partial x}+\mathbf{j}(\mathbf{i}\cdot\mathbf{i})D_{yx}\frac{\partial c}{\partial x}+\mathbf{k}(\mathbf{i}\cdot\mathbf{i})D_{zx}\frac{\partial c}{\partial x}\\
&\quad+\mathbf{i}(\mathbf{j}\cdot\mathbf{j})D_{xy}\frac{\partial c}{\partial y}+\mathbf{j}(\mathbf{j}\cdot\mathbf{j})D_{yy}\frac{\partial c}{\partial y}+\mathbf{k}(\mathbf{j}\cdot\mathbf{j})D_{zy}\frac{\partial c}{\partial y}\\
&\quad+\mathbf{i}(\mathbf{k}\cdot\mathbf{k})D_{xz}\frac{\partial c}{\partial z}+\mathbf{j}(\mathbf{k}\cdot\mathbf{k})D_{yz}\frac{\partial c}{\partial z}+\mathbf{k}(\mathbf{k}\cdot\mathbf{k})D_{zz}\frac{\partial c}{\partial z}\\
&= \mathbf{i}\left(D_{xx}\frac{\partial c}{\partial x}+D_{xy}\frac{\partial c}{\partial y}+D_{xz}\frac{\partial c}{\partial z}\right)\\
&\quad+\mathbf{j}\left(D_{yx}\frac{\partial c}{\partial x}+D_{yy}\frac{\partial c}{\partial y}+D_{yz}\frac{\partial c}{\partial z}\right)\\
&\quad+\mathbf{k}\left(D_{zx}\frac{\partial c}{\partial x}+D_{zy}\frac{\partial c}{\partial y}+D_{zz}\frac{\partial c}{\partial z}\right)
\end{aligned}\tag{A.1.6}$$

with dimensions $\mathbf{D} \Leftrightarrow 2$ (a second-order tensor), $\nabla c \Leftrightarrow 1$ (a vector), and therefore net dimensions of (from the dot product of Table A.1.2) $\Sigma - 2 = 2 + 1 - 2 = 1$ (a vector). We have made use of the property of the *dot product* between orthogonal unit vectors (e.g., **a** and **b**):

θ	$\mathbf{a}\cdot\mathbf{b} = \lvert a\rvert\lvert b\rvert\cos\theta$
0	1
$\pi/2$	0

Then,

$$\begin{aligned}
\nabla\cdot(\mathbf{D}\cdot\nabla c) &= \left(\mathbf{i}\frac{\partial}{\partial x}+\mathbf{j}\frac{\partial}{\partial y}+\mathbf{j}\frac{\partial}{\partial z}\right)\\
&\quad\cdot\left[\mathbf{i}\left(D_{xx}\frac{\partial c}{\partial x}+D_{xy}\frac{\partial c}{\partial y}+D_{xz}\frac{\partial c}{\partial z}\right)\right.\\
&\quad+\mathbf{j}\left(D_{yx}\frac{\partial c}{\partial x}+D_{yy}\frac{\partial c}{\partial y}+D_{yz}\frac{\partial c}{\partial z}\right)\\
&\quad\left.+\mathbf{k}\left(D_{zx}\frac{\partial c}{\partial x}+D_{zy}\frac{\partial c}{\partial y}+D_{zz}\frac{\partial c}{\partial z}\right)\right]
\end{aligned}$$

$$
\begin{aligned}
&= \frac{\partial}{\partial x}\left(D_{xx}\frac{\partial c}{\partial x} + D_{xy}\frac{\partial c}{\partial y} + D_{xz}\frac{\partial c}{\partial z}\right) \\
&+ \frac{\partial}{\partial y}\left(D_{yx}\frac{\partial c}{\partial x} + D_{yy}\frac{\partial c}{\partial y} + D_{yz}\frac{\partial c}{\partial z}\right) \\
&+ \frac{\partial}{\partial z}\left(D_{zx}\frac{\partial c}{\partial x} + D_{zy}\frac{\partial c}{\partial y} + D_{zz}\frac{\partial c}{\partial z}\right)
\end{aligned} \tag{A.1.7}
$$

which is the RHS of Eq. (A.1.3) with dimensions $1 + 1 - 2 = 0$ (a scalar as required by Eq. (A.1.3) since the LHS is a scalar $\partial c/\partial t$, and c and t are scalars).

If $\mathbf{D}$ is a constant tensor, Eq. (A.1.3) becomes

$$
\begin{aligned}
\frac{\partial c}{\partial t} &= D_{xx}\frac{\partial^2 c}{\partial x^2} + D_{xy}\frac{\partial^2 c}{\partial x\,\partial y} + D_{xz}\frac{\partial^2 c}{\partial x\,\partial z} \\
&+ D_{yx}\frac{\partial^2 c}{\partial y\,\partial x} + D_{yy}\frac{\partial^2 c}{\partial y^2} + D_{yz}\frac{\partial^2 c}{\partial y\,\partial z} \\
&+ D_{zx}\frac{\partial^2 c}{\partial z\,\partial x} + D_{zy}\frac{\partial^2 c}{\partial z\,\partial y} + D_{zz}\frac{\partial^2 c}{\partial z^2}
\end{aligned} \tag{A.1.8}
$$

Equation (A.1.8) can be written as

$$
\frac{\partial c}{\partial t} = \mathbf{D}^T : \nabla\nabla c \tag{A.1.9}
$$

where

$$
\begin{aligned}
\nabla\nabla c &= \left(\mathbf{i}\frac{\partial}{\partial x} + \mathbf{j}\frac{\partial}{\partial y} + \mathbf{k}\frac{\partial}{\partial z}\right)\left(\mathbf{i}\frac{\partial c}{\partial x} + \mathbf{j}\frac{\partial c}{\partial y} + \mathbf{k}\frac{\partial c}{\partial z}\right) \\
&= \mathbf{ii}\frac{\partial^2 c}{\partial x^2} + \mathbf{ij}\frac{\partial^2 c}{\partial x\,\partial y} + \mathbf{ik}\frac{\partial^2 c}{\partial x\,\partial z} \\
&+ \mathbf{ji}\frac{\partial^2 c}{\partial y\,\partial x} + \mathbf{jj}\frac{\partial^2 c}{\partial y^2} + \mathbf{jk}\frac{\partial^2 c}{\partial y\,\partial z} \\
&+ \mathbf{ki}\frac{\partial^2 c}{\partial z\,\partial x} + \mathbf{kj}\frac{\partial^2 c}{\partial z\,\partial y} + \mathbf{kk}\frac{\partial^2 c}{\partial z^2}
\end{aligned}
$$

with dimensions $1 + 1 - 0 = 2$ (a tensor).

$$
\begin{aligned}
\mathbf{D}^T : \nabla\nabla c = (&\mathbf{ii}D_{xx} + \mathbf{ij}D_{yx} + \mathbf{ik}D_{zx} \\
&+ \mathbf{ji}D_{xy} + \mathbf{jj}D_{yy} + \mathbf{jk}D_{zy} \\
&+ \mathbf{ki}D_{xz} + \mathbf{kj}D_{yz} + \mathbf{kk}D_{zz}) \\
&: \left(\mathbf{ii}\frac{\partial^2 c}{\partial x^2} + \mathbf{ij}\frac{\partial^2 c}{\partial x\,\partial y} + \mathbf{ik}\frac{\partial^2 c}{\partial x\,\partial z}\right. \\
&+ \mathbf{ji}\frac{\partial^2 c}{\partial y\,\partial x} + \mathbf{jj}\frac{\partial^2 c}{\partial y^2} + \mathbf{jk}\frac{\partial^2 c}{\partial y\,\partial z} \\
&\left.+ \mathbf{ki}\frac{\partial^2 c}{\partial z\,\partial x} + \mathbf{kj}\frac{\partial^2 c}{\partial z\,\partial y} + \mathbf{kk}\frac{\partial^2 c}{\partial z^2}\right)
\end{aligned}
$$

$$
\begin{aligned}
&= \mathbf{ii}:\mathbf{ii}D_{xx}\frac{\partial^2 c}{\partial x^2} + \mathbf{ij}:\mathbf{ji}D_{yx}\frac{\partial^2 c}{\partial y\,\partial x} + \mathbf{ik}:\mathbf{ki}D_{zx}\frac{\partial^2 c}{\partial z\,\partial x} \\
&+ \mathbf{ji}:\mathbf{ij}D_{xy}\frac{\partial^2 c}{\partial x\,\partial y} + \mathbf{jj}:\mathbf{jj}D_{yy}\frac{\partial^2 c}{\partial y^2} + \mathbf{jk}:\mathbf{kj}D_{zy}\frac{\partial^2 c}{\partial z\,\partial y} \\
&+ \mathbf{ki}:\mathbf{ik}D_{xz}\frac{\partial^2 c}{\partial x\,\partial z} + \mathbf{kj}:\mathbf{jk}D_{yz}\frac{\partial^2 c}{\partial y\,\partial z} + \mathbf{kk}:\mathbf{kk}D_{zz}\frac{\partial^2 c}{\partial z^2}
\end{aligned}
$$

with dimensions $2+2-4=0$ (a scalar).

Thus,

$$
\begin{aligned}
\frac{\partial c}{\partial t} &= D_{xx}\frac{\partial^2 c}{\partial x^2} + D_{yx}\frac{\partial^2 c}{\partial y\,\partial x} + D_{zx}\frac{\partial^2 c}{\partial z\,\partial x} \\
&+ D_{xy}\frac{\partial^2 c}{\partial x\,\partial y} + D_{yy}\frac{\partial^2 c}{\partial y^2} + D_{zy}\frac{\partial^2 c}{\partial z\,\partial y} \\
&+ D_{xz}\frac{\partial^2 c}{\partial x\,\partial z} + D_{yz}\frac{\partial^2 c}{\partial y\,\partial z} + D_{zz}\frac{\partial^2 c}{\partial z^2} \qquad \text{(A.1.10)}
\end{aligned}
$$

which is Eq. (A.1.8).

We can now apply the same analysis in cylindrical coordinates by starting with a mass balance on an incremental volume $(r\Delta\theta)(\Delta r)(\Delta z)$ ([1], p. 840):

$$
\begin{aligned}
r\Delta\theta\Delta r\Delta z\frac{\partial c}{\partial t} &= -r\Delta\theta\Delta z D_{rr}\frac{\partial c}{\partial r}|_r - \left(-r\Delta\theta\Delta z D_{rr}\frac{\partial c}{\partial r}|_{r+\Delta r}\right) \\
&- r\Delta\theta\Delta z D_{r\theta}\frac{\partial c}{r\,\partial\theta}|_r - \left(-r\Delta\theta\Delta z D_{r\theta}\frac{\partial c}{r\,\partial\theta}|_{r+\Delta r}\right) \\
&- r\Delta\theta\Delta z D_{rz}\frac{\partial c}{\partial z}|_r - \left(-r\Delta\theta\Delta z D_{rz}\frac{\partial c}{\partial z}|_{r+\Delta r}\right) \\
&- \Delta r\Delta z D_{\theta r}\frac{\partial c}{\partial r}|_\theta - \left(-\Delta r\Delta z D_{\theta r}\frac{\partial c}{\partial r}|_{\theta+\Delta\theta}\right) \\
&- \Delta r\Delta z D_{\theta\theta}\frac{\partial c}{r\,\partial\theta}|_\theta - \left(-\Delta r\Delta z D_{\theta\theta}\frac{\partial c}{r\,\partial\theta}|_{\theta+\Delta\theta}\right) \\
&- \Delta r\Delta z D_{\theta z}\frac{\partial c}{\partial z}|_\theta - \left(-\Delta r\Delta z D_{\theta z}\frac{\partial c}{\partial z}|_{\theta+\Delta\theta}\right) \\
&- r\Delta\theta\Delta r D_{zr}\frac{\partial c}{\partial r}|_z - \left(-r\Delta\theta\Delta r D_{zr}\frac{\partial c}{\partial r}|_{z+\Delta z}\right) \\
&- r\Delta\theta\Delta r D_{z\theta}\frac{\partial c}{r\,\partial\theta}|_z - \left(-r\Delta\theta\Delta r D_{z\theta}\frac{\partial c}{r\,\partial\theta}|_{z+\Delta z}\right) \\
&- r\Delta\theta\Delta r D_{zz}\frac{\partial c}{\partial z}|_z - \left(-r\Delta\theta\Delta r D_{zz}\frac{\partial c}{\partial z}|_{z+\Delta z}\right) \qquad \text{(A.1.11)}
\end{aligned}
$$

Rearrangement of Eq. (A.1.11) gives

$$
\begin{aligned}
\frac{\partial c}{\partial t} = {} & \frac{r\Delta\theta\Delta z D_{rr}\frac{\partial c}{\partial r}|_{r+\Delta r} - r\Delta\theta\Delta z D_{rr}\frac{\partial c}{\partial r}|_{r}}{r\Delta\theta\Delta r\Delta z} \\
& + \frac{r\Delta\theta\Delta z D_{r\theta}\frac{\partial c}{r\,\partial\theta}|_{r+\Delta r} - r\Delta\theta\Delta z D_{r\theta}\frac{\partial c}{r\,\partial\theta}|_{r}}{r\Delta\theta\Delta r\Delta z} \\
& + \frac{r\Delta\theta\Delta z D_{rz}\frac{\partial c}{\partial z}|_{r+\Delta r} - r\Delta\theta\Delta z D_{rz}\frac{\partial c}{\partial z}|_{r}}{r\Delta\theta\Delta r\Delta z} \\
& + \frac{\Delta r\Delta z D_{\theta r}\frac{\partial c}{\partial r}|_{\theta+\Delta\theta} - \Delta r\Delta z D_{\theta r}\frac{\partial c}{\partial r}|_{\theta}}{r\Delta\theta\Delta r\Delta z} \\
& + \frac{\Delta r\Delta z D_{\theta\theta}\frac{\partial c}{r\,\partial\theta}|_{\theta+\Delta\theta} - \Delta r\Delta z D_{\theta\theta}\frac{\partial c}{r\,\partial\theta}|_{\theta}}{r\Delta\theta\Delta r\Delta z} \\
& + \frac{\Delta r\Delta z D_{\theta z}\frac{\partial c}{\partial z}|_{\theta+\Delta\theta} - \Delta r\Delta z D_{\theta z}\frac{\partial c}{\partial z}|_{\theta}}{r\Delta\theta\Delta r\Delta z} \\
& + \frac{r\Delta\theta\Delta r D_{zr}\frac{\partial c}{\partial r}|_{z+\Delta z} - r\Delta\theta\Delta r D_{zr}\frac{\partial c}{\partial r}|_{z}}{r\Delta\theta\Delta r\Delta z} \\
& + \frac{r\Delta\theta\Delta r D_{z\theta}\frac{\partial c}{r\,\partial\theta}|_{z+\Delta z} - r\Delta\theta\Delta r D_{z\theta}\frac{\partial c}{r\,\partial\theta}|_{z}}{r\Delta\theta\Delta r\Delta z} \\
& + \frac{r\Delta\theta\Delta r D_{zz}\frac{\partial c}{\partial z}|_{z+\Delta z} - r\Delta\theta\Delta r D_{zz}\frac{\partial c}{\partial z}|_{z}}{r\Delta\theta\Delta r\Delta z}
\end{aligned}
\qquad (A.1.12)
$$

In the limit Δr, $\Delta\theta$, $\Delta z \to 0$, Eq. (A.1.12) becomes

$$
\begin{aligned}
\frac{\partial c}{\partial t} = {} & \frac{\partial\left(rD_{rr}\frac{\partial c}{\partial r}\right)}{r\,\partial r} + \frac{\partial\left(rD_{r\theta}\frac{\partial c}{r\,\partial\theta}\right)}{r\,\partial r} + \frac{\partial\left(rD_{rz}\frac{\partial c}{\partial z}\right)}{r\,\partial r} \\
& + \frac{\partial\left(D_{\theta r}\frac{\partial c}{\partial r}\right)}{r\,\partial\theta} + \frac{\partial\left(D_{\theta\theta}\frac{\partial c}{r\,\partial\theta}\right)}{r\,\partial\theta} + \frac{\partial\left(D_{\theta z}\frac{\partial c}{\partial z}\right)}{r\,\partial\theta} \\
& + \frac{\partial\left(D_{zr}\frac{\partial c}{\partial r}\right)}{\partial z} + \frac{\partial\left(D_{z\theta}\frac{\partial c}{r\,\partial\theta}\right)}{\partial z} + \frac{\partial\left(D_{zz}\frac{\partial c}{\partial z}\right)}{\partial z}
\end{aligned}
\qquad (A.1.13)
$$

For constant **D**, Eq. (A.1.13) becomes

$$\frac{\partial c}{\partial t} = D_{rr}\left(\frac{\partial^2 c}{\partial r^2} + \frac{1}{r}\frac{\partial c}{\partial r}\right) + \frac{D_{r\theta}}{r}\frac{\partial^2 c}{\partial r\,\partial\theta} + D_{rz}\left(\frac{\partial^2 c}{\partial r\,\partial z} + \frac{1}{r}\frac{\partial c}{\partial z}\right)$$
$$+ \frac{D_{\theta r}}{r}\frac{\partial^2 c}{\partial\theta\,\partial r} + \frac{D_{\theta\theta}}{r^2}\frac{\partial^2 c}{\partial\theta^2} + \frac{D_{\theta z}}{r}\frac{\partial^2 c}{\partial\theta\,\partial z}$$
$$+ D_{zr}\frac{\partial^2 c}{\partial z\,\partial r} + \frac{D_{z\theta}}{r}\frac{\partial^2 c}{\partial z\,\partial\theta} + D_{zz}\frac{\partial^2 c}{\partial z^2} \tag{A.1.14}$$

As a check on Eq. (A.1.14), we can consider the special case for which the main diagonal elements of the diffusivity tensor are equal ($D_{rr} = D_{\theta\theta} = D_{zz} = D$) and the off-diagonal diffusivities are zero. Equation (A.1.14) then becomes

$$\frac{\partial c}{\partial t} = D\left(\frac{\partial^2 c}{\partial r^2} + \frac{1}{r}\frac{\partial c}{\partial r} + \frac{1}{r^2}\frac{\partial^2 c}{\partial\theta^2} + \frac{\partial^2 c}{\partial z^2}\right)$$

which is the well-known form of the diffusion equation ([2], p. 3)

$$\frac{\partial c}{\partial t} = D\nabla^2 c$$

where ∇^2 is the Laplacian operator.

We now consider if Eq. (A.1.13) can be expressed in the vector–tensor notation of Eq. (A.1.4) with $\mathbf{i_r}$, $\mathbf{j}_\theta$, and $\mathbf{k_z}$ the unit vectors in cylindrical coordinates.

$$\nabla c = \mathbf{i}_r\frac{\partial c}{\partial r} + \mathbf{j}_\theta\frac{1}{r}\frac{\partial c}{\partial\theta} + \mathbf{k}_z\frac{\partial c}{\partial z} \tag{A.1.15}$$

$$\mathbf{D}\cdot\nabla c = (\mathbf{i}_r\mathbf{i}_r D_{rr} + \mathbf{i}_r\mathbf{j}_\theta D_{r\theta} + \mathbf{i}_r\mathbf{k}_z D_{rz}$$
$$+ \mathbf{j}_\theta\mathbf{i}_r D_{\theta r} + \mathbf{j}_\theta\mathbf{j}_\theta D_{\theta\theta} + \mathbf{j}_\theta\mathbf{k}_z D_{\theta z}$$
$$+ \mathbf{k}_z\mathbf{i}_r D_{zr} + \mathbf{k}_z\mathbf{j}_\theta D_{z\theta} + \mathbf{k}_z\mathbf{k}_z D_{zz})$$
$$\cdot\,\mathbf{i}_r\frac{\partial c}{\partial r} + \mathbf{j}_\theta\frac{1}{r}\frac{\partial c}{\partial\theta} + \mathbf{k}_z\frac{\partial c}{\partial z}$$
$$= \mathbf{i}_r(\mathbf{i}_r\cdot\mathbf{i}_r)D_{rr}\frac{\partial c}{\partial r} + \mathbf{j}_\theta(\mathbf{i}_r\cdot\mathbf{i}_r)D_{\theta r}\frac{\partial c}{\partial r} + \mathbf{k}_z(\mathbf{i}_r\cdot\mathbf{i}_r)D_{zr}\frac{\partial c}{\partial r}$$
$$+ \mathbf{i}_r(\mathbf{j}_\theta\cdot\mathbf{j}_\theta)D_{r\theta}\frac{1}{r}\frac{\partial c}{\partial\theta} + \mathbf{j}_\theta(\mathbf{j}_\theta\cdot\mathbf{j}_\theta)D_{\theta\theta}\frac{1}{r}\frac{\partial c}{\partial\theta} + \mathbf{k}_z(\mathbf{j}_\theta\cdot\mathbf{j}_\theta)D_{z\theta}\frac{1}{r}\frac{\partial c}{\partial\theta}$$
$$+ \mathbf{i}_r(\mathbf{k}_z\cdot\mathbf{k}_z)D_{rz}\frac{\partial c}{\partial z} + \mathbf{j}_\theta(\mathbf{k}_z\cdot\mathbf{k}_z)D_{\theta z}\frac{\partial c}{\partial z} + \mathbf{k}_z(\mathbf{k}_z\cdot\mathbf{k}_z)D_{zz}\frac{\partial c}{\partial z}$$
$$= \mathbf{i}_r\left(D_{rr}\frac{\partial c}{\partial r} + D_{r\theta}\frac{1}{r}\frac{\partial c}{\partial\theta} + D_{rz}\frac{\partial c}{\partial z}\right)$$
$$+ \mathbf{j}_\theta\left(D_{\theta r}\frac{\partial c}{\partial r} + D_{\theta\theta}\frac{1}{r}\frac{\partial c}{\partial\theta} + D_{\theta z}\frac{\partial c}{\partial z}\right)$$
$$+ \mathbf{k}_z\left(D_{zr}\frac{\partial c}{\partial r} + D_{z\theta}\frac{1}{r}\frac{\partial c}{\partial\theta} + D_{zz}\frac{\partial c}{\partial z}\right)$$

Finally, the RHS of Eq. (A.1.4) is ([2], p. 2)

$$\begin{aligned}
\nabla \cdot (\mathbf{D} \cdot \nabla c) &= \left(\mathbf{i}_r \frac{1}{r}\frac{\partial}{\partial r}(r) + \mathbf{j}_\theta \frac{1}{r}\frac{\partial}{\partial \theta} + \mathbf{k}_z \frac{\partial}{\partial z} \right) \\
&\quad \cdot \left[\mathbf{i}_r \left(D_{rr}\frac{\partial c}{\partial r} + D_{r\theta}\frac{1}{r}\frac{\partial c}{\partial \theta} + D_{rz}\frac{\partial c}{\partial z} \right) \right. \\
&\quad + \mathbf{j}_\theta \left(D_{\theta r}\frac{\partial c}{\partial r} + D_{\theta\theta}\frac{1}{r}\frac{\partial c}{\partial \theta} + D_{\theta z}\frac{\partial c}{\partial z} \right) \\
&\quad \left. + \mathbf{k}_z \left(D_{zr}\frac{\partial c}{\partial r} + D_{z\theta}\frac{1}{r}\frac{\partial c}{\partial y} + D_{zz}\frac{\partial c}{\partial z} \right) \right] \\
&= \frac{1}{r}\frac{\partial}{\partial r}\left(rD_{rr}\frac{\partial c}{\partial r} + rD_{r\theta}\frac{1}{r}\frac{\partial c}{\partial \theta} + rD_{rz}\frac{\partial c}{\partial z} \right) \\
&\quad + \frac{1}{r}\frac{\partial}{\partial \theta}\left(D_{\theta r}\frac{\partial c}{\partial r} + D_{\theta\theta}\frac{1}{r}\frac{\partial c}{\partial \theta} + D_{\theta z}\frac{\partial c}{\partial z} \right) \\
&\quad + \frac{\partial}{\partial z}\left(D_{zr}\frac{\partial c}{\partial r} + D_{z\theta}\frac{1}{r}\frac{\partial c}{\partial y} + D_{zz}\frac{\partial c}{\partial z} \right)
\end{aligned}$$

which is the RHS of Eq. (A.1.13). In other words, Eq. (A.1.4) is correct for Cartesian coordinates *and* cylindrical coordinates.

For constant **D** (Eq. (A.1.14)), we now determine if Eq. (A.1.9) applies. ∇c is given by Eq. (A.1.15). Then

$$\begin{aligned}
\nabla\nabla c &= \left(\mathbf{i}_r \frac{1}{r}\frac{\partial}{\partial r}(r) + \mathbf{j}_\theta \frac{1}{r}\frac{\partial}{\partial \theta} + \mathbf{k}_z \frac{\partial}{\partial z} \right)\left(\mathbf{i}_r \frac{\partial c}{\partial r} + \mathbf{j}_\theta \frac{1}{r}\frac{\partial c}{\partial \theta} + \mathbf{k}_z \frac{\partial c}{\partial z} \right) \\
&= \mathbf{i}_r\mathbf{i}_r \frac{1}{r}\frac{\partial}{\partial r}\left(r\frac{\partial c}{\partial r} \right) + \mathbf{i}_r\mathbf{j}_\theta \frac{1}{r}\frac{\partial^2 c}{\partial r\,\partial \theta} + \mathbf{i}_r\mathbf{k}_z \frac{1}{r}\frac{\partial}{\partial r}\left(r\frac{\partial c}{\partial z} \right) \\
&\quad + \mathbf{j}_\theta\mathbf{i}_r \frac{1}{r}\frac{\partial^2 c}{\partial \theta\,\partial r} + \mathbf{j}_\theta\mathbf{j}_\theta \frac{1}{r^2}\frac{\partial^2 c}{\partial \theta^2} + \mathbf{j}_\theta\mathbf{k}_z \frac{1}{r}\frac{\partial^2 c}{\partial \theta\,\partial z} \\
&\quad + \mathbf{k}_z\mathbf{i}_r \frac{\partial^2 c}{\partial z\,\partial r} + \mathbf{k}_z\mathbf{j}_\theta \frac{1}{r}\frac{\partial^2 c}{\partial z\,\partial \theta} + \mathbf{k}_z\mathbf{k}_z \frac{\partial^2 c}{\partial z^2}
\end{aligned}$$

The RHS of Eq. (A.1.9) is then

$$\begin{aligned}
\mathbf{D}^T : \nabla\nabla c &= (\mathbf{i}_r\mathbf{i}_r D_{rr} + \mathbf{i}_r\mathbf{j}_\theta D_{\theta r} + \mathbf{i}_r\mathbf{k}_z D_{zr} \\
&\quad + \mathbf{j}_\theta\mathbf{i}_r D_{r\theta} + \mathbf{j}_\theta\mathbf{j}_\theta D_{\theta\theta} + \mathbf{j}_\theta\mathbf{k}_z D_{z\theta} \\
&\quad + \mathbf{k}_z\mathbf{i}_r D_{rz} + \mathbf{k}_z\mathbf{j}_\theta D_{\theta z} + \mathbf{k}_z\mathbf{k}_z D_{zz}) \\
&\quad : \left[\mathbf{i}_r\mathbf{i}_r \frac{1}{r}\frac{\partial}{\partial r}\left(r\frac{\partial c}{\partial r} \right) + \mathbf{i}_r\mathbf{j}_\theta \frac{1}{r}\frac{\partial^2 c}{\partial r\,\partial \theta} + \mathbf{i}_r\mathbf{k}_z \frac{1}{r}\frac{\partial}{\partial r}\left(r\frac{\partial c}{\partial z} \right) \right. \\
&\quad + \mathbf{j}_\theta\mathbf{i}_r \frac{1}{r}\frac{\partial^2 c}{\partial \theta\,\partial r} + \mathbf{j}_\theta\mathbf{j}_\theta \frac{1}{r^2}\frac{\partial^2 c}{\partial \theta^2} + \mathbf{j}_\theta\mathbf{k}_z \frac{1}{r}\frac{\partial^2 c}{\partial \theta\,\partial z} \\
&\quad \left. + \mathbf{k}_z\mathbf{i}_r \frac{\partial^2 c}{\partial z\,\partial r} + \mathbf{k}_z\mathbf{j}_\theta \frac{1}{r}\frac{\partial^2 c}{\partial z\,\partial \theta} + \mathbf{k}_z\mathbf{k}_z \frac{\partial^2 c}{\partial z^2} \right]
\end{aligned}$$

$$
\begin{aligned}
&= \mathbf{i}_r\mathbf{i}_r : \mathbf{i}_r\mathbf{i}_r D_{rr}\frac{1}{r}\frac{\partial}{\partial r}\left(r\frac{\partial c}{\partial r}\right) + \mathbf{i}_r\mathbf{j}_\theta : \mathbf{j}_\theta\mathbf{i}_r D_{\theta r}\frac{1}{r}\frac{\partial^2 c}{\partial\theta\,\partial r} + \mathbf{i}_r\mathbf{k}_z : \mathbf{k}_z\mathbf{i}_r D_{zr}\frac{\partial^2 c}{\partial z\,\partial r} \\
&\quad + \mathbf{j}_\theta\mathbf{i}_r : \mathbf{i}_r\mathbf{j}_\theta D_{r\theta}\frac{1}{r}\frac{\partial^2 c}{\partial r\,\partial\theta} + \mathbf{j}_\theta\mathbf{j}_\theta : \mathbf{j}_\theta\mathbf{j}_\theta D_{\theta\theta}\frac{1}{r^2}\frac{\partial^2 c}{\partial\theta^2} + \mathbf{j}_\theta\mathbf{k}_z : \mathbf{k}_z\mathbf{j}_\theta D_{z\theta}\frac{1}{r}\frac{\partial^2 c}{\partial z\,\partial\theta} \\
&\quad + \mathbf{k}_z\mathbf{i}_r : \mathbf{i}_r\mathbf{k}_z D_{rz}\frac{1}{r}\frac{\partial}{\partial r}\left(r\frac{\partial c}{\partial z}\right) + \mathbf{k}_z\mathbf{j}_\theta : \mathbf{j}_\theta\mathbf{k}_z D_{\theta z}\frac{1}{r}\frac{\partial^2 c}{\partial\theta\,\partial z} + \mathbf{k}_z\mathbf{k}_z : \mathbf{k}_z\mathbf{k}_z D_{zz}\frac{\partial^2 c}{\partial z^2} \\
&= D_{rr}\left(\frac{\partial^2 c}{\partial r^2} + \frac{1}{r}\frac{\partial c}{\partial r}\right) + D_{\theta r}\frac{1}{r}\frac{\partial^2 c}{\partial\theta\,\partial r} + D_{zr}\frac{\partial^2 c}{\partial z\,\partial r} \\
&\quad + D_{r\theta}\frac{1}{r}\frac{\partial^2 c}{\partial r\,\partial\theta} + D_{\theta\theta}\frac{1}{r^2}\frac{\partial^2 c}{\partial\theta^2} + D_{z\theta}\frac{1}{r}\frac{\partial^2 c}{\partial z\,\partial\theta} \\
&\quad + D_{rz}\left(\frac{\partial^2 c}{\partial r\,\partial z} + \frac{1}{r}\frac{\partial c}{\partial z}\right) + D_{\theta z}\frac{1}{r}\frac{\partial^2 c}{\partial\theta\,\partial z} + D_{zz}\frac{\partial^2 c}{\partial z^2}
\end{aligned}
$$

which is the RHS of Eq. (A.1.14). Thus, Eq. (A.1.9) applies to Cartesian *and* cylindrical coordinates.

Finally, we apply the same analysis in spherical coordinates by starting with a mass balance on an incremental volume $(\Delta r)(r\Delta\theta)(r\sin\theta\Delta\phi)$ ([1], p. 840); θ is the angle of r with respect to the z axis and ϕ is the angle of r projected onto the x–y plane.

$$
\begin{aligned}
&(\Delta r)\,(r\Delta\theta)\,(r\sin\theta\Delta\phi)\,\frac{\partial c}{\partial t} \\
&= -\,(r\Delta\theta)\,(r\sin\theta\Delta\phi)\,D_{rr}\frac{\partial c}{\partial r}|_r - \left(-\,(r\Delta\theta)\,(r\sin\theta\Delta\phi)\,D_{rr}\frac{\partial c}{\partial r}|_{r+\Delta r}\right) \\
&\quad -\,(r\Delta\theta)\,(r\sin\theta\Delta\phi)\,D_{r\theta}\frac{\partial c}{r\,\partial\theta}|_r - \left(-\,(r\Delta\theta)\,(r\sin\theta\Delta\phi)\,D_{r\theta}\frac{\partial c}{r\,\partial\theta}|_{r+\Delta r}\right) \\
&\quad -\,(r\Delta\theta)\,(r\sin\theta\Delta\phi)\,D_{r\phi}\frac{\partial c}{r\sin\theta\,\partial\phi}|_r - \left(-\,(r\Delta\theta)\,(r\sin\theta\Delta\phi)\,D_{r\phi}\frac{\partial c}{r\sin\theta\,\partial\phi}|_{r+\Delta r}\right) \\
&\quad -\,(\Delta r)\,(r\sin\theta\Delta\phi)\,D_{\theta r}\frac{\partial c}{\partial r}|_\theta - \left(-\,(\Delta r)\,(r\sin\theta\Delta\phi)\,D_{\theta r}\frac{\partial c}{\partial r}|_{\theta+\Delta\theta}\right) \\
&\quad -\,(\Delta r)\,(r\sin\theta\Delta\phi)\,D_{\theta\theta}\frac{\partial c}{r\,\partial\theta}|_\theta - \left(-\,(\Delta r)\,(r\sin\theta\Delta\phi)\,D_{\theta\theta}\frac{\partial c}{r\,\partial\theta}|_{\theta+\Delta\theta}\right) \\
&\quad -\,(\Delta r)\,(r\sin\theta\Delta\phi)\,D_{\theta\phi}\frac{\partial c}{r\sin\theta\,\partial\phi}|_\theta - \left(-\,(\Delta r)\,(r\sin\theta\Delta\phi)\,D_{\theta\phi}\frac{\partial c}{r\sin\theta\,\partial\phi}|_{\theta+\Delta\theta}\right) \\
&\quad -\,(\Delta r)\,(r\Delta\theta)\,D_{\phi r}\frac{\partial c}{\partial r}|_\phi - \left(-\,(\Delta r)\,(r\Delta\theta)\,D_{\phi r}\frac{\partial c}{\partial r}|_{\phi+\Delta\phi}\right) \\
&\quad -\,(\Delta r)\,(r\Delta\theta)\,D_{\phi\theta}\frac{\partial c}{r\,\partial\theta}|_\phi - \left(-\,(\Delta r)\,(r\Delta\theta)\,D_{\phi\theta}\frac{\partial c}{r\,\partial\theta}|_{\phi+\Delta\phi}\right) \\
&\quad -\,(\Delta r)\,(r\Delta\theta)\,D_{\phi\phi}\frac{\partial c}{r\sin\theta\,\partial\phi}|_\phi - \left(-\,(\Delta r)\,(r\Delta\theta)\,D_{\phi\phi}\frac{\partial c}{r\sin\theta\partial\phi}|_{\phi+\Delta\phi}\right)
\end{aligned}
\tag{A.1.16}
$$

Rearrangement of Eq. (A.1.16) gives

$$
\begin{aligned}
\frac{\partial c}{\partial t} = {} & \frac{(r\Delta\theta)(r\sin\theta\Delta\phi)\,D_{rr}\dfrac{\partial c}{\partial r}|_{r+\Delta r} - (r\Delta\theta)(r\sin\theta\Delta\phi)\,D_{rr}\dfrac{\partial c}{\partial r}|_{r}}{(\Delta r)(r\Delta\theta)(r\sin\theta\Delta\phi)} \\
& + \frac{(r\Delta\theta)(r\sin\theta\Delta\phi)\,D_{r\theta}\dfrac{\partial c}{r\,\partial\theta}|_{r+\Delta r} - (r\Delta\theta)(r\sin\theta\Delta\phi)\,D_{r\theta}\dfrac{\partial c}{r\,\partial\theta}|_{r}}{(\Delta r)(r\Delta\theta)(r\sin\theta\Delta\phi)} \\
& + \frac{(r\Delta\theta)(r\sin\theta\Delta\phi)\,D_{rz}\dfrac{\partial c}{r\sin\theta\,\partial\phi}|_{r+\Delta r} - (r\Delta\theta)(r\sin\theta\Delta\phi)\,D_{rz}\dfrac{\partial c}{r\sin\theta\,\partial\phi}|_{r}}{(\Delta r)(r\Delta\theta)(r\sin\theta\Delta\phi)} \\
& + \frac{(\Delta r)(r\sin\theta\Delta\phi)\,D_{\theta r}\dfrac{\partial c}{\partial r}|_{\theta+\Delta\theta} - (\Delta r)(r\sin\theta\Delta\phi)\,D_{\theta r}\dfrac{\partial c}{\partial r}|_{\theta}}{(\Delta r)(r\Delta\theta)(r\sin\theta\Delta\phi)} \\
& + \frac{(\Delta r)(r\sin\theta\Delta\phi)\,D_{\theta\theta}\dfrac{\partial c}{r\,\partial\theta}|_{\theta+\Delta\theta} - (\Delta r)(r\sin\theta\Delta\phi)\,D_{\theta\theta}\dfrac{\partial c}{r\,\partial\theta}|_{\theta}}{(\Delta r)(r\Delta\theta)(r\sin\theta\Delta\phi)} \\
& + \frac{(\Delta r)(r\sin\theta\Delta\phi)\,D_{\theta\phi}\dfrac{\partial c}{r\sin\theta\,\partial\phi}|_{\theta+\Delta\theta} - (\Delta r)(r\sin\theta\Delta\phi)\,D_{\theta\phi}\dfrac{\partial c}{r\sin\theta\,\partial\phi}|_{\theta}}{(\Delta r)(r\Delta\theta)(r\sin\theta\Delta\phi)} \\
& + \frac{(\Delta r)(r\Delta\theta)\,D_{\phi r}\dfrac{\partial c}{\partial r}|_{\phi+\Delta\phi} - (\Delta r)(r\Delta\theta)\,D_{\phi r}\dfrac{\partial c}{\partial r}|_{\phi}}{(\Delta r)(r\Delta\theta)(r\sin\theta\Delta\phi)} \\
& + \frac{(\Delta r)(r\Delta\theta)\,D_{\phi\theta}\dfrac{\partial c}{r\,\partial\theta}|_{\phi+\Delta\phi} - (\Delta r)(r\Delta\theta)\,D_{\phi\theta}\dfrac{\partial c}{r\,\partial\theta}|_{\phi}}{(\Delta r)(r\Delta\theta)(r\sin\theta\Delta\phi)} \\
& + \frac{(\Delta r)(r\Delta\theta)\,D_{\phi\phi}\dfrac{\partial c}{r\sin\theta\,\partial\phi}|_{\phi+\Delta\phi} - (\Delta r)(r\Delta\theta)\,D_{\phi\phi}\dfrac{\partial c}{r\sin\theta\,\partial\phi}|_{\phi}}{(\Delta r)(r\Delta\theta)(r\sin\theta\Delta\phi)}
\end{aligned}
$$

or after some simplification,

$$
\begin{aligned}
\frac{\partial c}{\partial t} = {} & \frac{1}{r^2}\frac{r^2 D_{rr}\dfrac{\partial c}{\partial r}|_{r+\Delta r} - r^2 D_{rr}\dfrac{\partial c}{\partial r}|_{r}}{\Delta r} \\
& + \frac{1}{r^2}\frac{rD_{r\theta}\dfrac{\partial c}{\partial\theta}|_{r+\Delta r} - rD_{r\theta}\dfrac{\partial c}{\partial\theta}|_{r}}{\Delta r} \\
& + \frac{1}{r^2\,\sin\theta}\frac{rD_{r\phi}\dfrac{\partial c}{\partial\phi}|_{r+\Delta r} - rD_{r\phi}\dfrac{\partial c}{\partial\phi}|_{r}}{\Delta r} \\
& + \frac{1}{r\sin\theta}\frac{(\sin\theta)\,D_{\theta r}\dfrac{\partial c}{\partial r}|_{\theta+\Delta\theta} - (\sin\theta)\,D_{\theta r}\dfrac{\partial c}{\partial r}|_{\theta}}{\Delta\theta}
\end{aligned}
$$

$$+\frac{1}{r\sin\theta}\frac{\sin\theta D_{\theta\theta}\frac{\partial c}{r\,\partial\theta}|_{\theta+\Delta\theta}-\sin\theta D_{\theta\theta}\frac{\partial c}{r\,\partial\theta}|_{\theta}}{\Delta\theta}$$

$$+\frac{1}{r^2\sin\theta}\frac{D_{\theta\phi}\frac{\partial c}{\partial\phi}|_{\theta+\Delta\theta}-D_{\theta\phi}\frac{\partial c}{\partial\phi}|_{\theta}}{\Delta\theta}$$

$$+\frac{1}{r\sin\theta}\frac{D_{\phi r}\frac{\partial c}{\partial r}|_{\phi+\Delta\phi}-D_{\phi r}\frac{\partial c}{\partial r}|_{\phi}}{\Delta\phi}$$

$$+\frac{1}{r^2\sin\theta}\frac{D_{\phi\theta}\frac{\partial c}{\partial\theta}|_{\phi+\Delta\phi}-D_{\phi\theta}\frac{\partial c}{\partial\theta}|_{\phi}}{\Delta\phi}$$

$$+\frac{1}{r^2\sin^2\theta}\frac{D_{\phi\phi}\frac{\partial c}{\partial\phi}|_{\phi+\Delta\phi}-D_{\phi\phi}\frac{\partial c}{\partial\phi}|_{\phi}}{\Delta\phi} \tag{A.1.17}$$

In the limit Δr, $\Delta\theta$, $\Delta\phi \longrightarrow 0$, Eq. (A.1.17) becomes

$$\frac{\partial c}{\partial t}=\frac{1}{r^2}\frac{\partial\left(r^2D_{rr}\frac{\partial c}{\partial r}\right)}{\partial r}+\frac{1}{r^2}\frac{\partial\left(rD_{r\theta}\frac{\partial c}{\partial\theta}\right)}{\partial r}+\frac{1}{r^2\sin\theta}\frac{\partial\left(rD_{r\phi}\frac{\partial c}{\partial\phi}\right)}{\partial r}$$
$$+\frac{1}{r\sin\theta}\frac{\partial\left(\sin\theta D_{\theta r}\frac{\partial c}{\partial r}\right)}{\partial\theta}+\frac{1}{r^2\sin\theta}\frac{\partial\left(\sin\theta D_{\theta\theta}\frac{\partial c}{\partial\theta}\right)}{\partial\theta}+\frac{1}{r^2\sin\theta}\frac{\partial\left(D_{\theta\phi}\frac{\partial c}{\partial\phi}\right)}{\partial\theta}$$
$$+\frac{1}{r\sin\theta}\frac{\partial\left(D_{\phi r}\frac{\partial c}{\partial r}\right)}{\partial\phi}+\frac{1}{r^2\sin\theta}\frac{\partial\left(D_{\phi\theta}\frac{\partial c}{\partial\theta}\right)}{\partial\phi}+\frac{1}{r^2\sin^2\theta}\frac{\partial\left(D_{\phi\phi}\frac{\partial c}{\partial\phi}\right)}{\partial\phi} \tag{A.1.18}$$

For constant $\mathbf{D}$, Eq. (A.1.18) becomes

$$\frac{\partial c}{\partial t}=D_{rr}\left(\frac{\partial^2 c}{\partial r^2}+\frac{2}{r}\frac{\partial c}{\partial r}\right)+\frac{D_{r\theta}}{r}\left(\frac{\partial^2 c}{\partial r\,\partial\theta}+\frac{1}{r}\frac{\partial c}{\partial\theta}\right)+\frac{D_{r\phi}}{r\sin\theta}\left(\frac{\partial^2 c}{\partial r\,\partial\phi}+\frac{1}{r}\frac{\partial c}{\partial\phi}\right)$$
$$+\frac{D_{\theta r}}{r}\left(\frac{\partial^2 c}{\partial\theta\,\partial r}+\frac{\cos\theta}{\sin\theta}\frac{\partial c}{\partial r}\right)+\frac{D_{\theta\theta}}{r^2}\left(\frac{\partial^2 c}{\partial\theta^2}+\frac{\cos\theta}{\sin\theta}\frac{\partial c}{\partial\theta}\right)+\frac{D_{\theta\phi}}{r^2\sin\theta}\frac{\partial^2 c}{\partial\theta\,\partial\phi}$$
$$+\frac{D_{\phi r}}{r\sin\theta}\frac{\partial^2 c}{\partial\phi\,\partial r}+\frac{D_{\phi\theta}}{r^2\sin\theta}\frac{\partial^2 c}{\partial\phi\,\partial\theta}+\frac{D_{\phi\phi}}{r^2\sin^2\theta}\frac{\partial^2 c}{\partial\phi^2} \tag{A.1.19}$$

As a check on Eq. (A.1.19), we can consider the special case for which the main diagonal elements of the diffusivity tensor are equal ($D_{rr}=D_{\theta\theta}=D_{\phi\phi}=D$) and the off-diagonal diffusivities are zero. Equation (A.1.19) then becomes

$$\frac{\partial c}{\partial t}=D_{rr}\left(\frac{\partial^2 c}{\partial r^2}+\frac{2}{r}\frac{\partial c}{\partial r}\right)+\frac{D_{\theta\theta}}{r^2}\left(\frac{\partial^2 c}{\partial\theta^2}+\frac{\cos\theta}{\sin\theta}\frac{\partial c}{\partial\theta}\right)+\frac{D_{\phi\phi}}{r^2\sin^2\theta}\frac{\partial^2 c}{\partial\phi^2}$$

which is ([2], p. 3)

$$\frac{\partial c}{\partial t} = D\nabla^2 c$$

in spherical coordinates.

We now consider if Eq. (A.1.18) can be expressed in the vector–tensor notation of Eq. (A.1.4).

$$\nabla c = \mathbf{i}_r \frac{\partial c}{\partial r} + \mathbf{j}_\theta \frac{1}{r}\frac{\partial c}{\partial \theta} + \mathbf{k}_\phi \frac{1}{r\sin\theta}\frac{\partial c}{\partial \phi} \tag{A.1.20}$$

$$\begin{aligned}
\mathbf{D}\cdot\nabla c = {} & (\mathbf{i}_r\mathbf{i}_r D_{rr} + \mathbf{i}_r\mathbf{j}_\theta D_{r\theta} + \mathbf{i}_r\mathbf{k}_\phi D_{r\phi} \\
& + \mathbf{j}_\theta\mathbf{i}_r D_{\theta r} + \mathbf{j}_\theta\mathbf{j}_\theta D_{\theta\theta} + \mathbf{j}_\theta\mathbf{k}_\phi D_{\theta\phi} \\
& + \mathbf{k}_\phi\mathbf{i}_r D_{\phi r} + \mathbf{k}_\phi\mathbf{j}_\theta D_{\phi\theta} + \mathbf{k}_\phi\mathbf{k}_\phi D_{\phi\phi}) \\
& \cdot\left(\mathbf{i}_r \frac{\partial c}{\partial r} + \mathbf{j}_\theta \frac{1}{r}\frac{\partial c}{\partial \theta} + \mathbf{k}_\phi \frac{1}{r\sin\theta}\frac{\partial c}{\partial \phi}\right) \\
= {} & \mathbf{i}_r(\mathbf{i}_r\cdot\mathbf{i}_r)D_{rr}\frac{\partial c}{\partial r} + \mathbf{j}_\theta(\mathbf{i}_r\cdot\mathbf{i}_r)D_{\theta r}\frac{\partial c}{\partial r} + \mathbf{k}_\phi(\mathbf{i}_r\cdot\mathbf{i}_r)D_{\phi r}\frac{\partial c}{\partial r} \\
& + \mathbf{i}_r(\mathbf{j}_\theta\cdot\mathbf{j}_\theta)D_{r\theta}\frac{1}{r}\frac{\partial c}{\partial \theta} + \mathbf{j}_\theta(\mathbf{j}_\theta\cdot\mathbf{j}_\theta)D_{\theta\theta}\frac{1}{r}\frac{\partial c}{\partial \theta} + \mathbf{k}_\phi(\mathbf{j}_\theta\cdot\mathbf{j}_\theta)D_{\phi\theta}\frac{1}{r}\frac{\partial c}{\partial \theta} \\
& + \mathbf{i}_r(\mathbf{k}_\phi\cdot\mathbf{k}_\phi)D_{r\phi}\frac{1}{r\sin\theta}\frac{\partial c}{\partial \phi} + \mathbf{j}_\theta(\mathbf{k}_\phi\cdot\mathbf{k}_\phi)D_{\theta\phi}\frac{1}{r\sin\theta}\frac{\partial c}{\partial \phi} \\
& + \mathbf{k}_\phi(\mathbf{k}_\phi\cdot\mathbf{k}_\phi)D_{\phi\phi}\frac{1}{r\sin\theta}\frac{\partial c}{\partial \phi} \\
= {} & \mathbf{i}_r\left(D_{rr}\frac{\partial c}{\partial r} + D_{r\theta}\frac{1}{r}\frac{\partial c}{\partial \theta} + D_{r\phi}\frac{1}{r\sin\theta}\frac{\partial c}{\partial \phi}\right) \\
& + \mathbf{j}_\theta\left(D_{\theta r}\frac{\partial c}{\partial r} + D_{\theta\theta}\frac{1}{r}\frac{\partial c}{\partial \theta} + D_{\theta\phi}\frac{1}{r\sin\theta}\frac{\partial c}{\partial \phi}\right) \\
& + \mathbf{k}_\phi\left(D_{\phi r}\frac{\partial c}{\partial r} + D_{\phi\theta}\frac{1}{r}\frac{\partial c}{\partial \theta} + D_{\phi\phi}\frac{1}{r\sin\theta}\frac{\partial c}{\partial \phi}\right)
\end{aligned}$$

Finally, the RHS of Eq. (A.1.4) is ([2], p. 2)

$$\begin{aligned}
\nabla\cdot(\mathbf{D}\cdot\nabla c) = {} & \left(\mathbf{i}_r\frac{1}{r^2}\frac{\partial}{\partial r}(r^2) + \mathbf{j}_\theta\frac{1}{r\sin\theta}\frac{\partial}{\partial\theta}(\sin\theta) + \mathbf{k}_\phi\frac{1}{r\sin\theta}\frac{\partial}{\partial\phi}\right) \\
& \cdot\left[\mathbf{i}_r\left(D_{rr}\frac{\partial c}{\partial r} + D_{r\theta}\frac{1}{r}\frac{\partial c}{\partial \theta} + D_{r\phi}\frac{1}{r\sin\theta}\frac{\partial c}{\partial \phi}\right)\right. \\
& + \mathbf{j}_\theta\left(D_{\theta r}\frac{\partial c}{\partial r} + D_{\theta\theta}\frac{1}{r}\frac{\partial c}{\partial \theta} + D_{\theta\phi}\frac{1}{r\sin\theta}\frac{\partial c}{\partial \phi}\right) \\
& \left. + \mathbf{k}_\phi\left(D_{\phi r}\frac{\partial c}{\partial r} + D_{\phi\theta}\frac{1}{r}\frac{\partial c}{\partial \theta} + D_{\phi\phi}\frac{1}{r\sin\theta}\frac{\partial c}{\partial \phi}\right)\right]
\end{aligned}$$

$$
\begin{aligned}
&= \frac{1}{r^2}\frac{\partial}{\partial r}\left(r^2 D_{rr}\frac{\partial c}{\partial r}\right) + \frac{1}{r^2}\frac{\partial}{\partial r}\left(r^2 D_{r\theta}\frac{1}{r}\frac{\partial c}{\partial \theta}\right) + \frac{1}{r^2}\frac{\partial}{\partial r}\left(r^2 D_{r\phi}\frac{1}{r\sin\theta}\frac{\partial c}{\partial \phi}\right) \\
&+ \frac{1}{r\sin\theta}\frac{\partial}{\partial \theta}\left(\sin\theta\, D_{\theta r}\frac{\partial c}{\partial r}\right) + \frac{1}{r\sin\theta}\frac{\partial}{\partial \theta}\left(\sin\theta\, D_{\theta\theta}\frac{1}{r}\frac{\partial c}{\partial \theta}\right) \\
&+ \frac{1}{r\sin\theta}\frac{\partial}{\partial \theta}\left(\sin\theta\, D_{\theta\phi}\frac{1}{r\sin\theta}\frac{\partial c}{\partial \phi}\right) + \frac{1}{r\sin\theta}\frac{\partial}{\partial \phi}\left(D_{\phi r}\frac{\partial c}{\partial r}\right) \\
&+ \frac{1}{r\sin\theta}\frac{\partial}{\partial \phi}\left(D_{\phi\theta}\frac{1}{r}\frac{\partial c}{\partial \theta}\right) + \frac{1}{r\sin\theta}\frac{\partial}{\partial \phi}\left(D_{\phi\phi}\frac{1}{r\sin\theta}\frac{\partial c}{\partial \phi}\right)\Bigg)
\end{aligned}
\tag{A.1.21}
$$

which is the RHS of Eq. (A.1.18). In other words, Eq. (A.1.4) is correct for Cartesian, cylindrical, *and* spherical coordinates. The final result is therefore Eq. (A.1.18).

For $\mathbf{D}$ constant (Eq. (A.1.19)), we now determine if Eq. (A.1.9) applies. ∇c is given by Eq. (A.1.20). Then

$$
\begin{aligned}
\nabla\nabla c &= \left(\mathbf{i}_r\frac{1}{r^2}\frac{\partial}{\partial r}(r^2) + \mathbf{j}_\theta\frac{1}{r\sin\theta}\frac{\partial}{\partial \theta}(\sin\theta) + \mathbf{k}_\phi\frac{1}{r\sin\theta}\frac{\partial}{\partial \phi}\right)\left(\mathbf{i}_r\frac{\partial c}{\partial r} + \mathbf{j}_\theta\frac{1}{r}\frac{\partial c}{\partial \theta} + \mathbf{k}_\phi\frac{1}{r\sin\theta}\frac{\partial c}{\partial \phi}\right) \\
&= \mathbf{i}_r\mathbf{i}_r\frac{1}{r^2}\frac{\partial}{\partial r}\left(r^2\frac{\partial c}{\partial r}\right) + \mathbf{i}_r\mathbf{j}_\theta\frac{1}{r^2}\frac{\partial}{\partial r}\left(r^2\frac{1}{r}\frac{\partial c}{\partial \theta}\right) + \mathbf{i}_r\mathbf{k}_\phi\frac{1}{r^2}\frac{\partial}{\partial r}\left(r^2\frac{1}{r\sin\theta}\frac{\partial c}{\partial \phi}\right) \\
&+ \mathbf{j}_\theta\mathbf{i}_r\frac{1}{r\sin\theta}\frac{\partial}{\partial \theta}\left(\sin\theta\frac{\partial c}{\partial r}\right) + \mathbf{j}_\theta\mathbf{j}_\theta\frac{1}{r\sin\theta}\frac{\partial}{\partial \theta}\left(\sin\theta\frac{1}{r}\frac{\partial c}{\partial \theta}\right) \\
&+ \mathbf{j}_\theta\mathbf{k}_\phi\frac{1}{r\sin\theta}\frac{\partial}{\partial \theta}\left(\sin\theta\frac{1}{r\sin\theta}\frac{\partial c}{\partial \phi}\right) \\
&+ \mathbf{k}_\phi\mathbf{i}_r\frac{1}{r\sin\theta}\frac{\partial}{\partial \phi}\frac{\partial c}{\partial r} + \mathbf{k}_\phi\mathbf{j}_\theta\frac{1}{r\sin\theta}\frac{\partial}{\partial \phi}\frac{1}{r}\frac{\partial c}{\partial \theta} + \mathbf{k}_\phi\mathbf{k}_\phi\frac{1}{r\sin\theta}\frac{\partial}{\partial \phi}\frac{1}{r\sin\theta}\frac{\partial c}{\partial \phi}
\end{aligned}
$$

The RHS of Eq. (A.1.9) is then

$$
\begin{aligned}
\mathbf{D}^T : \nabla\nabla c &= (\mathbf{i}_r\mathbf{i}_r D_{rr} + \mathbf{i}_r\mathbf{j}_\theta D_{\theta r} + \mathbf{i}_r\mathbf{k}_\phi D_{\phi r} \\
&+ \mathbf{j}_\theta\mathbf{i}_r D_{r\theta} + \mathbf{j}_\theta\mathbf{j}_\theta D_{\theta\theta} + \mathbf{j}_\theta\mathbf{k}_\phi D_{\phi\theta} \\
&+ \mathbf{k}_\phi\mathbf{i}_r D_{r\phi} + \mathbf{k}_\phi\mathbf{j}_\theta D_{\theta\phi} + \mathbf{k}_\phi\mathbf{k}_\phi D_{\phi\phi}) \\
&: \Bigg[\mathbf{i}_r\mathbf{i}_r\frac{1}{r^2}\frac{\partial}{\partial r}\left(r^2\frac{\partial c}{\partial r}\right) + \mathbf{i}_r\mathbf{j}_\theta\frac{1}{r^2}\frac{\partial}{\partial r}\left(r^2\frac{1}{r}\frac{\partial c}{\partial \theta}\right) + \mathbf{i}_r\mathbf{k}_\phi\frac{1}{r^2}\frac{\partial}{\partial r}\left(r^2\frac{1}{r\sin\theta}\frac{\partial c}{\partial \phi}\right) \\
&+ \mathbf{j}_\theta\mathbf{i}_r\frac{1}{r\sin\theta}\frac{\partial}{\partial \theta}\left(\sin\theta\frac{\partial c}{\partial r}\right) + \mathbf{j}_\theta\mathbf{j}_\theta\frac{1}{r\sin\theta}\frac{\partial}{\partial \theta}\left(\sin\theta\frac{1}{r}\frac{\partial c}{\partial \theta}\right) \\
&+ \mathbf{j}_\theta\mathbf{k}_\phi\frac{1}{r\sin\theta}\frac{\partial}{\partial \theta}\left(\sin\theta\frac{1}{r\sin\theta}\frac{\partial c}{\partial \phi}\right) \\
&+ \mathbf{k}_\phi\mathbf{i}_r\frac{1}{r\sin\theta}\frac{\partial}{\partial \phi}\frac{\partial c}{\partial r} + \mathbf{k}_\phi\mathbf{j}_\theta\frac{1}{r\sin\theta}\frac{\partial}{\partial \phi}\frac{1}{r}\frac{\partial c}{\partial \theta} + \mathbf{k}_\phi\mathbf{k}_\phi\frac{1}{r\sin\theta}\frac{\partial}{\partial \phi}\frac{1}{r\sin\theta}\frac{\partial c}{\partial \phi}\Bigg]
\end{aligned}
$$

$$
\begin{aligned}
&= \mathbf{i}_r\mathbf{i}_r : \mathbf{i}_r\mathbf{i}_r D_{rr}\frac{1}{r^2}\frac{\partial}{\partial r}\left(r^2\frac{\partial c}{\partial r}\right) + \mathbf{i}_r\mathbf{j}_\theta : \mathbf{j}_\theta\mathbf{i}_r D_{\theta r}\frac{1}{r\sin\theta}\frac{\partial}{\partial\theta}\left(\sin\theta\frac{\partial c}{\partial r}\right)\\
&\quad + \mathbf{i}_r\mathbf{k}_\phi : \mathbf{k}_\phi\mathbf{i}_r D_{\phi r}\frac{1}{r\sin\theta}\frac{\partial}{\partial\phi}\frac{\partial c}{\partial r}\\
&\quad + \mathbf{j}_\theta\mathbf{i}_r : \mathbf{i}_r\mathbf{j}_\theta D_{r\theta}\frac{1}{r^2}\frac{\partial}{\partial r}\left(r^2\frac{1}{r}\frac{\partial c}{\partial\theta}\right) + \mathbf{j}_\theta\mathbf{j}_\theta : \mathbf{j}_\theta\mathbf{j}_\theta D_{\theta\theta}\frac{1}{r\sin\theta}\frac{\partial}{\partial\theta}\left(\sin\theta\frac{1}{r}\frac{\partial c}{\partial\theta}\right)\\
&\quad + \mathbf{j}_\theta\mathbf{k}_\phi : \mathbf{k}_\phi\mathbf{j}_\theta D_{\phi\theta}\frac{1}{r\sin\theta}\frac{\partial}{\partial\phi}\frac{1}{r}\frac{\partial c}{\partial\theta}\\
&\quad + \mathbf{k}_\phi\mathbf{i}_r : \mathbf{i}_r\mathbf{k}_\phi D_{r\phi}\frac{1}{r^2}\frac{\partial}{\partial r}\left(r^2\frac{1}{r\sin\theta}\frac{\partial c}{\partial\phi}\right)\\
&\quad + \mathbf{k}_\phi\mathbf{j}_\theta : \mathbf{j}_\theta\mathbf{k}_\phi D_{\theta\phi}\frac{1}{r\sin\theta}\frac{\partial}{\partial\theta}\left(\sin\theta\frac{1}{r\sin\theta}\frac{\partial c}{\partial\phi}\right)\\
&\quad + \mathbf{k}_\phi\mathbf{k}_\phi : \mathbf{k}_\phi\mathbf{k}_\phi D_{\phi\phi}\frac{1}{r\sin\theta}\frac{\partial}{\partial\phi}\frac{1}{r\sin\theta}\frac{\partial c}{\partial\phi}\\
&= D_{rr}\frac{1}{r^2}\frac{\partial}{\partial r}\left(r^2\frac{\partial c}{\partial r}\right) + D_{\theta r}\frac{1}{r\sin\theta}\frac{\partial}{\partial\theta}\left(\sin\theta\frac{\partial c}{\partial r}\right) + D_{\phi r}\frac{1}{r\sin\theta}\frac{\partial}{\partial\phi}\frac{\partial c}{\partial r}\\
&\quad + D_{r\theta}\frac{1}{r^2}\frac{\partial}{\partial r}\left(r^2\frac{1}{r}\frac{\partial c}{\partial\theta}\right) + D_{\theta\theta}\frac{1}{r\sin\theta}\frac{\partial}{\partial\theta}\left(\sin\theta\frac{1}{r}\frac{\partial c}{\partial\theta}\right) + D_{\phi\theta}\frac{1}{r\sin\theta}\frac{\partial}{\partial\phi}\frac{1}{r}\frac{\partial c}{\partial\theta}\\
&\quad + D_{r\phi}\frac{1}{r^2}\frac{\partial}{\partial r}\left(r^2\frac{1}{r\sin\theta}\frac{\partial c}{\partial\phi}\right) + D_{\theta\phi}\frac{1}{r\sin\theta}\frac{\partial}{\partial\theta}\left(\sin\theta\frac{1}{r\sin\theta}\frac{\partial c}{\partial\phi}\right)\\
&\quad + D_{\phi\phi}\frac{1}{r\sin\theta}\frac{\partial}{\partial\phi}\frac{1}{r\sin\theta}\frac{\partial c}{\partial\phi}\\
&= D_{rr}\left(\frac{\partial^2 c}{\partial r^2} + \frac{2}{r}\frac{\partial c}{\partial r}\right) + \frac{D_{\theta r}}{r}\left(\frac{\partial^2 c}{\partial\theta\,\partial r} + \frac{\cos\theta}{\sin\theta}\frac{\partial c}{\partial r}\right) + \frac{D_{\phi r}}{r\sin\theta}\frac{\partial^2 c}{\partial\phi\,\partial r}\\
&\quad + \frac{D_{r\theta}}{r}\left(\frac{\partial^2 c}{\partial r\,\partial\theta} + \frac{1}{r}\frac{\partial c}{\partial\theta}\right) + \frac{D_{\theta\theta}}{r^2}\left(\frac{\partial^2 c}{\partial\theta^2} + \frac{\cos\theta}{\sin\theta}\frac{\partial c}{\partial\theta}\right) + \frac{D_{\phi\theta}}{r^2\sin\theta}\frac{\partial^2 c}{\partial\phi\,\partial\theta}\\
&\quad + \frac{D_{r\phi}}{r\sin\theta}\left(\frac{\partial^2 c}{\partial r\,\partial\phi} + \frac{1}{r}\frac{\partial c}{\partial\phi}\right) + \frac{D_{\theta\phi}}{r^2\sin\theta}\frac{\partial^2 c}{\partial\theta\,\partial\phi} + \frac{D_{\phi\phi}}{r^2\sin^2\theta}\frac{\partial^2 c}{\partial\phi^2}
\end{aligned}
$$

which is the RHS of Eq. (A.1.19). Thus, Eq. (A.1.9) applies to Cartesian, cylindrical, *and* spherical coordinates.

As a few concluding points:

1. The preceding analysis (mass balance on an incremental volume) can be applied in any *orthogonal coordinate system* ([3], pp. 26, 115).
2. The analysis can be extended to include convective terms, and thereby arrive at tensor forms of *convective–diffusive* (*hyperbolic–parabolic*) PDE systems.
3. In principle, the method of lines (MOL) can be applied to these various PDE systems. In practice, experience has indicated that, in particular, when mixed

partial derivatives are included, as in most of the preceding PDEs, numerical difficulties can develop.

4. Mixed partials can be calculated within the MOL by using stagewise differentiation. For example, to compute the mixed partial $\partial^2 u/\partial x\, \partial y$, we can differentiate u first with respect to x and then with respect to y.
5. Because numerical difficulties can occur with each new problem, generally some trial and error is required to arrive at a numerical solution with acceptable accuracy and computational effort. In other words, the MOL is not necessarily a straightforward, mechanical procedure, and success in using it is generally dependent, to some degree, on the experience and creativity of the analyst. Our hope is that the preceding examples can serve as a guide to the solution of new problems.

REFERENCES

[1] Bird, R. B., W. E. Stewart, and E. N. Lightfoot (2002), *Transport Phenomena*, 2nd ed., Wiley, New York

[2] vande Wouwer, A., P. Saucez, and W. E. Schiesser (Eds.) (2001), *Adaptive Method of Lines*, CRC Press, Boca Raton, FL

[3] Morse, P. and H. Feshbach (1953), *Methods of Theoretical Physics*, McGraw-Hill, New York

APPENDIX 2

Order Conditions for Finite-Difference Approximations

We have used the DSS (Differentiation in Space Subroutines[1]) to compute finite-difference (FD) approximations of spatial (boundary-value) derivatives in partial differential equations (PDEs). Also, we have mentioned the order of these approximations, for example, second-order approximations in `dss002`. In this appendix, we explain briefly what is meant by the order of an FD.

If we consider the derivative of a polynomial of order p, for example, of second order with $p = 2$,

$$p_2(x) = a_0 + a_1 x + a_2 x^2 \tag{A.2.1}$$

then the pth derivative is a constant $(2a_2)$ and the $(p + 1)$st derivative is zero.

Then a *pth-order FD differentiates a pth-order polynomial exactly*. Thus, if we apply `dss002` (with three-point, second-order FDs) to a second-order polynomial, the numerical derivatives should be exact. But the numerical derivatives of third- and higher-order polynomials will not be exact.

To illustrate these ideas, consider the test program given in Listing A.2.1.

```
%
  clear all
  clc
%
% Varying order of differentiator
  for ndss=2:2:10
%
% Varying order of test polynomial
  for norder=1:12
```

[1] These routines were first written in Fortran, and thus the name "subroutines." They were subsequently translated into Matlab, for which the corresponding name is "function." Also, we use just "routines" in referring to them.

```
%
% Grid in x
  xl=0.0;
  xu=1.0;
  n=11;
  dx=(xu-xl)/(n-1);
%
% Test polynomial
  for i=1:n
    x(i)=xl+(i-1)*dx;
    u(i)=x(i)^norder;
  end
%
% ux
  if ndss==2   ux=dss002(xl,xu,n,u); end
  if ndss==4   ux=dss004(xl,xu,n,u); end
  if ndss==6   ux=dss006(xl,xu,n,u); end
  if ndss==8   ux=dss008(xl,xu,n,u); end
  if ndss==10  ux=dss010(xl,xu,n,u); end
%
% Comparison of numerical and analytical ux
  if(norder-ndss)>-1 & (norder-ndss)< 3
  fprintf('\n\n ndss = %2d   order = %2d\n\n',ndss,norder);
  for i=1:n
    uxa(i)=norder*x(i)^(norder-1);
    err(i)=ux(i)-uxa(i);
    fprintf(' i = %3d  ux(i) = %8.5f  uxa(i) = %8.5f
             err(i) = %8.5f\n',i,ux(i),uxa(i),err(i));
  end
  end
%
% Next order
  end
%
% Next differentiator
  end
```

Listing A.2.1. Program ux_poly_1 for the numerical differentiation of polynomials

We can note the following points about this program:

1. The pair of nested `for` loops varies the order of the FD approximation (outer loop), and for each FD order, the order of the test polynomial is varied (inner loop).

```
%
  clear all
  clc
%
% Varying order of differentiator
  for ndss=2:2:10
%
% Varying order of test polynomial
  for norder=1:12
```

2. A grid in x is defined and the polynomial of order `norder` is evaluated as `u(i)`.

```
%
% Grid in x
  xl=0.0;
  xu=1.0;
  n=11;
  dx=(xu-xl)/(n-1);
%
% Test polynomial
  for i=1:n
    x(i)=xl+(i-1)*dx;
    u(i)=x(i)^norder;
  end
```

3. The numerical first derivative `ux` of `u` is then computed for FD orders two (`dss002`) to ten (`dss010`).

```
%
% ux
  if ndss==2   ux=dss002(xl,xu,n,u); end
  if ndss==4   ux=dss004(xl,xu,n,u); end
  if ndss==6   ux=dss006(xl,xu,n,u); end
  if ndss==8   ux=dss008(xl,xu,n,u); end
  if ndss==10  ux=dss010(xl,xu,n,u); end
```

4. The exact derivative (from analytical differentiation) is put in array `uxa(i)` and the difference between the exact and numerical derivatives is computed as `err(i)`.

```
%
% Comparison of numerical and analytical ux
  if(norder-ndss)>-1 & (norder-ndss)< 3
  fprintf('\n\n ndss = %2d   order = %2d\n\n',ndss,norder);
  for i=1:n
    uxa(i)=norder*x(i)^(norder-1);
    err(i)=ux(i)-uxa(i);
    fprintf(' i = %3d  ux(i) = %8.5f  uxa(i) = %8.5f
                err(i) = %8.5f\n',i,ux(i),uxa(i),err(i));
  end
  end
```

The first `if` statement limits the comparison of the analytical and exact derivatives (in the output) to the cases `norder = ndss` to `norder = ndss +2` (just to display the most interesting conditions when the order of the polynomial equals or exceeds by 2 the order of the differentiator).

A selected portion of the output from this test program is given in Table A.2.1. We can note the following details about the output given in Table A.2.1:

1. For `ndss = norder = 2`, the numerical (FD) derivative is exact (to machine accuracy, which is about 15 figures for Matlab running on a 32-bit machine).

```
ndss =  2   order =  2

 i =  1  ux(i) = 0.00000  uxa(i) = 0.00000  err(i) =  0.00000
 i =  2  ux(i) = 0.20000  uxa(i) = 0.20000  err(i) =  0.00000
 i =  3  ux(i) = 0.40000  uxa(i) = 0.40000  err(i) =  0.00000
 i =  4  ux(i) = 0.60000  uxa(i) = 0.60000  err(i) =  0.00000
 i =  5  ux(i) = 0.80000  uxa(i) = 0.80000  err(i) = -0.00000
 i =  6  ux(i) = 1.00000  uxa(i) = 1.00000  err(i) =  0.00000
 i =  7  ux(i) = 1.20000  uxa(i) = 1.20000  err(i) =  0.00000
 i =  8  ux(i) = 1.40000  uxa(i) = 1.40000  err(i) =  0.00000
 i =  9  ux(i) = 1.60000  uxa(i) = 1.60000  err(i) = -0.00000
 i = 10  ux(i) = 1.80000  uxa(i) = 1.80000  err(i) = -0.00000
 i = 11  ux(i) = 2.00000  uxa(i) = 2.00000  err(i) = -0.00000
```

2. For `ndss = 2, norder = 3`, the numerical (FD) derivatives are not exact. In fact, the difference between the exact and numerical derivatives is a constant that is given by the *principal error term* (with $p = 2$).

$$\texttt{err} = c\frac{d^{p+1}f(x)}{dx^{p+1}} = c\frac{d^3f(x)}{dx^3} \tag{A.2.2a}$$

where c is a constant.

Table A.2.1. Partial output from `ux_poly_1` of Listing A.2.1

```
ndss =  2   order =  2

i =   1  ux(i) =  0.00000  uxa(i) =  0.00000  err(i) =  0.00000
i =   2  ux(i) =  0.20000  uxa(i) =  0.20000  err(i) =  0.00000
i =   3  ux(i) =  0.40000  uxa(i) =  0.40000  err(i) =  0.00000
i =   4  ux(i) =  0.60000  uxa(i) =  0.60000  err(i) =  0.00000
i =   5  ux(i) =  0.80000  uxa(i) =  0.80000  err(i) = -0.00000
i =   6  ux(i) =  1.00000  uxa(i) =  1.00000  err(i) =  0.00000
i =   7  ux(i) =  1.20000  uxa(i) =  1.20000  err(i) =  0.00000
i =   8  ux(i) =  1.40000  uxa(i) =  1.40000  err(i) =  0.00000
i =   9  ux(i) =  1.60000  uxa(i) =  1.60000  err(i) = -0.00000
i =  10  ux(i) =  1.80000  uxa(i) =  1.80000  err(i) = -0.00000
i =  11  ux(i) =  2.00000  uxa(i) =  2.00000  err(i) = -0.00000

ndss =  2   order =  3

i =   1  ux(i) = -0.02000  uxa(i) =  0.00000  err(i) = -0.02000
i =   2  ux(i) =  0.04000  uxa(i) =  0.03000  err(i) =  0.01000
i =   3  ux(i) =  0.13000  uxa(i) =  0.12000  err(i) =  0.01000
i =   4  ux(i) =  0.28000  uxa(i) =  0.27000  err(i) =  0.01000
i =   5  ux(i) =  0.49000  uxa(i) =  0.48000  err(i) =  0.01000
i =   6  ux(i) =  0.76000  uxa(i) =  0.75000  err(i) =  0.01000
i =   7  ux(i) =  1.09000  uxa(i) =  1.08000  err(i) =  0.01000
i =   8  ux(i) =  1.48000  uxa(i) =  1.47000  err(i) =  0.01000
i =   9  ux(i) =  1.93000  uxa(i) =  1.92000  err(i) =  0.01000
i =  10  ux(i) =  2.44000  uxa(i) =  2.43000  err(i) =  0.01000
i =  11  ux(i) =  2.98000  uxa(i) =  3.00000  err(i) = -0.02000

ndss =  2   order =  4

i =   1  ux(i) = -0.00600  uxa(i) =  0.00000  err(i) = -0.00600
i =   2  ux(i) =  0.00800  uxa(i) =  0.00400  err(i) =  0.00400
i =   3  ux(i) =  0.04000  uxa(i) =  0.03200  err(i) =  0.00800
i =   4  ux(i) =  0.12000  uxa(i) =  0.10800  err(i) =  0.01200
i =   5  ux(i) =  0.27200  uxa(i) =  0.25600  err(i) =  0.01600
i =   6  ux(i) =  0.52000  uxa(i) =  0.50000  err(i) =  0.02000
i =   7  ux(i) =  0.88800  uxa(i) =  0.86400  err(i) =  0.02400
i =   8  ux(i) =  1.40000  uxa(i) =  1.37200  err(i) =  0.02800
i =   9  ux(i) =  2.08000  uxa(i) =  2.04800  err(i) =  0.03200
i =  10  ux(i) =  2.95200  uxa(i) =  2.91600  err(i) =  0.03600
i =  11  ux(i) =  3.92600  uxa(i) =  4.00000  err(i) = -0.07400

                 .                                             .
                 .                                             .
                 .                                             .
```

```
                Output for ndss = 4 to 8 removed

          .                                         .
          .                                         .
          .                                         .
ndss = 10   order = 10

i =   1  ux(i) = -0.00000  uxa(i) =  0.00000  err(i) = -0.00000
i =   2  ux(i) =  0.00000  uxa(i) =  0.00000  err(i) =  0.00000
i =   3  ux(i) =  0.00001  uxa(i) =  0.00001  err(i) =  0.00000
i =   4  ux(i) =  0.00020  uxa(i) =  0.00020  err(i) =  0.00000
i =   5  ux(i) =  0.00262  uxa(i) =  0.00262  err(i) = -0.00000
i =   6  ux(i) =  0.01953  uxa(i) =  0.01953  err(i) =  0.00000
i =   7  ux(i) =  0.10078  uxa(i) =  0.10078  err(i) = -0.00000
i =   8  ux(i) =  0.40354  uxa(i) =  0.40354  err(i) = -0.00000
i =   9  ux(i) =  1.34218  uxa(i) =  1.34218  err(i) =  0.00000
i =  10  ux(i) =  3.87420  uxa(i) =  3.87420  err(i) = -0.00000
i =  11  ux(i) = 10.00000  uxa(i) = 10.00000  err(i) =  0.00000

ndss = 10   order = 11

i =   1  ux(i) = -0.00036  uxa(i) =  0.00000  err(i) = -0.00036
i =   2  ux(i) =  0.00004  uxa(i) =  0.00000  err(i) =  0.00004
i =   3  ux(i) = -0.00001  uxa(i) =  0.00000  err(i) = -0.00001
i =   4  ux(i) =  0.00007  uxa(i) =  0.00006  err(i) =  0.00000
i =   5  ux(i) =  0.00115  uxa(i) =  0.00115  err(i) = -0.00000
i =   6  ux(i) =  0.01074  uxa(i) =  0.01074  err(i) =  0.00000
i =   7  ux(i) =  0.06651  uxa(i) =  0.06651  err(i) = -0.00000
i =   8  ux(i) =  0.31073  uxa(i) =  0.31072  err(i) =  0.00000
i =   9  ux(i) =  1.18111  uxa(i) =  1.18112  err(i) = -0.00001
i =  10  ux(i) =  3.83550  uxa(i) =  3.83546  err(i) =  0.00004
i =  11  ux(i) = 10.99964  uxa(i) = 11.00000  err(i) = -0.00036

ndss = 10   order = 12

i =   1  ux(i) = -0.00200  uxa(i) =  0.00000  err(i) = -0.00200
i =   2  ux(i) =  0.00020  uxa(i) =  0.00000  err(i) =  0.00020
i =   3  ux(i) = -0.00005  uxa(i) =  0.00000  err(i) = -0.00005
i =   4  ux(i) =  0.00004  uxa(i) =  0.00002  err(i) =  0.00002
i =   5  ux(i) =  0.00049  uxa(i) =  0.00050  err(i) = -0.00001
i =   6  ux(i) =  0.00587  uxa(i) =  0.00586  err(i) =  0.00001
i =   7  ux(i) =  0.04353  uxa(i) =  0.04354  err(i) = -0.00001
i =   8  ux(i) =  0.23730  uxa(i) =  0.23728  err(i) =  0.00002
i =   9  ux(i) =  1.03074  uxa(i) =  1.03079  err(i) = -0.00005
i =  10  ux(i) =  3.76596  uxa(i) =  3.76573  err(i) =  0.00023
i =  11  ux(i) = 11.99764  uxa(i) = 12.00000  err(i) = -0.00236
```

An alternate form of the principal error term of Eq. (A.2.2a) is

$$\texttt{err} = c\Delta x^p \tag{A.2.2b}$$

where Δx is the grid spacing in x (Note that the constant c is different in each case).

```
ndss =  2   order =  3

 i =  1  ux(i) = -0.02000 uxa(i) = 0.00000  err(i) = -0.02000
 i =  2  ux(i) =  0.04000 uxa(i) = 0.03000  err(i) =  0.01000
 i =  3  ux(i) =  0.13000 uxa(i) = 0.12000  err(i) =  0.01000
 i =  4  ux(i) =  0.28000 uxa(i) = 0.27000  err(i) =  0.01000
 i =  5  ux(i) =  0.49000 uxa(i) = 0.48000  err(i) =  0.01000
 i =  6  ux(i) =  0.76000 uxa(i) = 0.75000  err(i) =  0.01000
 i =  7  ux(i) =  1.09000 uxa(i) = 1.08000  err(i) =  0.01000
 i =  8  ux(i) =  1.48000 uxa(i) = 1.47000  err(i) =  0.01000
 i =  9  ux(i) =  1.93000 uxa(i) = 1.92000  err(i) =  0.01000
 i = 10  ux(i) =  2.44000 uxa(i) = 2.43000  err(i) =  0.01000
 i = 11  ux(i) =  2.98000 uxa(i) = 3.00000  err(i) = -0.02000
```

Thus, for a third-order polynomial, $d^3f(x)/dx^3$ is constant. Note, however, that c in Eqs. (A.2.2a) and (A.2.2b) at the boundaries ($x = xl =$ `x(1)`, $xu =$ `x(11)`) or `-0.02000` is twice the value at the interior points, `0.01000`. This is usually the case. That is, although the order of the FD is the same throughout x, the *multiplying constant c in Eqs. (A.2.2a) and (A.2.2b) is larger at the boundaries than at the interior points.*

3. For `ndss = 2, norder = 4`, the numerical (FD) derivative is not exact and increases with x as expected from Eq. (A.2.2a) (since for the fourth-order polynomial, the derivative $d^3f(x)/dx^3$ increases with x).

```
ndss =  2   order =  4

 i =  1  ux(i) = -0.00600 uxa(i) = 0.00000  err(i) = -0.00600
 i =  2  ux(i) =  0.00800 uxa(i) = 0.00400  err(i) =  0.00400
 i =  3  ux(i) =  0.04000 uxa(i) = 0.03200  err(i) =  0.00800
 i =  4  ux(i) =  0.12000 uxa(i) = 0.10800  err(i) =  0.01200
 i =  5  ux(i) =  0.27200 uxa(i) = 0.25600  err(i) =  0.01600
 i =  6  ux(i) =  0.52000 uxa(i) = 0.50000  err(i) =  0.02000
 i =  7  ux(i) =  0.88800 uxa(i) = 0.86400  err(i) =  0.02400
 i =  8  ux(i) =  1.40000 uxa(i) = 1.37200  err(i) =  0.02800
 i =  9  ux(i) =  2.08000 uxa(i) = 2.04800  err(i) =  0.03200
 i = 10  ux(i) =  2.95200 uxa(i) = 2.91600  err(i) =  0.03600
 i = 11  ux(i) =  3.92600 uxa(i) = 4.00000  err(i) = -0.07400
```

4. The same general results are observed for a tenth-order FD (`ndss = 10`) applied to tenth- to twelfth-order polynomials. For the tenth-order polynomial, the numerical derivatives are exact.

```
ndss = 10    order = 10

 i =  1  ux(i) = -0.00000 uxa(i) =  0.00000 err(i) = -0.00000
 i =  2  ux(i) =  0.00000 uxa(i) =  0.00000 err(i) =  0.00000
 i =  3  ux(i) =  0.00001 uxa(i) =  0.00001 err(i) =  0.00000
 i =  4  ux(i) =  0.00020 uxa(i) =  0.00020 err(i) =  0.00000
 i =  5  ux(i) =  0.00262 uxa(i) =  0.00262 err(i) = -0.00000
 i =  6  ux(i) =  0.01953 uxa(i) =  0.01953 err(i) =  0.00000
 i =  7  ux(i) =  0.10078 uxa(i) =  0.10078 err(i) = -0.00000
 i =  8  ux(i) =  0.40354 uxa(i) =  0.40354 err(i) = -0.00000
 i =  9  ux(i) =  1.34218 uxa(i) =  1.34218 err(i) =  0.00000
 i = 10  ux(i) =  3.87420 uxa(i) =  3.87420 err(i) = -0.00000
 i = 11  ux(i) = 10.00000 uxa(i) = 10.00000 err(i) =  0.00000
```

5. For `ndss = 10, norder = 11`, the numerical (FD) derivatives are not exact. The difference between the exact and numerical derivatives is a constant that is given by the principal error term (with $p = 10$).

```
ndss = 10    order = 11

 i =  1  ux(i) = -0.00036 uxa(i) =  0.00000 err(i) = -0.00036
 i =  2  ux(i) =  0.00004 uxa(i) =  0.00000 err(i) =  0.00004
 i =  3  ux(i) = -0.00001 uxa(i) =  0.00000 err(i) = -0.00001
 i =  4  ux(i) =  0.00007 uxa(i) =  0.00006 err(i) =  0.00000
 i =  5  ux(i) =  0.00115 uxa(i) =  0.00115 err(i) = -0.00000
 i =  6  ux(i) =  0.01074 uxa(i) =  0.01074 err(i) =  0.00000
 i =  7  ux(i) =  0.06651 uxa(i) =  0.06651 err(i) = -0.00000
 i =  8  ux(i) =  0.31073 uxa(i) =  0.31072 err(i) =  0.00000
 i =  9  ux(i) =  1.18111 uxa(i) =  1.18112 err(i) = -0.00001
 i = 10  ux(i) =  3.83550 uxa(i) =  3.83546 err(i) =  0.00004
 i = 11  ux(i) = 10.99964 uxa(i) = 11.00000 err(i) = -0.00036
```

As before, the numerical derivatives are less accurate near the boundaries in x (because of the variation in c in Eqs. (A.2.2a) and (A.2.2b) with x, but the FD approximations are tenth order throughout x).

6. For `ndss = 10, norder = 12`, the errors are greater than those for the eleventh-order polynomial.

```
ndss = 10   order = 12

 i =  1  ux(i) = -0.00200 uxa(i) =  0.00000 err(i) = -0.00200
 i =  2  ux(i) =  0.00020 uxa(i) =  0.00000 err(i) =  0.00020
 i =  3  ux(i) = -0.00005 uxa(i) =  0.00000 err(i) = -0.00005
 i =  4  ux(i) =  0.00004 uxa(i) =  0.00002 err(i) =  0.00002
 i =  5  ux(i) =  0.00049 uxa(i) =  0.00050 err(i) = -0.00001
 i =  6  ux(i) =  0.00587 uxa(i) =  0.00586 err(i) =  0.00001
 i =  7  ux(i) =  0.04353 uxa(i) =  0.04354 err(i) = -0.00001
 i =  8  ux(i) =  0.23730 uxa(i) =  0.23728 err(i) =  0.00002
 i =  9  ux(i) =  1.03074 uxa(i) =  1.03079 err(i) = -0.00005
 i = 10  ux(i) =  3.76596 uxa(i) =  3.76573 err(i) =  0.00023
 i = 11  ux(i) = 11.99764 uxa(i) = 12.00000 err(i) = -0.00236
```

The preceding general conclusions (mainly, a pth-order FD differentiates a pth-order polynomial exactly) do not necessarily apply to nonpolynomial functions. For example, if we consider an exponential function, e^{ax}, the FD numerical derivatives will in general not be exact. One way to anticipate this result is to consider the exponential function as an infinite-order polynomial expressed as its Taylor series expansion. In fact, the principal error term of Eq. (A.2.2a) is the first term of a Taylor series expansion beyond the point of truncation of the series that gives the FD approximation; thus, this error term is called the *truncation error* of the FD.

These conclusions are illustrated with the test program given in Listing A.2.2, which is a minor variation of ux_poly_1 in Listing A.2.1.

```
%
  clear all
  clc
%
% Varying order of differentiator
  for ndss=2:2:10
%
% Grid in x
  xl=0.0;
  xu=1.0;
  n=11;
  dx=(xu-xl)/(n-1);
%
% Test exponential
  a=1.0;
  for i=1:n
    x(i)=xl+(i-1)*dx;
    u(i)=exp(a*x(i));
  end
```

```
%
% ux
  if ndss==2   ux=dss002(xl,xu,n,u); end
  if ndss==4   ux=dss004(xl,xu,n,u); end
  if ndss==6   ux=dss006(xl,xu,n,u); end
  if ndss==8   ux=dss008(xl,xu,n,u); end
  if ndss==10  ux=dss010(xl,xu,n,u); end
%
% Comparison of numerical and analytical ux
  fprintf('\n\n ndss = %2d   a = %4.2f\n\n',ndss,a);
  for i=1:n
    uxa(i)=a*exp(x(i));
    err(i)=ux(i)-uxa(i);
    fprintf(' i = %3d  ux(i) = %8.5f  uxa(i) = %8.5f
                err(i) = %8.5f\n',i,ux(i),uxa(i),err(i));
  end
%
% Next differentiator
  end
```

Listing A.2.2. Program ux_exp_1 for the numerical differentiation of an exponential function

The only essential difference is the use of `exp(a*x(i))` and its analytical derivative `a*exp(a*x(i))`. A portion of the output from this program is given in Table A.2.2.

We can note the following points about the output given in Table A.2.2:

1. The second-order FDs give a first derivative that is accurate to less than three figures.

```
ndss =  2   a = 1.00

i =  1  ux(i) = 0.99640  uxa(i) = 1.00000  err(i) = -0.00360
i =  2  ux(i) = 1.10701  uxa(i) = 1.10517  err(i) =  0.00184
i =  3  ux(i) = 1.22344  uxa(i) = 1.22140  err(i) =  0.00204
i =  4  ux(i) = 1.35211  uxa(i) = 1.34986  err(i) =  0.00225
i =  5  ux(i) = 1.49431  uxa(i) = 1.49182  err(i) =  0.00249
i =  6  ux(i) = 1.65147  uxa(i) = 1.64872  err(i) =  0.00275
i =  7  ux(i) = 1.82516  uxa(i) = 1.82212  err(i) =  0.00304
i =  8  ux(i) = 2.01711  uxa(i) = 2.01375  err(i) =  0.00336
i =  9  ux(i) = 2.22925  uxa(i) = 2.22554  err(i) =  0.00371
i = 10  ux(i) = 2.46370  uxa(i) = 2.45960  err(i) =  0.00410
i = 11  ux(i) = 2.70987  uxa(i) = 2.71828  err(i) = -0.00841
```

Table A.2.2. Partial output from ux_exp_1 of Listing A.2.2

```
ndss =  2   a = 1.00

i =   1  ux(i) =  0.99640  uxa(i) =  1.00000  err(i) = -0.00360
i =   2  ux(i) =  1.10701  uxa(i) =  1.10517  err(i) =  0.00184
i =   3  ux(i) =  1.22344  uxa(i) =  1.22140  err(i) =  0.00204
i =   4  ux(i) =  1.35211  uxa(i) =  1.34986  err(i) =  0.00225
i =   5  ux(i) =  1.49431  uxa(i) =  1.49182  err(i) =  0.00249
i =   6  ux(i) =  1.65147  uxa(i) =  1.64872  err(i) =  0.00275
i =   7  ux(i) =  1.82516  uxa(i) =  1.82212  err(i) =  0.00304
i =   8  ux(i) =  2.01711  uxa(i) =  2.01375  err(i) =  0.00336
i =   9  ux(i) =  2.22925  uxa(i) =  2.22554  err(i) =  0.00371
i =  10  ux(i) =  2.46370  uxa(i) =  2.45960  err(i) =  0.00410
i =  11  ux(i) =  2.70987  uxa(i) =  2.71828  err(i) = -0.00841

ndss =  4   a = 1.00

i =   1  ux(i) =  0.99998  uxa(i) =  1.00000  err(i) = -0.00002
i =   2  ux(i) =  1.10518  uxa(i) =  1.10517  err(i) =  0.00001
i =   3  ux(i) =  1.22140  uxa(i) =  1.22140  err(i) = -0.00000
i =   4  ux(i) =  1.34985  uxa(i) =  1.34986  err(i) = -0.00000
i =   5  ux(i) =  1.49182  uxa(i) =  1.49182  err(i) = -0.00000
i =   6  ux(i) =  1.64872  uxa(i) =  1.64872  err(i) = -0.00001
i =   7  ux(i) =  1.82211  uxa(i) =  1.82212  err(i) = -0.00001
i =   8  ux(i) =  2.01375  uxa(i) =  2.01375  err(i) = -0.00001
i =   9  ux(i) =  2.22553  uxa(i) =  2.22554  err(i) = -0.00001
i =  10  ux(i) =  2.45961  uxa(i) =  2.45960  err(i) =  0.00001
i =  11  ux(i) =  2.71824  uxa(i) =  2.71828  err(i) = -0.00005

            .                                            .
            .                                            .
            .                                            .

               Output for ndss = 6 and 8 removed

            .                                            .
            .                                            .
            .                                            .

ndss = 10   a = 1.00

i =   1  ux(i) =  1.00000  uxa(i) =  1.00000  err(i) = -0.00000
i =   2  ux(i) =  1.10517  uxa(i) =  1.10517  err(i) =  0.00000
i =   3  ux(i) =  1.22140  uxa(i) =  1.22140  err(i) = -0.00000
i =   4  ux(i) =  1.34986  uxa(i) =  1.34986  err(i) =  0.00000
i =   5  ux(i) =  1.49182  uxa(i) =  1.49182  err(i) = -0.00000
i =   6  ux(i) =  1.64872  uxa(i) =  1.64872  err(i) =  0.00000
i =   7  ux(i) =  1.82212  uxa(i) =  1.82212  err(i) = -0.00000
i =   8  ux(i) =  2.01375  uxa(i) =  2.01375  err(i) =  0.00000
i =   9  ux(i) =  2.22554  uxa(i) =  2.22554  err(i) = -0.00000
i =  10  ux(i) =  2.45960  uxa(i) =  2.45960  err(i) =  0.00000
i =  11  ux(i) =  2.71828  uxa(i) =  2.71828  err(i) = -0.00000
```

2. The fourth-order FDs give a substantial improvement in the accuracy of the derivative.

```
ndss =  4    a = 1.00

i =  1  ux(i) = 0.99998  uxa(i) = 1.00000  err(i) = -0.00002
i =  2  ux(i) = 1.10518  uxa(i) = 1.10517  err(i) =  0.00001
i =  3  ux(i) = 1.22140  uxa(i) = 1.22140  err(i) = -0.00000
i =  4  ux(i) = 1.34985  uxa(i) = 1.34986  err(i) = -0.00000
i =  5  ux(i) = 1.49182  uxa(i) = 1.49182  err(i) = -0.00000
i =  6  ux(i) = 1.64872  uxa(i) = 1.64872  err(i) = -0.00001
i =  7  ux(i) = 1.82211  uxa(i) = 1.82212  err(i) = -0.00001
i =  8  ux(i) = 2.01375  uxa(i) = 2.01375  err(i) = -0.00001
i =  9  ux(i) = 2.22553  uxa(i) = 2.22554  err(i) = -0.00001
i = 10  ux(i) = 2.45961  uxa(i) = 2.45960  err(i) =  0.00001
i = 11  ux(i) = 2.71824  uxa(i) = 2.71828  err(i) = -0.00005
```

3. The tenth-order FDs appear to give an exact first derivative. However, the output is deceptive and this conclusion is not correct.

```
ndss = 10    a = 1.00

i =  1  ux(i) = 1.00000  uxa(i) = 1.00000  err(i) = -0.00000
i =  2  ux(i) = 1.10517  uxa(i) = 1.10517  err(i) =  0.00000
i =  3  ux(i) = 1.22140  uxa(i) = 1.22140  err(i) = -0.00000
i =  4  ux(i) = 1.34986  uxa(i) = 1.34986  err(i) =  0.00000
i =  5  ux(i) = 1.49182  uxa(i) = 1.49182  err(i) = -0.00000
i =  6  ux(i) = 1.64872  uxa(i) = 1.64872  err(i) =  0.00000
i =  7  ux(i) = 1.82212  uxa(i) = 1.82212  err(i) = -0.00000
i =  8  ux(i) = 2.01375  uxa(i) = 2.01375  err(i) =  0.00000
i =  9  ux(i) = 2.22554  uxa(i) = 2.22554  err(i) = -0.00000
i = 10  ux(i) = 2.45960  uxa(i) = 2.45960  err(i) =  0.00000
i = 11  ux(i) = 2.71828  uxa(i) = 2.71828  err(i) = -0.00000
```

Rather, the errors in the numerical derivative are not displayed by the `%8.5f` format. This can easily be demonstrated by increasing `a` (i.e., $a > 1.0$) so that the exponential has greater curvature and therefore the numerical differentiation gives larger errors.

Since exponential functions are frequently solutions to ODE/PDE systems, the preceding discussion implies that in general we cannot expect the numerical solutions to be exact. In fact, numerical solutions computed with approximations such as FDs will generally not be exact since the approximations will have a *nonzero truncation error* (as expressed by the principal error term, such as in Eq. (A.2.2a) for polynomials, which defines the apparent order of the approximation).

This completes the discussion of numerical first derivatives. We next consider second derivatives, again with polynomial and exponential test functions. A program for the second-order derivative of a polynomial with Dirichlet boundary conditions (BCs) computed by FDs of orders 2–10 is given in Listing A.2.3.

```
%
  clear all
  clc
%
% Varying order of differentiator
  for ndss=42:2:50
%
% Varying order of test polynomial
  for norder=1:12
%
% Grid in x
  xl=0.0;
  xu=1.0;
  n=12;
  dx=(xu-xl)/(n-1);
%
% Test polynomial
  for i=1:n
    x(i)=xl+(i-1)*dx;
    u(i)=x(i)^norder;
  end
%
% uxx, Dirichlet BC
  nl=1;
  nu=1;
  ux=zeros(1,n);
  if ndss==42  uxx=dss042(xl,xu,n,u,ux,nl,nu); end
  if ndss==44  uxx=dss044(xl,xu,n,u,ux,nl,nu); end
  if ndss==46  uxx=dss046(xl,xu,n,u,ux,nl,nu); end
  if ndss==48  uxx=dss048(xl,xu,n,u,ux,nl,nu); end
  if ndss==50  uxx=dss050(xl,xu,n,u,ux,nl,nu); end
%
% Comparison of numerical and analytical uxx
  if(norder-(ndss-40))>-1 & (norder-(ndss-40))< 3
  fprintf('\n\n ndss = %2d    nl = 1    nu = 1
         order = %2d\n\n',ndss,norder);
  for i=1:n
```

```
      uxxa(i)=norder*(norder-1)*x(i)^(norder-2);
      err(i)=uxx(i)-uxxa(i);
      fprintf(' i = %3d  uxx(i) = %8.5f  uxxa(i) = %8.5f
                 err(i) = %8.5f\n',i,uxx(i),uxxa(i),err(i));
    end
    end
%
% Next order
    end
%
% Next differentiator
    end
```

Listing A.2.3. Program uxx_poly_1 for the numerical second derivative of polynomials

We can note the following points about this program:

1. As with the two preceding programs, two `for` loops cycle through FD routines for second derivatives, `ndss = 42` to `50` corresponding to `dss042` to `dss050`, for polynomials of order 1–12. In this case, 12 grid points are used because the tenth-order FDs of `dss050` use 12 grid points.

```
%
    clear all
    clc
%
% Varying order of differentiator
    for ndss=42:2:50
%
% Varying order of test polynomial
    for norder=1:12
%
% Grid in x
    xl=0.0;
    xu=1.0;
    n=12;
    dx=(xu-xl)/(n-1);
```

2. The polynomial test function in u is differentiated by the five FD routines for Dirichlet BCs (`nl = nu = 1`). The call to `zeros` is used to assign values to the first derivative `ux` that is an input argument to the FD routines. In the case of Dirichlet BCs, this first derivative is not actually used by the FD routines, but Matlab requires that all input arguments for a function must have at least one assigned value.

```
%
% Test polynomial
  for i=1:n
    x(i)=xl+(i-1)*dx;
    u(i)=x(i)^norder;
  end
%
% uxx, Dirichlet BC
  nl=1;
  nu=1;
  ux=zeros(1,n);
  if ndss==42  uxx=dss042(xl,xu,n,u,ux,nl,nu); end
  if ndss==44  uxx=dss044(xl,xu,n,u,ux,nl,nu); end
  if ndss==46  uxx=dss046(xl,xu,n,u,ux,nl,nu); end
  if ndss==48  uxx=dss048(xl,xu,n,u,ux,nl,nu); end
  if ndss==50  uxx=dss050(xl,xu,n,u,ux,nl,nu); end
```

3. The difference between the analytical and numerical derivatives is computed and the numerical output is displayed.

```
%
% Comparison of numerical and analytical uxx
  if(norder-(ndss-40))>-1 & (norder-(ndss-40))< 3
  fprintf('\n\n ndss = %2d    nl = 1    nu = 1
           order = %2d\n\n',ndss,norder);
  for i=1:n
    uxxa(i)=norder*(norder-1)*x(i)^(norder-2);
    err(i)=uxx(i)-uxxa(i);
    fprintf(' i = %3d  uxx(i) = %8.5f  uxxa(i) = %8.5f
             err(i) = %8.5f\n',i,uxx(i),uxxa(i),err(i));
  end
  end
```

The output from this test program has the same essential properties as in the case of the first derivative: (a) the pth-order FD is exact for pth-order polynomials, (b) the error in the $(p+1)$th-order numerical derivative is constant (except for the multiplying constant as in Eqs. (A.2.2a) and (A.2.2b) that increases near the boundaries in x), and (c) the error in the $(p+2)$th numerical derivative is not constant. Therefore, because of the similarity with the output in Table A.2.1, we will not list this output here.

A program for the numerical second derivative with Neumann BCs can be constructed directly from the program of Listing A.2.3. The only required difference is the programming of the BCs.

```
%
% uxx, Neumann BC
  nl=2;
  nu=2;
  ux(1)=norder*x(1)^(norder-1);
  ux(n)=norder*x(n)^(norder-1);
  if ndss==42  uxx=dss042(xl,xu,n,u,ux,nl,nu); end
  if ndss==44  uxx=dss044(xl,xu,n,u,ux,nl,nu); end
  if ndss==46  uxx=dss046(xl,xu,n,u,ux,nl,nu); end
  if ndss==48  uxx=dss048(xl,xu,n,u,ux,nl,nu); end
  if ndss==50  uxx=dss050(xl,xu,n,u,ux,nl,nu); end
```

Here we have used the analytical derivatives of the polynomial at the boundaries, `ux(1), ux(n)`. Again the output has the same essential properties as before and is therefore not listed here.

Test programs for the numerical second derivative of e^{ax} for Dirichlet and Neumann BCs can easily be constructed from the preceding programs (they are not discussed here to save some space). The general conclusions for the output are the same as before: (a) the FDs are not exact, (b) the error decreases with increasing order of the FDs, and (c) the error increases with increasing curvature of e^{ax} (from larger a).

In summary, our intention in presenting this appendix is to demonstrate the order conditions for the FDs in the DSS routines through application of the routines to some test problems. Basically, we have considered *p-refinement* (for the improvement of numerical derivative accuracy by increasing the order p of the FD approximations). In general, we could also improve the accuracy of the numerical derivatives by increasing the number of grid points. This approach is suggested by Eq. (A.2.2b); if the number of grid points is increased, the grid spacing Δx is decreased (and thus, the truncation error is decreased). Since the grid spacing is frequently given the symbol h in the numerical analysis literature, this procedure of changing the grid spacing is termed *h-refinement*.

Finally, the accuracy of the numerical derivatives is dependent on the properties of the function that is differentiated. Polynomials can be differentiated exactly by FDs. Nonpolynomial functions generally cannot (and FDs are not recommended for discontinuous functions).

APPENDIX 3

Analytical Solution of Nonlinear, Traveling Wave Partial Differential Equations

Partial differential equations (PDEs) that are nonlinear and have traveling wave solutions (to be explained) are of wide interest for their mathematical properties and their application to a broad spectrum of physical systems. We have used the traveling wave analytical solution of the Korteweg-de Vries (KdV) equation [1] in Chapter 7. Here we consider the analytical solution of a class of nonlinear, traveling wave PDEs whose analytical solutions can provide useful test problems for numerical PDE methods.

The starting point for the analysis is the following nonlinear, third-order PDE, which is a minor extension of the KdV equation with a coefficient a multiplying the *nonlinear convective* term, uu_x, and a coefficient b multiplying the *linear, dispersive* term, u_{xxx} (here we use subscript notation for partial derivatives, e.g., $\partial u/\partial t = u_t$).

$$u_t + auu_x + bu_{xxx} = 0 \tag{A.3.1}$$

The analytical method that follows is an extension of the method given by Knobel [2]. We look for a *traveling wave solution* of the form $u(x, t) = f(z)$, $z = x - ct$, where $c > 0$, and $f(z)$, $f'(z)$, and $f''(z)$ tend to 0 as $z \to \pm\infty$ ($'$ denotes differentiation with respect to z).

Substituting $u(x, t) = f(x - ct)$ into Eq. (A.3.1) gives a third-order nonlinear ordinary differential equation (ODE) for $f(z)$:

$$-cf' + aff' + bf''' = 0 \tag{A.3.2}$$

In arriving at Eq. (A.3.2), we have used

$$\frac{\partial z}{\partial t} = \frac{\partial(x - ct)}{\partial t} = -c$$

$$\frac{\partial z}{\partial x} = \frac{\partial(x - ct)}{\partial t} = 1$$

$$u_t = \frac{\partial f(x - ct)}{\partial t} = \frac{df(z)}{dz}\frac{\partial z}{\partial t} = -cf'$$

$$u_x = \frac{\partial f(x-ct)}{\partial x} = \frac{df(z)}{dz}\frac{\partial z}{\partial x} = f'$$

$$u_{xx} = \frac{\partial^2 f(x-ct)}{\partial x^2} = \frac{\partial\left[\dfrac{\partial f(x-ct)}{\partial x}\right]}{\partial x} = \frac{df'}{dz}\frac{\partial z}{\partial x} = f''$$

$$u_{xxx} = \frac{\partial^3 f(x-ct)}{\partial x^3} = \frac{\partial\left[\dfrac{\partial^2 f(x-ct)}{\partial x^2}\right]}{\partial x} = \frac{df''}{dz}\frac{\partial z}{\partial x} = f'''$$

Equation (A.3.2) can be integrated once to give

$$-cf + \frac{1}{2}af^2 + bf'' = c_1 \tag{A.3.3}$$

where c_1 is a constant of integration. From the assumption that $f(z)$ and $f'' \rightarrow 0$ as $z \rightarrow \infty$, the value of c_1 is zero.

Multiplying Eq. (A.3.3) by f'

$$-cff' + \frac{1}{2}af^2f' + bf'f'' = 0 \tag{A.3.4}$$

and integrating Eq. (A.3.4) results in the first-order equation

$$-\frac{1}{2}cf^2 + \frac{1}{6}af^3 + \frac{1}{2}b(f')^2 = c_2 \tag{A.3.5}$$

Since $f(z), f'(z) \rightarrow 0$ as $z \rightarrow \infty$, the constant of integration c_2 is zero. Solving for $(f')^2$ gives

$$3b(f')^2 = (3c - af)f^2 \tag{A.3.6}$$

From here we will require $0 < f(z) < 3c/a$ in order to have a positive RHS in Eq. (A.3.6). Taking the positive square root of Eq. (A.3.6) gives

$$\frac{\sqrt{3b}}{\sqrt{3c - af}\,f}f' = 1 \tag{A.3.7}$$

To integrate the LHS of Eq. (A.3.7), we make the rationalizing substitution $g^2 = 3c - af$; substituting $f = 3c/a - g^2/a$ and $f' = -(2/a)gg'$ gives

$$\frac{\sqrt{3b}}{g\,(3c/a - g^2/a)}[-(2/a)gg'] = 1$$

or

$$\frac{2\sqrt{3b}}{3c - g^2}g' = -1 \tag{A.3.8}$$

By the method of partial fractions, we have

$$\frac{1}{3c - g^2} = \frac{A}{\sqrt{3c} - g} + \frac{B}{\sqrt{3c} + g} = \frac{A(\sqrt{3c} + g) + B(\sqrt{3c} - g)}{3c - g^2}$$

and therefore (equating numerators)

$$A(\sqrt{3c}+g)+B(\sqrt{3c}-g)=1$$

or

$$A=B=\frac{1}{2\sqrt{3c}}$$

Then, from Eq. (A.3.8),

$$\frac{2\sqrt{3b}}{3c-g^2}g'=\frac{\sqrt{b/c}}{\sqrt{3c}-g}g'+\frac{\sqrt{b/c}}{\sqrt{3c}+g}g'=-1 \tag{A.3.9}$$

Integration of both sides of Eq. (A.3.9) with respect to z gives

$$\ln(\sqrt{3c}+g)-\ln(\sqrt{3c}-g)=-\sqrt{c/b}\,z+d$$

or

$$\ln\left(\frac{\sqrt{3c}+g}{\sqrt{3c}-g}\right)=-\sqrt{c/b}\,z+d \tag{A.3.10}$$

with a constant of integration d. Solving Eq. (A.3.10) for g, we have

$$\frac{\sqrt{3c}+g}{\sqrt{3c}-g}=e^{-\sqrt{c/b}z+d}$$

$$\sqrt{3c}+g=(\sqrt{3c}-g)e^{-\sqrt{c/b}z+d}$$

$$g(1+e^{-\sqrt{c/b}z+d})=\sqrt{3c}(e^{-\sqrt{c/b}z+d}-1)$$

$$g(z)=\sqrt{3c}\frac{\exp(-\sqrt{c/b}z+d)-1}{\exp(-\sqrt{c/b}z+d)+1} \tag{A.3.11}$$

Since

$$\tanh(x)=\frac{\sinh(x)}{\cosh(x)}=\frac{\frac{e^x-e^{-x}}{2}}{\frac{e^x+e^{-x}}{2}}=\frac{e^0-e^{-2x}}{e^0+e^{-2x}}=-\frac{e^{-2x}-1}{e^{-2x}+1}$$

Eq. (A.3.11) can be written as

$$g(z)=\sqrt{3c}\frac{\exp(-\sqrt{c/b}z+d)-1}{\exp(-\sqrt{c/b}z+d)+1}=-\sqrt{3c}\,\tanh\left[\frac{1}{2}(\sqrt{c/b}z-d)\right] \tag{A.3.12}$$

Then from $f=3c/a-g^2/a$, we obtain

$$af=3c-\left\{-\sqrt{3c}\,\tanh\left[\frac{1}{2}(\sqrt{c/b}z-d)\right]\right\}^2$$

$$af=3c-3c\,\tanh^2\left[\frac{1}{2}(\sqrt{c/b}z-d)\right]$$

$$af = 3c\frac{\cosh^2\left[\frac{1}{2}(\sqrt{c/b}z - d)\right] - \sinh^2\left[\frac{1}{2}(\sqrt{c/b}z - d)\right]}{\cosh^2\left[\frac{1}{2}(\sqrt{c/b}z - d)\right]} \tag{A.3.13}$$

Since

$$\cosh^2(x) - \sinh^2(x) = \left(\frac{e^x + e^{-x}}{2}\right)^2 - \left(\frac{e^x - e^{-x}}{2}\right)^2$$

$$= \frac{e^{2x} + 2e^x e^{-x} + e^{-2x} - e^{2x} + 2e^x e^{-x} - e^{-2x}}{4} = 1$$

Eq. (A.3.13) reduces to

$$f(z) = \frac{3c}{a}\operatorname{sech}^2\left[\frac{1}{2}(\sqrt{c/b}z - d)\right] \tag{A.3.14}$$

Since the arbitrary constant d is simply a shift of the sech, it does not have a substantial effect on the solution and we therefore take $d = 0$.

$$u(x, t) = f(z) = \frac{3c}{a}\operatorname{sech}^2\left[\frac{1}{2}\sqrt{\frac{c}{b}}z\right] = \frac{3c}{a}\operatorname{sech}^2\left[\frac{1}{2}\sqrt{\frac{c}{b}}(x - ct)\right] \tag{A.3.15}$$

We can get an idea of what this traveling wave looks like as follows. The traveling wave solution of the KdV equation (see also Chapter 7) is

$$u(x, t) = f(z) = 3c\operatorname{sech}^2\left[\frac{\sqrt{c}}{2}(x - ct)\right] \tag{A.3.16}$$

A comparison of Eqs. (A.3.15) and (A.3.16) indicates the following.

1. The amplitude of the sech soliton of Eq. (A.3.16) is changed from $3c$ to $3c/a$ in Eq. (A.3.15) (thus, the soliton of Eq. (A.3.15) can be scaled vertically).
2. The argument of the soliton of Eq. (A.3.16) is changed from $(\sqrt{c}/2)(x - ct)$ to $(1/2)\sqrt{c/b}(x - ct)$ in Eq. (A.3.15) (thus, the soliton of Eq. (A.3.15) can be made "sharper" using $b < 1$).

This method of solution for Eq. (A.3.1) can be readily extended to other PDEs. For example, we can consider PDEs with mixed partials; the analytical solutions (derived next) provide tests of numerical solutions for this important class of PDEs (with mixed partials).

To start,

$$u_t + auu_x - bu_{xxt} = 0 \tag{A.3.17}$$

is the so-called *equal-width* equation. Again, we look for a traveling wave solution $u(x, t) = f(x - ct)$. Following the previous approach, we arrive at the ODE

$$-cf' + aff' + df''' = 0 \tag{A.3.18}$$

with $d = bc$. Note that this equation is essentially the same as Eq. (A.3.2), so the solution follows immediately from Eq. (A.3.15):

$$u(x,t) = f(z) = \frac{3c}{a}\,\text{sech}^2\left[\frac{\sqrt{1/b}}{2}(x-ct)\right] \tag{A.3.19}$$

Equation (A.3.19) indicates that the *width* of the soliton is independent of c (thus, the name "equal-width" equation). Also, the *Benjamin-Bona-Mahony* equation ([3], p. 583) is a special case of Eq. (A.3.17) for $a = -1$.

For the equation

$$u_t + auu_x + bu_{xtt} = 0 \tag{A.3.20}$$

the solution follows immediately from Eq. (A.3.15)

$$u(x,t) = f(z) = \frac{3c}{a}\text{sech}^2\left[\frac{\sqrt{1/(bc)}}{2}(x-ct)\right] \tag{A.3.21}$$

so that the *width* now increases with increasing c.

For the equation

$$u_t + auu_x - bu_{ttt} = 0 \tag{A.3.22}$$

the solution follows immediately from Eq. (A.3.15)

$$u(x,t) = f(z) = \frac{3c}{a}\text{sech}^2\left[\frac{\sqrt{1/(bc^2)}}{2}(x-ct)\right] \tag{A.3.23}$$

so that again the *width* increases with increasing c. Equations (A.3.19), (A.3.21), and (A.3.23) could be used to test numerical solutions of Eqs. (A.3.17), (A.3.20), and (A.3.22), respectively.

The following variation of Eq. (A.3.17) also has a traveling wave solution ([3], pp. 583–584):

$$u_t + au^k u_x - bu_{xxt} = 0 \tag{A.3.24}$$

We could also consider equations in which x and t are interchanged, for example,

$$u_x + auu_t + bu_{ttt} = 0 \tag{A.3.25}$$

Development of traveling wave solutions based on minor variations of Eq. (A.3.2) and its solution, Eq. (A.3.15), would be possible. Many other PDEs with traveling wave solutions are discussed in [3].

In summary, this discussion of a method for deriving analytical, traveling wave, PDE solutions is intended to illustrate a rather general approach in terms of some specific example PDEs. The essential features of this method are

1. A traveling wave solution of the form $u(x,t) = f(x-ct)$ is assumed that is then substituted in the PDE.
2. A (generally nonlinear) ODE for $f(z)$ results from this substitution that is then integrated with respect to the composite independent variable $z = x - ct$.

3. In the process of integrating the ODE, the assumption that $f(z)$ and its derivatives tend to zero as $|z| \to \infty$ is employed.
4. Basically, this assumption requires that the initial condition for the PDE, $u(x, t = 0) = f(x)$, is nonzero only over a finite interval in x such that the solution and its derivatives remain at zero for large $|z|$. Recall, also, that we require $0 < f(z) < 3c/a$ from Eq. (A.3.6).
5. The generality of the method is limited largely by the integration of the ODE in z. This integration is performed analytically, possibly with the assistance of a symbolic (computer) algebra system.

REFERENCES

[1] Korteweg, D. J. and F. de Vries (1895), On the Change of Form of Long Waves Advancing in a Rectangular Canal, and on a New Type of Long Stationary Waves, *Phil. Mag.*, **39**: 422–443.

[2] Knobel, R. (2000), Korteweg–de Vries Equation, In: *An Introduction to the Mathematical Theory of Waves*, American Mathematical Society, Institute for Advanced Study, Providence, RI Chapter 5, pp. 31–35

[3] Polyanin, A. D. and V. F. Zaitsev (2004), *Handbook of Nonlinear Partial Differential Equations*, Chapman & Hall/CRC, Boca Raton, FL

APPENDIX 4

Implementation of Time-Varying Boundary Conditions

We consider here several methods for the implementation of time-varying boundary conditions (BCs). To illustrate the basic problem, consider the one-dimensional (1D) diffusion equation

$$\frac{\partial u}{\partial t} = D\frac{\partial^2 u}{\partial x^2} \tag{A.4.1a}$$

Equation (A.4.1a) is first order in t and second order in x. It therefore requires one initial condition (IC) and two boundary conditions (BCs), which are taken as

$$u(x, t = 0) = 0 \tag{A.4.1b}$$

$$u(x = 0, t) = f(t), \tag{A.4.1c}$$

$$\frac{\partial u(x = 1, t)}{\partial x} = 0 \tag{A.4.1d}$$

Equation (A.4.1c) is the time-varying Dirichlet BC; $f(t)$ is a function to be specified.

In considering this problem, it would seem that the implementation of BC (A.4.1c) should be straightforward. For example, in the ordinary differential equation (ODE) routine, since t is an input argument, a statement to implement Eq. (A.4.1c) might be

```
u(1)=f(t);
```

However, most of the Matlab ODE integrators, for example, `ode15s`, do not accept the setting of the dependent variables (such as `u(1)`). Rather, if the dependent variables are defined as in Eq. (A.4.1c), they must be programmed either through a corresponding ODE or by explicit programming of the dependent variables outside of the ODE routine.

These alternatives are illustrated with the Matlab routines given in Listing A.4.1, in which $f(t) = 1 - \cos(a\pi t)$ in Eq. (A.4.1c). First, a main program for Eqs. (A.4.1a)–(A.4.1d) follows:

```
%
% Clear previous files
  clear all
  clc
%
% Parameters shared with the ODE routine
  global a D n xl xu dx ncall ndss
%
% Parameters
  a=1.0;
  D=0.3;
  xl=0.0;
  xu=1.0;
%
% Cases with the time varying Dirichlet boundary condition
% (TVDBC) in or outside the spatial grid
  mf=5;
%
% mf = 1,2,3; TVDBC not part of the spatial grid
%
%   mf = 1 - explicit programming of the FDs with boundary
%            value calculated for the ODE at x = dx in pde_1.m
%
%   mf = 2 - ux routines dss002 to dss010 with boundary value
%            calculated for MOL ODEs in pde_2.m
%
%   mf = 3 - uxx routines dss042 to dss050 with boundary value
%            calculated for MOL ODEs in pde_3.m
%
%   For mf = 1,2,3, the boundary value must be calculated
%   after the return from the ODE integrator if it is to be
%   displayed and/or plotted.
    if(mf<=3)
      n=20;
      dx=(xu-xl)/n;
    end
%
% mf = 4,5,6,7; TVDBC part of the spatial grid
%
%   mf = 4 - explicit programming of the FDs with boundary
%            value calculated at boundary in pde_4.m; this
%            boundary value is not returned by the ODE
```

```
%            integrator and it must therefore be restored
%            after the return from the integrator if it is
%            to be displayed and/or plotted
%
%   mf = 5 - explicit programming of the FDs with boundary
%            value calculated and derivative of boundary value
%            also calculated (for ODE at boundary) in pde_5.m;
%            the ODE integrator returns the correct boundary
%            value
%
%   mf = 6 - ux routines dss002 to dss010 with boundary value
%            calculated for MOL ODEs in pde_6.m; the ODE
%            integrator returns the correct boundary value
%
%   mf = 7 - uxx routines dss042 to dss050 with boundary value
%            calculated for MOL ODEs in pde_7.m; the ODE
%            integrator returns the correct boundary value
    if(mf>3)
      n=21;
      dx=(xu-xl)/(n-1);
    end
%
% Spatial grid
  x=[xl:dx:xu]';
%
% Initial condition
  u0=zeros(n,1);
%
% Independent variable for ODE integration
  t0=0.0;
  tf=2.5;
  tout=[0.0:0.25:tf]';
  nout=11;
  ncall=0;
%
% ODE integration
  reltol=1.0e-04; abstol=1.0e-04;
  options=odeset('RelTol',reltol,'AbsTol',abstol);
  if(mf==1) % explicit FDs
    [t,u]=ode15s(@pde_1,tout,u0,options); end
  if(mf==2) ndss=4; % ndss = 2, 4, 6, 8 or 10 required
    [t,u]=ode15s(@pde_2,tout,u0,options); end
  if(mf==3) ndss=44; % ndss = 42, 44, 46, 48 or 50 required
    [t,u]=ode15s(@pde_3,tout,u0,options); end
  if(mf==4) % explicit FDs
    [t,u]=ode15s(@pde_4,tout,u0,options); end
  if(mf==5) % explicit FDs
```

```
    [t,u]=ode15s(@pde_5,tout,u0,options); end
  if(mf==6) ndss=4; % ndss = 2, 4, 6, 8 or 10 required
    [t,u]=ode15s(@pde_6,tout,u0,options); end
  if(mf==7) ndss=44; % ndss = 42, 44, 46, 48 or 50 required
    [t,u]=ode15s(@pde_7,tout,u0,options); end
%
% Display selected output
  if(mf==1)
    n2=n/2;
    fprintf('\n             mf = %2d   n = %3d',mf,n);
  elseif(mf==2|mf==3)
    n2=n/2;
    fprintf('\n      mf = %2d   n = %3d   ndss = %2d', ...
            mf,n,ndss);
  elseif(mf==4|mf==5)
    n2=(n-1)/2+1;
    fprintf('\n             mf = %2d   n = %3d',mf,n);
  elseif(mf==6|mf==7)
    n2=(n-1)/2+1;
    fprintf('\n       mf = %2d    n = %3d   ndss = %2d', ...
            mf,n,ndss);
  end
  fprintf('\n   abstol = %8.1e    reltol = %8.1e', ...
          abstol,reltol);
  if(mf<=3)
    fprintf('\n\n    t   u(x=dx,t)  u(x=0.5,t)   u(x=1,t)\n');
  else
    fprintf('\n\n    t    u(x=0,t)  u(x=0.5,t)   u(x=1,t)\n');
  end
  for it=1:nout
    if(mf==4)u(it,1)=1-cos(a*pi*t(it));end
    fprintf('%6.2f%12.6f%12.6f%12.6f\n',t(it),u(it,1), ...
            u(it,n2),u(it,n));
  end
  fprintf('\n ncall = %4d\n',ncall);
%
% Plot numerical solution as u(x,t) vs x with t as parameter
  if(mf<=3)
    uplot(:,2:n+1)=u(:,1:n);
%
%   Include boundary value in plot
    for it=1:nout
      uplot(it,1)=1-cos(a*pi*t(it));
    end
  elseif(mf>3)
    uplot=u;
%
```

```
%   Restore boundary value for plot (mf = 4)
    if(mf==4)
      for it=1:nout
        uplot(it,1)=u(it,1);
      end
    end
  end
  figure(1);
  plot(x,uplot,'-k'); axis tight
  title('u(x,t) vs x, t = 0,0.25,...,2.5');
  xlabel('x');
  ylabel('u(x,t)')
  figure(2);
  surf(x,tout,uplot);
  shading interp; axis tight
  colormap jet
  title('u(x,t) vs x, t = 0,0.25,...,2.5');
  xlabel('x');
  ylabel('t');
  zlabel('u(x,t)');
  rotate3d on
% print -deps pde.eps; print -dps pde.ps
```

Listing A.4.1. Main program pde_1_main for the solution of Eqs. (A.4.1a)–(A.4.1d)

We can note the following points about this program:

1. A *global* area is set up to share parameters between this main program and the ODE routines. The parameters pertaining to BCs (A.4.1c) and (A.4.1d) are then defined.

```
%
% Clear previous files
  clear all
  clc
%
% Parameters shared with the ODE routine
  global a D n xl xu dx ncall ndss
%
% Parameters
  a=1.0;
  D=0.3;
  xl=0.0;
  xu=1.0;
```

2. As the comments explain, for the first three cases (`mf<=3`), BC (A.4.1c) is implemented at a grid point for $x = 0$ that is *algebraic* (does not involve an ODE), as will be explained when ODE routines `pde_1, pde_2, pde_3` are discussed.

```
%
% Cases with the time varying Dirichlet boundary condition
% (TVDBC) in or outside the spatial grid
  mf=1;
%
% mf = 1,2,3; TVDBC not part of the spatial grid
%
%   mf = 1 - explicit programming of the FDs with boundary
%            value calculated for the ODE at x = dx in pde_1.m
%
%   mf = 2 - ux routines dss002 to dss010 with boundary value
%            calculated for MOL ODEs in pde_2.m
%
%   mf = 3 - uxx routines dss042 to dss050 with boundary value
%            calculated for MOL ODEs in pde_3.m
%
%   For mf = 1,2,3, the boundary value must be calculated
%   after the return from the ODE integrator if it is to be
%   displayed and/or plotted.
    if(mf<=3)
      n=20;
      dx=(xu-xl)/n;
    end
```

Note that 20 grid points in x are used for these three cases.

3. For the next four cases (`mf = 4` to `mf = 7`), BC (A.4.1c) is implemented at a grid point for $x = 0$ that is *differential* (involves an ODE), as will be explained when ODE routines pde_4 to pde_7 are discussed. Note that 21 grid points are now used.

```
%
% mf = 4,5,6,7; TVDBC part of the spatial grid
%
%   mf = 4 - explicit programming of the FDs with boundary
%            value calculated at boundary in pde_4.m; this
%            boundary value is not returned by the ODE
%            integrator and it must therefore be restored
%            after the return from the integrator if it is to
%            be displayed and/or plotted
```

```
%
%   mf = 5 - explicit programming of the FDs with boundary
%            value calculated and derivative of boundary value
%            also calculated (for ODE at boundary) in pde_5.m;
%            the ODE integrator returns the correct boundary
%            value
%
%   mf = 6 - ux routines dss002 to dss010 with boundary value
%            calculated for MOL ODEs in pde_6.m; the ODE
%            integrator returns the correct boundary value
%
%   mf = 7 - uxx routines dss042 to dss050 with boundary value
%            calculated for MOL ODEs in pde_7.m; the ODE
%            integrator returns the correct boundary value
    if(mf>3)
      n=21;
      dx=(xu-xl)/(n-1);
    end
```

The comments largely explain how BC (A.4.1c) is handled within each of the seven ODE routines.

4. A spatial grid in x and IC (A.4.1b) are specified.

```
%
% Spatial grid
  x=[xl:dx:xu]';
%
% Initial condition
  u0=zeros(n,1);
```

5. A timescale $0 \leq t \leq 5$ is set with outputs displayed every 0.5 for a total of 11 output points.

```
%
% Independent variable for ODE integration
  t0=0.0;
  tf=2.5;
  tout=[0.0:0.25:tf]';
  nout=11;
  ncall=0;
%
% ODE integration
  reltol=1.0e-04; abstol=1.0e-04;
```

```
  options=odeset('RelTol',reltol,'AbsTol',abstol);
  if(mf==1) % explicit FDs
    [t,u]=ode15s(@pde_1,tout,u0,options); end
  if(mf==2) ndss=4; % ndss = 2, 4, 6, 8 or 10 required
    [t,u]=ode15s(@pde_2,tout,u0,options); end
  if(mf==3) ndss=44; % ndss = 42, 44, 46, 48 or 50 required
    [t,u]=ode15s(@pde_3,tout,u0,options); end
  if(mf==4) % explicit FDs
    [t,u]=ode15s(@pde_4,tout,u0,options); end
  if(mf==5) % explicit FDs
    [t,u]=ode15s(@pde_5,tout,u0,options); end
  if(mf==6) ndss=4; % ndss = 2, 4, 6, 8 or 10 required
    [t,u]=ode15s(@pde_6,tout,u0,options); end
  if(mf==7) ndss=44; % ndss = 42, 44, 46, 48 or 50 required
    [t,u]=ode15s(@pde_7,tout,u0,options); end
```

Integration of the ODEs for the seven cases is by ode15s.

6. Output is displayed for each of the seven cases at $x = 0.5$ according to what is available and pertains to the case. Finally, the number of calls to the ODE routine, ncall, is displayed.

```
%
% Display selected output
  if(mf==1)
    n2=n/2;
    fprintf('\n                mf = %2d   n = %3d',mf,n);
  elseif(mf==2|mf==3)
    n2=n/2;
    fprintf('\n       mf = %2d   n = %3d   ndss = %2d', ...
            mf,n,ndss);
  elseif(mf==4|mf==5)
    n2=(n-1)/2+1;
    fprintf('\n                mf = %2d   n = %3d',mf,n);
  elseif(mf==6|mf==7)
    n2=(n-1)/2+1;
    fprintf('\n       mf = %2d   n = %3d   ndss = %2d', ...
            mf,n,ndss);
  end
  fprintf('\n   abstol = %8.1e   reltol = %8.1e', ...
          abstol,reltol);
  if(mf<=3)
    fprintf('\n\n    t   u(x=dx,t)  u(x=0.5,t)   u(x=1,t)\n');
  else
    fprintf('\n\n    t    u(x=0,t)  u(x=0.5,t)   u(x=1,t)\n');
  end
```

```
for it=1:nout
  if(mf==4)u(it,1)=1-cos(a*pi*t(it));end
  fprintf('%6.2f%12.6f%12.6f%12.6f\n',t(it),u(it,1), ...
          u(it,n2),u(it,n));
end
fprintf('\n ncall = %4d\n',ncall);
```

7. Plotting of the solution is included according to the particular case.

```
%
% Plot numerical solution as u(x,t) vs x with t as parameter
  if(mf<=3)
    uplot(:,2:n+1)=u(:,1:n);
%
%   Include boundary value in plot
    for it=1:nout
      uplot(it,1)=1-cos(a*pi*t(it));
    end
  elseif(mf>3)
    uplot=u;
%
%   Restore boundary value for plot (mf = 4)
    if(mf==4)
      for it=1:nout
        uplot(it,1)=u(it,1);
      end
    end
  end
  figure(1);
  plot(x,uplot,'-k'); axis tight
  title('u(x,t) vs x, t = 0,0.25,...,2.5');
  xlabel('x');
  ylabel('u(x,t)')
  figure(2);
  surf(x,tout,uplot);
  shading interp; axis tight
  colormap jet
  title('u(x,t) vs x, t = 0,0.25,...,2.5');
  xlabel('x');
  ylabel('t');
  zlabel('u(x,t)');
  rotate3d on
% print -deps pde.eps; print -dps pde.ps
```

Note that the plots include the boundary value at $x = 0$ through the use of the BC function $u(x = 0, t) = f(t) = 1 - \cos(a\pi t)$ according to BC (A.4.1c).

We now consider the seven ODE routines. The first, pde_1, called by ode15s with mf=1, is given in Listing A.4.2.

```
  function ut=pde_1(t,u)
%
% Problem parameters
  global a D n xl xu dx ncall ndss
%
% PDE
  dx2=dx^2;
  for i=1:n
    if(i==1)
      ubc=1-cos(a*pi*t);
      ut(1)=(u(2)-2.0*u(1)+ubc)/dx2;
    elseif(i==n)
      ut(i)=2.0*(u(i-1)-u(i))/dx2;
    else
      ut(i)=(u(i+1)-2.0*u(i)+u(i-1))/dx2;
    end
  end
  ut=D*ut';
%
% Increment calls to pde_1
  ncall=ncall+1;
```

Listing A.4.2. ODE routine pde_1

We can note the following points about pde_1:

1. The function is defined and a global area is included to share parameter values with the main program, pde_1_main.
2. The for loop steps through the $n = 20$ grid points in the method of lines (MOL) solution. For the first grid point at x=dx (i=1), the function $f(t)$ in BC (A.4.1c) is computed and then used in the three-point FD approximation of $\partial^2 u/\partial x^2$

$$\frac{\partial^2 u(x = \Delta x, t)}{\partial x^2} \approx \frac{u(x = 2\Delta x, t) - 2u(x = \Delta x, t) + u(x = 0, t)}{\Delta x^2} \tag{A.4.2}$$

to give the derivative $\partial u/\partial t$ according to Eq. (A.4.1a) (with $a = 1, D = 0.3$).

```
%
% PDE
  dx2=dx^2;
  for i=1:n
    if(i==1)
      ubc=1-cos(a*pi*t);
      ut(1)=(u(2)-2.0*u(1)+ubc)/dx2;
```

Again, note that the first ODE is at $x = dx$, but the required boundary value `ubc` is used, which in effect "drives" the solution at the grid points `i=1` to `20`.

3. The ODE at $x = 1$ includes BC (A.4.1d) approximated (from Eq. (A.4.2)) as

$$\frac{\partial^2 u(x=1,t)}{\partial x^2} \approx 2\frac{u(x=1-\Delta x,t) - u(x=1,t)}{\Delta x^2}$$

where BC (A.4.1d) is approximated as

$$\frac{\partial u(x=1,t)}{\partial x} \approx \frac{u(x=1+\Delta x,t) - u(x=1-\Delta x,t)}{2\Delta x} = 0$$

or

$$u(x=1+\Delta x,t) = u(x=1-\Delta x,t)$$

```
    elseif(i==n)
      ut(i)=2.0*(u(i-1)-u(i))/dx2;
```

4. At the interior points (`i=2` to `i=n-1`), $\partial^2 u/\partial x^2$ is approximated by Eq. (A.4.2) and $\partial u/\partial t$ is coded as

```
    else
      ut(i)=(u(i+1)-2.0*u(i)+u(i-1))/dx2;
    end
  end
```

5. A transpose is required for `ode15s` and the counter for the calls to `pde_1` is incremented.

```
  ut=D*ut';
%
% Increment calls to pde_1
  ncall=ncall+1;
```

Table A.4.1. Output from pde_1_main and pde_1 (mf=1)

```
        mf =  1    n =  20
abstol = 1.0e-004   reltol = 1.0e-004

   t    u(x=dx,t)  u(x=0.5,t)    u(x=1,t)
 0.00    0.000000    0.000000    0.000000
 0.25    0.222812    0.012251    0.000372
 0.50    0.834292    0.127294    0.018234
 0.75    1.499688    0.380408    0.105870
 1.00    1.838224    0.691448    0.290116
 1.25    1.657580    0.930678    0.533912
 1.50    1.068480    1.000662    0.754220
 1.75    0.419962    0.895990    0.871942
 2.00    0.095138    0.707407    0.859815
 2.25    0.287020    0.569835    0.759544
 2.50    0.885440    0.584187    0.658633

ncall =  126
```

The output from pde_1_main (Listing A.4.1) and pde_1 (Listing A.4.2) (for mf=1) is given in Table A.4.1. Note that the first column of u(x,t) values corresponds to x=dx and therefore does not correspond to BC (A.4.1c) $u(x = 0, t) = f(t) = 1 - \cos(a\pi t)$. The values of this BC are (for selected values of t from Table A.4.1 with $a = 1$) as follows:

t	$1 - \cos(a\pi t)$
0	0
0.5	1
1	2
1.5	1
2	0
2.5	1

These values of $u(x = 0, t)$ were added in pde_1_main of Listing A.4.1 so that the plot shown in Figure A.4.1 includes these boundary values.

The interpretation of the line plot of Figure A.4.1 is not obvious (note the $u(x = 0, t)$ values) and is facilitated by the 3D plot of Figure A.4.2. In order to gain satisfactory resolution of the 3D plot, additional output values for $t = 0, 0.05, \ldots, 2.5$ were specified with the following code (used in pde_1_main):

```
%
% Independent variable for ODE integration
  t0=0.0;
  tf=2.5;
  tout=[0.0:0.05:tf]';
  nout=51;
```

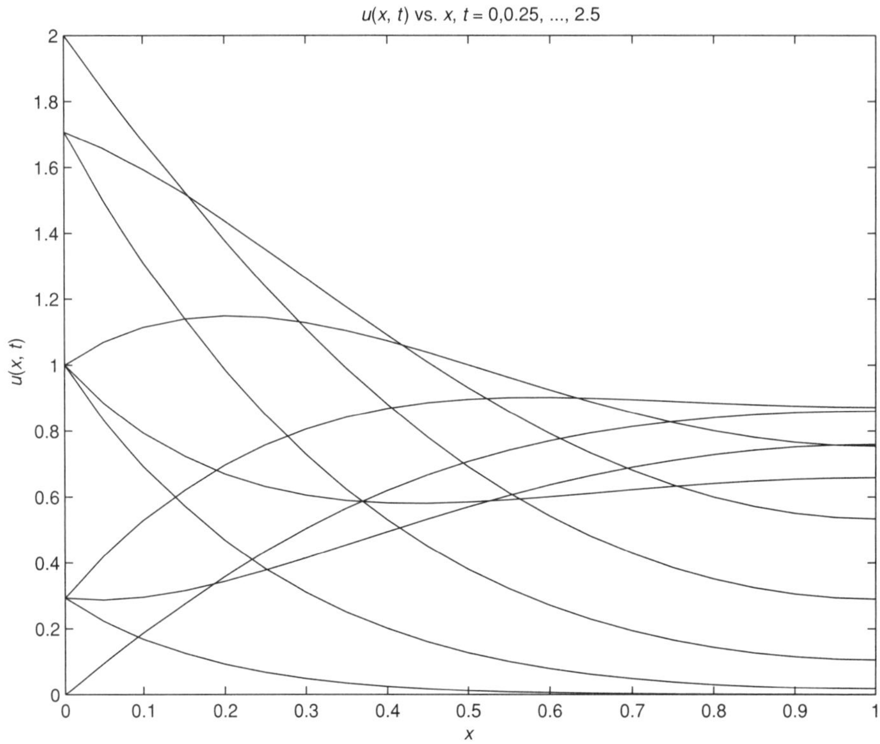

Figure A.4.1. Solution of Eqs. (A.4.1a)–(A.4.1d) from `pde_1_main` and `pde_1` (`mf=1`)

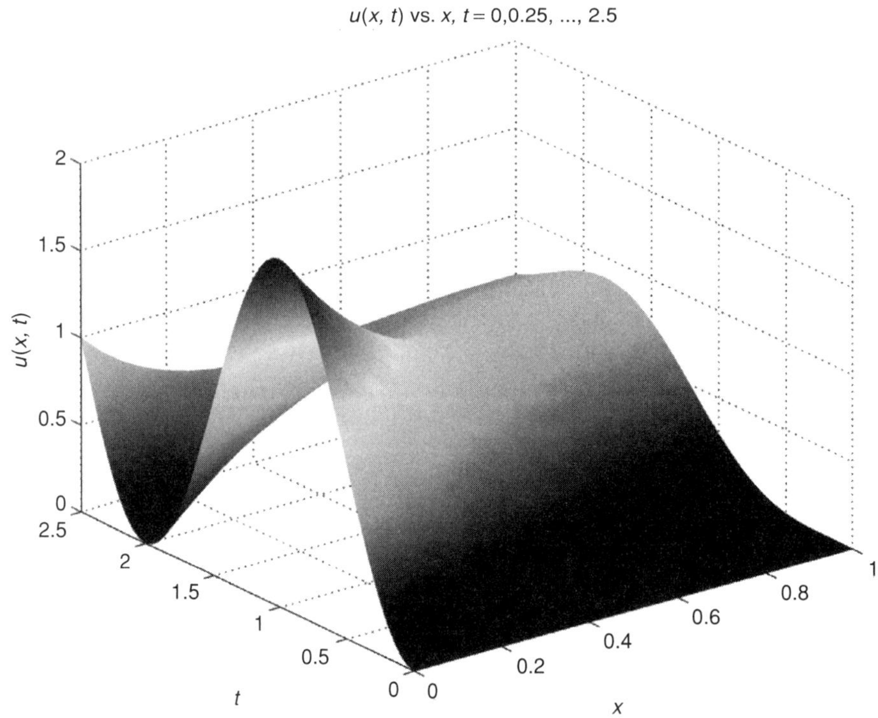

Figure A.4.2. Solution of Eqs. (A.4.1a)–(A.4.1d) from `pde_1_main` and `pde_1` (`mf=1`)

We note from Figure A.4.1

1. BC (A.4.1c) at $x = 0$
2. BC (A.4.1d) (zero slope) at $x = 1$

An analytical solution could easily be derived to test the numerical solution, but we did not include this test to save some space. The next ODE routine, pde_2, called by ode15s with mf=2, is given in Listing A.4.3.

```
  function ut=pde_2(t,u)
%
% Problem parameters
  global a D n xl xu ncall ndss
%
% Calculate ux
  ue(2:n+1)=u;
  ue(1)=1-cos(a*pi*t);
  if     (ndss== 2) uex=dss002(xl,xu,n+1,ue); % second order
  elseif(ndss== 4) uex=dss004(xl,xu,n+1,ue); % fourth order
  elseif(ndss== 6) uex=dss006(xl,xu,n+1,ue); % sixth order
  elseif(ndss== 8) uex=dss008(xl,xu,n+1,ue); % eighth order
  elseif(ndss==10) uex=dss010(xl,xu,n+1,ue); % tenth order
  end
%
% BC at x = 1 (Neumann)
  uex(n+1)=0.0;
%
% Calculate uxx
  if     (ndss== 2) uexx=dss002(xl,xu,n+1,uex); % second order
  elseif(ndss== 4) uexx=dss004(xl,xu,n+1,uex); % fourth order
  elseif(ndss== 6) uexx=dss006(xl,xu,n+1,uex); % sixth order
  elseif(ndss== 8) uexx=dss008(xl,xu,n+1,uex); % eighth order
  elseif(ndss==10) uexx=dss010(xl,xu,n+1,uex); % tenth order
  end
%
% PDE
  uxx=uexx(2:n+1);
  ut=D*uxx';
%
% Increment calls to pde_2
  ncall=ncall+1;
```

Listing A.4.3. ODE routine pde_2

We can note the following points about pde_2:

1. After the definition of the function and inclusion of a global area, an additional grid point (corresponding to $x = 0$) is added and the BC function of Eq. (A.4.1c) is added. This is done so that the differentiation routine `dss004` (from `ndss=4` set before `ode15s` is called in `pde_1_main` with `mf=2`) can operate on the expanded grid with dependent variable `ue`.

```
% Calculate ux
  ue(2:n+1)=u;
  ue(1)=1-cos(a*pi*t);
  if     (ndss== 2) uex=dss002(xl,xu,n+1,ue); % second order
  elseif(ndss== 4) uex=dss004(xl,xu,n+1,ue); % fourth order
  elseif(ndss== 6) uex=dss006(xl,xu,n+1,ue); % sixth order
  elseif(ndss== 8) uex=dss008(xl,xu,n+1,ue); % eighth order
  elseif(ndss==10) uex=dss010(xl,xu,n+1,ue); % tenth order
  end
```

2. BC (A.4.1d) is used to reset the $x = 1$ value of `uex`. Then the second derivative in x, in array `uexx`, is computed by differentiating the first derivative (*stagewise differentiation* is used).

```
%
% BC at x = 1 (Neumann)
  uex(n+1)=0.0;
%
% Calculate uxx
  if     (ndss== 2) uexx=dss002(xl,xu,n+1,uex); % second order
  elseif(ndss== 4) uexx=dss004(xl,xu,n+1,uex); % fourth order
  elseif(ndss== 6) uexx=dss006(xl,xu,n+1,uex); % sixth order
  elseif(ndss== 8) uexx=dss008(xl,xu,n+1,uex); % eighth order
  elseif(ndss==10) uexx=dss010(xl,xu,n+1,uex); % tenth order
  end
```

3. Equation (A.4.1a) is then programmed, followed by the transpose (for `ode15s`) and the incrementing of the counter `ncall`. Note that Eq. (A.4.1a) is programmed with $n = 20$ ODEs (not $n = 21$ ODEs).

```
%
% PDE
  uxx=uexx(2:n+1);
  ut=D*uxx';
%
% Increment calls to pde_2
  ncall=ncall+1;
```

The output from pde_1_main and pde_2 is essentially the same as in Table A.4.1 and Figures A.4.1 and A.4.2, so it is not reproduced here.

The next ODE routine, pde_3, called by ode15s with mf=3, is given in Listing A.4.4.

```
  function ut=pde_3(t,u)
%
% Problem parameters
  global a D n xl xu ncall ndss
%
% Calculate ux
  ue(2:n+1)=u;
  ue(1)=1-cos(a*pi*t);
  uex(n+1)=0.0;
  nl=1; % Dirichlet
  nu=2; % Neumann
  if    (ndss==42) uexx=dss042(xl,xu,n+1,ue,uex,nl,nu);
  % second order
  elseif(ndss==44) uexx=dss044(xl,xu,n+1,ue,uex,nl,nu);
  % fourth order
  elseif(ndss==46) uexx=dss046(xl,xu,n+1,ue,uex,nl,nu);
  % sixth order
  elseif(ndss==48) uexx=dss048(xl,xu,n+1,ue,uex,nl,nu);
  % eighth order
  elseif(ndss==50) uexx=dss050(xl,xu,n+1,ue,uex,nl,nu);
  % tenth order
  end
%
% PDE
  uxx=uexx(2:n+1);
  ut=D*uxx';
%
% Increment calls to pde_3
  ncall=ncall+1;
```

Listing A.4.4. ODE routine pde_3

This routine is similar to pde_2 (Listing A.4.3). The only essential difference is the call to dss044 (with mf=3, ndss=44 in pde_1_main rather than two calls to dss004); dss044 computes the second derivative directly from the PDE dependent variable ue. Note that both BC (A.4.1c) and BC (A.4.1d) are included. The output from pde_1_main and pde_3 is essentially the same as in Table A.4.1 and Figures A.4.1 and A.4.2, so it is not reproduced here.

The next ODE routine, pde_4, called by ode15s with mf=4, is given in Listing A.4.5.

```
  function ut=pde_4(t,u)
%
% Problem parameters
  global a D n xl xu dx ncall ndss
%
% PDE
  dx2=dx^2;
  for i=1:n
    if(i==1)
      u(1)=1-cos(a*pi*t);
      ut(1)=0.0;
    elseif(i==n)
      ut(i)=2.0*(u(i-1)-u(i))/dx2;
    else
      ut(i)=(u(i+1)-2.0*u(i)+u(i-1))/dx2;
    end
  end
  ut=D*ut';
%
% Increment calls to pde_4
  ncall=ncall+1;
```

Listing A.4.5. ODE routine pde_4

pde_4 is similar to pde_1 in Listing A.4.1. Here are a few additional points:

1. The essential difference between pde_1 and pde_4 is that the latter uses an ODE grid point at $x = 0$ (rather than an algebraic grid point as in pde_1).
2. Note that the programming of the BC function $f(t)$ in Eq. (A.4.1c) is stored as the dependent variable at the first grid point, u(1)=1-cos(a*pi*t);. While this value of u(1) is used to "drive" the other $n - 1 = 20$ ODEs, it does not actually set u(1) so that it can be returned from pde_1 (just because that is the way ode15s works).
3. In order for u(1) to be available for plotting, it is reset in the main program pde_1_main. Here is the code in pde_1_main.

```
%
%   Restore boundary value for plot (mf = 4)
    if(mf==4)
      for it=1:nout
        uplot(it,1)=u(it,1);
      end
    end
```

4. A similar resetting of u(1) is also included for the tabulated numerical output:

```
  for it=1:nout
    if(mf==4)u(it,1)=1-cos(a*pi*t(it));end
    fprintf('%6.2f%12.6f%12.6f%12.6f\n', ...
            t(it),u(it,1),u(it,n2),u(it,n));
  end
```

The output from pde_1_main and pde_4 is essentially the same as in Table A.4.1 and Figures A.4.1 and A.4.2, so it is not reproduced here.

The next ODE routine, pde_5, called by ode15s with mf=5, is given in Listing A.4.6.

```
  function ut=pde_5(t,u)
%
% Problem parameters
  global a D n xl xu dx ncall ndss
%
% PDE
  dx2=dx^2;
  for i=1:n
    if(i==1)
      u(1)=1-cos(a*pi*t);
      ut(1)=a*pi*sin(a*pi*t)/D;
    elseif(i==n)
      ut(i)=2.0*(u(i-1)-u(i))/dx2;
    else
      ut(i)=(u(i+1)-2.0*u(i)+u(i-1))/dx2;
    end
  end
  ut=D*ut';
%
% Increment calls to pde_5
  ncall=ncall+1;
```

Listing A.4.6. ODE routine pde_5

pde_5 is similar to pde_4 (in Listing A.4.6). The essential difference is that the ODE at $x = 0$ is programmed as the derivative of the function $f(t)$ in BC (A.4.1c).

```
      ut(1)=a*pi*sin(a*pi*t)/D;
```

The division by D is required to maintain the time-varying BC identity $\{du(x = 0, t)\}/dt$ = (ut(1) $= [d\{1 - \cos(a\pi t)\}]/dt = a\pi\sin(a\pi t)$, which is the first element in the vector ut ($= u_t$). This is because, once it is assembled, the whole vector ut is multiplied by D (through ut=D*ut' in the Matlab code) after which ut(1) has the correct value.

This approach illustrates the basic requirement of ode15s that the dependent variables such as u(1) are changed by their corresponding ODEs, and not by algebraic statements such as in pde_4. While this approach may seem completely logical, it has the drawback that the derivative of $f(t)$ is required, and this may not be readily available (depending on the particular $f(t)$ used in an application). The output from pde_1_main and pde_5 is essentially the same as in Table A.4.1 and Figures A.4.1 and A.4.2, so it is not reproduced here.

The next ODE routine, pde_6, called by ode15s with mf=6, is given in Listing A.4.7.

```
  function ut=pde_6(t,u)
%
% Problem parameters
  global a D n xl xu ncall ndss
%
% Calculate ux
  u(1)=1-cos(a*pi*t);
  if     (ndss== 2) ux=dss002(xl,xu,n,u); % second order
  elseif(ndss== 4) ux=dss004(xl,xu,n,u); % fourth order
  elseif(ndss== 6) ux=dss006(xl,xu,n,u); % sixth order
  elseif(ndss== 8) ux=dss008(xl,xu,n,u); % eighth order
  elseif(ndss==10) ux=dss010(xl,xu,n,u); % tenth order
  end
%
% BC at x = 1 (Neumann)
  ux(n)=0.0;
%
% Calculate uxx
  if     (ndss== 2) uxx=dss002(xl,xu,n,ux); % second order
  elseif(ndss== 4) uxx=dss004(xl,xu,n,ux); % fourth order
  elseif(ndss== 6) uxx=dss006(xl,xu,n,ux); % sixth order
  elseif(ndss== 8) uxx=dss008(xl,xu,n,ux); % eighth order
  elseif(ndss==10) uxx=dss010(xl,xu,n,ux); % tenth order
  end
%
% PDE
  ut=D*uxx';
%
% Increment calls to pde_6
  ncall=ncall+1;
```

Listing A.4.7. ODE routine pde_6

pde_6 is similar to pde_2 (in Listing A.4.3). The essential difference is that the grid point at the boundary $x = 0$ is included in the spatial grid (rather than using a separate value outside the grid for $u(x = 0, t)$ as in pde_2).

Again, u(1)=1-cos(a*pi*t); sets the value of u(1), so it "drives" the other $n - 1 = 20$ ODEs, but it does not actually set u(1) so that it can be returned from pde_1. Rather the coding

```
%
% PDE
  ut=uxx';
```

includes the ODE at $x = 0$. In other words, this approach is both algebraic (from u(1)=1-cos(a*pi*);) and differential (from ut=uxx';). The output from pde_1_main and pde_6 is essentially the same as in Table A.4.1 and Figures A.4.1 and A.4.2, so it is not reproduced here.

The final ODE routine, pde_7, called by ode15s with mf=7, is given in Listing A.4.8.

```
  function ut=pde_7(t,u)
%
% Problem parameters
  global a D n xl xu ncall ndss
%
% Calculate ux
  u(1)=1-cos(a*pi*t);
  ux(n)=0.0;
  nl=1; % Dirichlet
  nu=2; % Neumann
  if    (ndss==42) uxx=dss042(xl,xu,n,u,ux,nl,nu);
  % second order
  elseif(ndss==44) uxx=dss044(xl,xu,n,u,ux,nl,nu);
  % fourth order
  elseif(ndss==46) uxx=dss046(xl,xu,n,u,ux,nl,nu);
  % sixth order
  elseif(ndss==48) uxx=dss048(xl,xu,n,u,ux,nl,nu);
  % eighth order
  elseif(ndss==50) uxx=dss050(xl,xu,n,u,ux,nl,nu);
  % tenth order
  end
%
% PDE
  ut=D*uxx';
%
% Increment calls to pde_7
  ncall=ncall+1;
```

Listing A.4.8. ODE routine pde_7

pde_7 is similar to pde_3 (in Listing A.4.4). The essential difference is that the grid point at the boundary $x = 0$ is included in the spatial grid (rather than using a separate value outside the grid for $u(x = 0, t)$ as in pde_3).

Again, u(1)=1-cos(a*pi*t); sets the value of u(1), so it "drives" the other $n - 1 = 20$ ODEs, but it does not actually set u(1) so that it can be returned from pde_1. Rather the coding

```
%
% PDE
  ut=uxx';
```

includes the ODE at $x = 0$. In other words, this approach is both algebraic (from u(1)=1-cos(a*pi*t);) and differential (from ut=uxx';). The output from pde_1_main and pde_7 is essentially the same as in Table A.4.1 and Figures A.4.1 and A.4.2, so it is not reproduced here.

In summary, our intent for this discussion of a time-varying Dirichlet BC is to show several (seven) different ways the same BC can be programmed. The output in all cases was essentially the same. We do not have a preferred method and consider all seven to be about equivalent. The computational requirement for this problem was not significantly different (e.g., ncall = 126 and ncall = 133 for pde_1 and pde_7, respectively). Perhaps the most important point to retain is that the Matlab ODE integrators do not return dependent variables that are set algebraically in the ODE routine. Rather, they must be set via ODEs if they are to be returned by the integrator.

APPENDIX 5

The Differentiation in Space Subroutines Library

FIRST-DERIVATIVE ROUTINES

The Differentiation in Space Subroutines (DSS) library contains five Matlab routines for the numerical calculation of first derivatives. They are as follows:

`dss002()`	3-point stencil, 2nd-order approximation
`dss004()`	5-point stencil, 4th-order approximation
`dss006()`	7-point stencil, 6th-order approximation
`dss008()`	9-point stencil, 8th-order approximation
`dss010()`	11-point stencil, 10th-order approximation

The call format is the same for each routine, and the call definition of `dss004()` is given next by way of example as

```
function [ux]=dss004(xl,xu,n,u)
```

where function `dss004` computes the first derivative, u_x, of a variable $u(x)$ over the spatial domain $x_l \le x \le x_u$ from a classical *five-point, fourth-order finite-difference* approximation. To provide the necessary number of grid points in x, we recommend that the minimum number of grid points used for a particular problem should be equal to the *stencil size plus 10.*

Argument List

Input Variables

`xl` left value of the spatial independent variable, x

`xu` right value of the spatial independent variable, x

n number of spatial grid points in the x domain including the boundary points

u one-dimensional (1D) array of n values of dependent variable u at the n points for which the derivative is to be computed

Output Variables

ux one-dimensional array of the *first derivative* of u at the n grid points

SECOND-DERIVATIVE ROUTINES

The DSS library contains five Matlab routines for the numerical calculation of second derivatives. They are as follows:

`dss042()`	4-point stencil, 2nd-order approximation
`dss044()`	6-point stencil, 4th-order approximation
`dss046()`	8-point stencil, 6th-order approximation
`dss048()`	10-point stencil, 8th-order approximation
`dss050()`	12-point stencil, 10th-order approximation

The call format is the same for each routine, and the call definition of `dss044()` is given next by way of example as

```
function [ux]=dss044(xl,xu,n,u,ux,nl,nu)
```

where function `dss044` computes the second derivative, u_{xx} of a variable $u(x)$ over the spatial domain $x_l \leq x \leq x_u$ from a classical *six-point, fourth-order finite-difference* approximation. To provide the necessary number of grid points in x, we recommend that the minimum number of grid points used for a particular problem should be equal to the *stencil size plus 10*.

Argument List

Input Variables

xl left value of the spatial independent variable, x

xu right value of the spatial independent variable, x

n number of spatial grid points in the x domain including the boundary points

u one-dimensional array of n values of dependent variable u at the n grid points for which the derivative is to be computed

ux one-dimensional array with the *first derivative* of u; the end values of u_x, $u_x(x = x_l)$ and $u_x(x = x_u)$, are used in Dirichlet or Neumann boundary conditions (BCs) at $x = x_l$ and $x = x_u$, depending on the arguments `nl` and `nu`

`nl` integer index for the type of BC at $x = x_l$. The allowable values are

1. Dirichlet BC at $x = x_l$ ($u_x(x = x_l)$ is not used)
2. Neumann BC at $x = x_l$ ($u_x(x = x_l)$ is used)

`nu` integer index for the type of BC at $x = x_u$ (input). The allowable values are

1. Dirichlet BC at $x = x_u$ ($u_x(x = x_u)$ is not used)
2. Neumann BC at $x = x_u$ ($u_x(x = x_u)$ is used)

Output Variables

`uxx` one-dimensional array with the *second derivative* of u at the n grid points

HIGHER-ORDER AND MIXED DERIVATIVES

Higher-order and mixed derivatives can be calculated by calling the appropriate DSS routines in sequence. This is known as *stagewise differentiation*. For instance, a finite-difference approximation to a third derivative can be obtained by three successive calls to the first-derivative library routine `dss004()`. As we are dealing with a third derivative, this would require three BCs that would be determined by the physical nature of the problem.

Note: In general, p boundary conditions (BCs) are required for each pth derivative with respect to a spatial variable and q initial conditions (ICs) are required for each qth derivative with respect to a temporal variable. Thus, for example, the 1D Korteweg–deVries (KdV) equation $\partial u/\partial t = -u(\partial u/\partial x) + \mu(\partial^3 u/\partial x^3)$ would require three spatial BCs and one IC.

This is illustrated by the Matlab code fragment given in Listing A.5.1 for a dependent variable $u(x, t)$ with a Dirichlet BC at the left end of the spatial domain $u(x = x_l, t) = u_{bl}$, a time-varying Neumann BC at the right end of the spatial domain $u_x(x = x_u, t) = f_1(t)$, and a time-varying second derivative BC at the right end of the spatial domain $u_{xx}(x = x_u, t) = f_2(t)$.

```
function [uxxx] = thirdDerivEx1(t,u)
global n ubl xl xu          % Global variables
 .
 .
u(1) = ubl;                 % Dirichlet BC at left end of
                            % domain
ux   = dss004(xl,xu,n,u);   % First derivative of u
ux(n)= f1(t);               % Neumann BC at right end of
                            % domain
uxx  = dss004(xl,xu,n,ux);  % Second derivative of u
```

```
uxx(n)=f2(t);                 % Second derivative BC at right
                              % end of domain
uxxx = dss004(xl,xu,n,uxx);  % Third derivative of u
return;
```

Listing A.5.1. Calculating a third derivative by three successive calls to dss004()

An equally valid approach would be to call the second-derivative function dss044() followed by the first-derivative function dss004(), as illustrated in Listing A.5.2.

```
function [uxxx]=thirdDerivEx2(t,u)
global n ubl xl xu            % Global variables
.
.
u(1) = ubl;                   % Dirichlet BC at left end of
                              % domain
ux(n)= f1(t);                 % Neumann BC at right end of
                              % domain
nl   = 1;                     % Left BC is Dirichlet (do
                              % nothing)
nu   = 2;                     % Right BC is Neumann
uxx  = dss044(xl,xu,n,u,ux,nl,nu); % Second derivative of u
uxx(n)=f2(t);                 % Second derivative BC at right
                              % end of domain
uxxx = dss004(xl,xu,n,uxx);  % Third derivative of u
return;
```

Listing A.5.2. Calculating a third derivative by calls to dss004() and dss044()

In the code fragments given in Listings A.5.1 and A.5.2 both calls are made to fourth-order finite-difference approximations. In formulating the solution to any PDE problem, it is important to select numerical approximations of the same order for all partial derivatives. This will, in general, ensure that truncation errors (for each derivative) vary by the same order of magnitude as the grid spacing is increased or decreased. An example of the use of the DSS library to calculate mixed partial derivatives is given in Chapter 12.

OBTAINING THE DSS LIBRARY

Additional information about the DSS library, including the PDE routines, is available from http://www.scholarpedia.org/article/method_of_lines. Enquiries can be directed to wes1@lehigh.edu.

APPENDIX 6

Animating Simulation Results

GENERAL

The results obtained from running the simulations described in this book are generally provided in the form of a matrix. This matrix includes the solutions for each time step and provides the means of generating a graphical sequence of images that can be used to create an animation. An animation creates the effect of motion and often provides the viewer with a clearer insight into the physical process being analyzed than purely static images. In this appendix we demonstrate some Matlab methods and the corresponding code for creating *AVI movies* and *animated GIF files*. For additional information the reader is referred to general Matlab references [1–3].

MATLAB MOVIE

The creation of a movie is quite straightforward using the built-in tools provided with Matlab. The process consists of the following steps, assuming a type `surfl` plot with fixed viewpoint:

- Define problem mathematically.

```
u=f(x,y);                    % Create initial 2D reference data
```

- Plot reference frame with desired specification – only basic option shown; for additional options refer to Matlab help.

```
surfl(x,y,u);                % Surface plot with lighting
axis([x1 x2 y1 y2 u1 u2]);  % Ensures scaling is fixed
set(gca,'nextplot', ...     % Set where to draw next plot
        'replacechildren');
```

```
colormap cool;                % Select colormap to enhance plot
shading interp;               % Interpolated color shading
view(elevation azimuth)       % Select desired viewpoint
                              % for frames
```

■ Define number of frames say, `N`.

```
myMovie=moviein(N);           % Define frame matrix of size N
```

■ Perform `N` calculations, plotting the result of each calculation and saving as a movie frame.

```
for i=1:N                     % for loop to generate N frames
  .
  .
  perform calculation(s)      % Calculations as necessary
  .
  .
  surfl(x,y,u);               % New surface plot
  myMovie(:,i)=getframe;      % Add new plot to frame matrix
end
```

■ Display movie within Matlab environment.

```
movie(myMovie)                % Play movie
```

Note:

1. If desired, the viewpoint can also be changed during the movie to reflect the changing solution by including an appropriate `view` statement within the `for` loop after the calculations and prior to the `surfl` statement.
2. The Matlab function `movie` also has additional options, for example, to control the number of frames displayed per second; refer to Matlab help for additional information.

BASIC EXAMPLE

The *peaks example* that uses the `surf` function, given in Listing A.6.1 has been taken from the Matlab help. It shows with just a few lines of code how effective this facility can be in providing clarity to a physical or mathematical process.

```
  Z=peaks; surf(Z);
  axis tight
  set(gca,'nextplot','replacechildren')
%
% Record the movie
  for j=1:20
    surf(sin(2*pi*j/20)*Z,Z)
    F(j)=getframe;
  end
%
% Play the movie twenty times
  movie(F,20)
```

Listing A.6.1. *Peaks example* from Matlab help

AVI MOVIES

The *AVI movie* creation facility within Matlab enables the user to create a standalone *Audio/Video Interleaved* (AVI) file that can be viewed in media players available on most personal computers. The process for the generation of an AVI file is a simple extension to creating a Matlab movie, as described earlier. The peaks example code given in Listing A.6.2 has been modified to include an extra line to convert the frames to an AVI movie and save the result to the file `peaks.avi`.

```
  Z=peaks; surf(Z);
  axis tight
  set(gca,'nextplot','replacechildren');
%
% Record the movie
  for j=1:20
    surf(sin(2*pi*j/20)*Z,Z)
    F(j)=getframe;
  end
%
% Play the movie twenty times
  movie(F,20)
  movie2avi(F,'peaks.avi','fps',4) % Save movie to file
                                   % 'peaks.avi' with the
                                   % Frames per second option
                                   % set to 4
```

Listing A.6.2. Peaks example with conversion to *AVI format* and *save to file*

We will now provide AVI movie examples that can be used in conjunction with the code detailed in the *Burgers equation*, *Schrödinger equation*, and *Korteweg–deVries equation* chapters. These examples use the Matlab `avifile` function that is slightly more complex than `movie2avi`, but is more flexible. The differences will be apparent from the coding and more information is available from the Matlab help facility.

EXAMPLE – BURGERS EQUATION MOVIE

The following code generates an animation movie file in AVI format and utilizes Matlab workspace data populated by a previous simulation run; that is, this code should be run following the Burgers equation simulation code detailed in Chapter 5.

The file name to save the movie to is initially set to `BurgersEqn_01.avi`. However, if this file already exists, then the user is prompted to choose to either supply a new file name or overwrite the existing file.

The user is then presented with a dialog box from which to choose:

1. Two-dimensional (2D) plot
2. Two-dimensional plus analytical solution plots
3. Three-dimensional surface plot

The movie-generating process then proceeds and the final frame is saved as an image at a resolution of 300 dpi to file `BurgersEqnPlot.png`. A detailed description of the code will not be provided as ample comments are included (see Listing A.6.3).

```
%
% Code to generate Burgers equation Movie
% Run after pde_1_main.m
% Generates animation movies
% User should select from choices presented in dialog boxes!
%
% Get screen size
  scrsz=get(0,'ScreenSize');
%
% Define required position for figure
  rect=[0 scrsz(4)/2 scrsz(3)/2 scrsz(4)/2];
%
% Create figure at required position
  h1=figure('Position',rect);
  set(h1,'DoubleBuffer','on');
%
% Define required dimensions of image  to be saved within
% figure
  frameRect=[0 0 scrsz(3)/2 scrsz(4)/2];
%
```

```
% Set up animation parameters
  frameSteps=3;                   % Plot image every frameSteps
  nframes=ceil(mm/frameSteps);    % Number of frames in the
                                  % movie
  Frames=moviein(nframes);        % Initialize the matrix
                                  % 'Frames'
  avi_fName='BurgersEqn_01.avi'; % Movie file name
  if exist(avi_fName) ~= 0
    fprintf('File: %s, already exists!\n\n', avi_fName);
    fileMsg=sprintf('File: %s, already exists - Choose to:', ...
                    avi_fName);
    namChoice=menu(fileMsg,'Overwrite','New Name');
    if namChoice==1
    delete(avi_fName);
  else
    newName=inputdlg('Movie file name:','File Name Request', ...
                     1,{avi_fName});
    if isempty(newName{1})
      delete(avi_fName);
    else
      [pathstr,name,ext,versn]=fileparts(newName{1});
      if strcmp('.avi',ext)==0
        avi_fName=strcat(newName{1},'.avi');
      else
        avi_fName=newName{1};
      end
    end
    if exist(avi_fName)
      delete(avi_fName);
    end
  end
 %
 % avi movie parameters
   fps=1;                      % Frames per second
   aviQuality=100;             % Image quality, 0-100 (100 best)
   aviCompression='Cinepak'; % Image compression
                               % mode - see help
   aviMov=avifile(avi_fName,'compression',aviCompression, ...
          'quality', aviQuality);
 %
 % Movie calculations
   plotType=1;        % Set plot type, 1=2D, 2=3D
   plotType=menu('Choose Plot Type:','2D', ...
             '2D+Analytical Solution','3D');
   qq=1;              % Set frame skip counter
   frameCount=0;      % Initialise frame counter
   u1=0.1*ones(mm,n);
```

```
    for it=1:mm
      qq=qq-1;
      if qq==0,
        frameCount=frameCount+1;
        if plotType==1 | plotType==2
%
%       2D Plot
          if plotType==1
            h2=plot(x,u(it,:));
          else
            plot(x,u(it,:),'o',x,u_anal(it,:),'-')
            legend('Numerical solution', ...
                   'Analytical solution', ...
                   'location', 'southwest');
          end
        hold on
        xlabel('x')
        ylabel('u(x,t)')
        set(gca,'XLim',[0 1],'YLim',[0 1]);
        title('Burgers equation');
        grid on
      else
%
%     Update u1 for each frame
      if it <= frameSteps      % First frame
        u1(it,:)=u(it,:);
      else                     % Subsequent frames
        u1(it-frameSteps:it,:)=u(it-frameSteps:it,:);
      end
      surfl(x,t,u1,'light');
      shading interp
      title('Burgers equation');
      set(get(gca,'XLabel'),'String','space, x')
      set(get(gca,'YLabel'),'String','time, t')
      set(get(gca,'ZLabel'),'String',...
                          'dependent variable, u(t,x)')
      set(gca,'XLim',[0 1],'YLim',[0 1],'ZLim', [0 1]);
      axis tight
      view(40,29);
      colormap('cool');   % set color map
      end
      Frames=getframe(h1, frameRect); % Create image in Frames
                                      % array
      aviMov=addframe(aviMov,Frames); % Add frame to avi movie
                                      % file
      disp(['    Plot number: ', num2str(frameCount), ...
            ' created']);
```

```
      qq=frameSteps;  % reset qq
    end
    aviMov=close(aviMov); % Close avi movie file
    fprintf('\navi movie created in file: %s \n', avi_fName);
%
% Print image to file
% print -dpng -r300 BurgersEqnPlot
```

Listing A.6.3. Code to be used following a Burgers equation simulation run to produce a variety of 2D and 3D plots

An example of 3D plot is shown in Chapter 5.

EXAMPLE – SCHRÖDINGER EQUATION MOVIE

The following code generates an animation movie file in AVI format and utilizes Matlab workspace data populated by a previous simulation run; that is, this code should be run following the Schrödinger simulation code detailed in Chapter 6.

The file name to save the movie to is initially set to `SchrodingersEqn_01.avi`. However, if this file already exists, then the user is prompted to choose to either supply a new file name or overwrite the existing file.

The user is then presented with a dialog box from which to choose:

1. Two-dimensional plot
2. Two-dimensional plus analytical solution plots
3. Three-dimensional surface plot

The movie-generating process then proceeds and the final frame is saved as an image at a resolution of 300 dpi to file `SchrodingersEqnPlot.png`. A detailed description of the code will not be provided as ample comments are included (see Listing A.6.4).

```
%
% Code to generate Schrodingers equation movie
% Run after pde_1_main.m
% Generates animation movies
% User should select from choices presented in dialog boxes!
%
  avi_fName='SchrodingerEqn_01.avi'; % Movie file name
  if exist(avi_fName) ~= 0
    fprintf('File: %s, already exists!\n\n', avi_fName);
    fileMsg=sprintf('File: %s, already exists - Choose to:', ...
                    avi_fName);
    namChoice=menu(fileMsg,'Overwrite','New Name');
    if namChoice==1
```

```
      delete(avi_fName);
    else
      newName=inputdlg('Movie file name:', ...
                       'File Name Request',1,{avi_fName});
      if isempty(newName{1})
        delete(avi_fName);
      else
        [pathstr,name,ext,versn]=fileparts(newName{1});
        if strcmp('.avi',ext)==0
          avi_fName=strcat(newName{1},'.avi');
        else
          avi_fName=newName{1};
        end
      end
      if exist(avi_fName)
        delete(avi_fName);
      end
    end
  end
%
% Set Plot Type: 1=2D, 2=2D + Analytic, 3=3D
  plotType=menu([{'Cubic Schrodinger Equation Solution.'}, ...
           {'Choose Plot Type:'}],'2D', ...
            '2D+Analytical Solution','3D');
%
% Get screen size
  scrsz=get(0,'ScreenSize');
%
% Define required position for figure
  rect=[0 scrsz(4)/2 scrsz(3)/2 scrsz(4)/2];
%
% Create figure at required position
  h1=figure('Position',rect);
  set(h1,'DoubleBuffer','on');
%
% Define required dimensions of image  to be saved
% within figure
  frameRect=[0 0 scrsz(3)/2 scrsz(4)/2];
%
% avi movie parameters
  fps=2;                        % Frames per second
  aviQuality=100;               % Image quality, 0-100 (100 best)
  aviCompression='Cinepak';  % Image compression mode - see
                                % help
  aviMov=avifile(avi_fName,'compression',aviCompression, ...
        'quality', aviQuality);
%
```

```
% Movie calculations
% Plot image every frameStep
  if plotType==1 | plotType==2
    frameStep=10;
  else
    frameStep=10;
  end
  qq=frameStep;                % Set frame skip counter
%
% Set up animation parameters
  [mm nn]=size(v2w2);
  nframes=ceil(mm/frameStep); % Number of frames in the movie
  Frames=moviein(nframes);    % Initialize the matrix 'Frames'
  frameCount=0;               % Initialise frame counter
  u1=zeros(mm,n);
  for it=1:mm
    if qq==0
      frameCount=frameCount+1;
      if plotType==1 | plotType==2
%
%       2D Plot
        if plotType==1
          h2=plot(x,v2w2(it,:));
          title('CSE - numerical solution.');
        else
          h2=plot(x,v2w2(it,:),'o',x,v2w2_anal(it,:),'-');
          title('CSE: o - numerical; solid - analytical.');
        end
        xlabel('x')
        ylabel('u(x,t)')
        set(gca,'XLim',[-10 40],'YLim', [0 2]);
        grid on
      else
%
%     Update u1 for each frame
      if it <= frameStep       % First frame
        u1(it,:)=v2w2(it,:);
      else                     % Subsequent frames
        u1(it-frameStep:it,:)=v2w2(it-frameStep:it,:);
      end
      h2=surfl(x,t,u1, 'light');
      shading interp
      title('Cubic Schrodinger Equation');
      set(get(gca,'XLabel'),'String','space, x')
      set(get(gca,'YLabel'),'String','time, t')
      set(get(gca,'ZLabel'),'String','dependent variable,
          u(t,x)')
```

```
        set(gca,'XLim',[-10 40],'YLim',[0 30],'ZLim', [0 2]);
        axis tight
        view(-10,45)
        colormap('cool');   % set color map
      end
      Frames=getframe(h1, frameRect); % Create image in Frames
                                      % array
      aviMov=addframe(aviMov,Frames); % Add frame to avi movie
                                      % file
      fprintf('\n    Plot number: %d created', frameCount);
      qq=frameStep;  % Reset qq
    end
      qq=qq-1;
      aviMov=close(aviMov); % Close avi movie file
      fprintf('\navi movie created in file: %s \n', avi_fName);
%
% Print image to file
  print -dpng -r300 SchrodingersEqnPlot
```

Listing A.6.4. Code to be used following a Schrödinger equation simulation run to produce a variety of 2D and 3D plots

An example 3D plot is shown in Chapter 6.

EXAMPLE – KdV EQUATION MOVIE

The following code generates an animation movie file in AVI format and utilizes Matlab workspace data populated by a previous simulation run; that is, this code should be run following the KdV code detailed in Chapter 7.

Depending on whether the initial simulation run was based on a single- or two-soliton solution, the user is presented with a dialog box from which to choose either

1. Single-soliton solution, `ncase=1`
   (a) Two-dimensional plot
   (b) Two-dimensional plus analytical solution plots
   (c) Three-dimensional surface plot
2. Two-soliton solution, `ncase=2`
   (a) Two-dimensional plot
   (b) Three-dimensional waterfall plot
   (c) Three-dimensional surface plot

The file name to save the movie to is initially set to `KdV_Eqn_01.avi`. However, if this file already exists, then the user is prompted to choose to either supply a new file name or overwrite the existing file. The movie-generating process then proceeds and the final frame is saved as an image at a resolution of 300 dpi to file `KdVEqnPlot.png`. A detailed description of the code will not be provided as ample comments are included (see Listing A.6.5).

```
%
% Code to generate Korteweg de Vries equation movie
% Run after pde_1_main.m
% Generates animation movies
% User should select from choices presented in dialog boxes!
%
% Get screen size
  scrsz=get(0,'ScreenSize');
%
% Define required position for figure
  rect=[0 scrsz(4) scrsz(3)/2 scrsz(4)/3];
%
% Create figure at required position
  h1=figure('Position',rect);
  set(h1,'DoubleBuffer','on');
%
% Define required dimensions of image  to be saved within
% figure
  frameRect=[0 0 scrsz(3)/2 scrsz(4)/3];
  plotType=1;       % Set plot type, 1=2D, 2=3D
  if ncase==1
  plotType=menu([{'KdV Equation - Single soliton solution.'}, ...
           {'Choose Plot Type:'}], ...
           '2D','2D+Analytical Solution','3D');
  else
    plotType=menu([{'KdV Equation - Two soliton solution.'}, ...
             {'Choose Plot Type:'}], ...
             '2D','3D Surface', '3D waterfall');
  end
%
% Set up animation parameters
  [mm nn]=size(u);
  if ncase==2 && plotType==3
    frameSteps=floor(mm/20);
  else
    frameSteps=2;
  end
%
% Plot image every frameSteps
  nframes=ceil(mm/frameSteps);   % Number of frames in the
                                 % movie
  Frames=moviein(nframes);       % Initialize the matrix
                                 % 'Frames'
  avi_fName='KdV_Eqn_01.avi';    % Movie file name
  if exist(avi_fName, 'file') ~= 0
    fprintf('File: %s, already exists!\n\n', avi_fName);
```

```
      fileMsg=sprintf('File: %s, already exists - Choose to:', ...
                      avi_fName);
      namChoice=menu(fileMsg,'Overwrite','New Name');
      if namChoice==1
        delete(avi_fName);
      else
      newName=inputdlg('Movie file name:', ...
                       'File Name Request',1,{avi_fName});
      if isempty(newName{1})
        delete(avi_fName);
      else
        [pathstr,name,ext,versn]=fileparts(newName{1});
        if strcmp('.avi',ext)==0
          avi_fName=strcat(newName{1},'.avi');
        else
          avi_fName=newName{1};
        end
      end
      if exist(avi_fName, 'file')
        delete(avi_fName);
      end
    end
  %
  % avi movie parameters
    fps=1;                        % Frames per second
    aviQuality=100;               % Image quality, 0-100
                                  % (100 best)
  %
  % aviCompression='Cinepak';    % Image compression mode - see
                                  % help
                                  % Options On Windows:
                                  % 'Indeo3'
                                  % 'Indeo5'
                                  % 'Cinepak'
                                  % 'MSVC'
                                  % 'None'
    aviCompression='Indeo5'; % Image compression mode - Large
                             % file!
    aviMov=avifile(avi_fName,'compression',aviCompression, ...
           'quality', aviQuality);
  %
  % Movie calculations
    qq=1;              % Set frame skip counter
    frameCount=0;      % Initialise frame counter
    u1=zeros(mm,n);
    hbar=waitbar(0,'please wait...');
    for it=1:mm
```

```
    qq=qq-1;
    if qq==0
      frameCount=frameCount+1;
      if plotType==1 || (plotType==2 && ncase==1)
%
%       2D Plot
        if plotType==1
          h2D=plot(x,u(it,:));
        else
          h2D=plot(x,u(it,:),'o',x,u_anal(it,:),'-');
          legend('Numerical solution', ...
                 'Analytical solution', ...
                 'location', 'northeast');
        end
        xlabel('x')
        ylabel('dependent variable, u(x,t)')
        if ncase==1
          axis([-30 70 0 0.6]);
        else
          axis([-30 70 0 1.0]);
        end
          set(h2D,'LineWidth',2)
          title('Korteweg de Vries (KdV) equation');
          grid on
      else
%
%     Update u1 for each frame
      if it <= frameSteps       % First frame
        u1(it,:)=u(it,:);
      else                      % Subsequent frames
        u1(it-frameSteps:it,:)=u(it-frameSteps:it,:);
      end
      if plotType==2
        h3D=surfl(x,t,u1, 'light');
        shading interp
        axis tight
      else
        h3D=waterfall(x,t(it),u1(it,:));
        set(h3D,'LineWidth',2);
        axis([xl xu t0 tf 0 1])
        hold on;
      end
      title('Korteweg deVries (KdV) equation');
      set(get(gca,'XLabel'),'String','space, x')
      set(get(gca,'YLabel'),'String','time, t')
      set(get(gca,'ZLabel'),'String', ...
          'dependent variable,u(t,x)')
```

```
      view(-10,70);
      colormap('cool');   % set color map
    end
    Frames=getframe(h1, frameRect);   % Create image in Frames
                                      % array
    aviMov=addframe(aviMov,Frames);   % Add frame to avi movie
                                      % file
    disp(['    Plot number: ', num2str(frameCount),...
          ' created']);
    qq=frameSteps;  % Reset qq
    waitbar(it/mm);
  end
  close(hbar);
  aviMov=close(aviMov); % Close avi movie file
  fprintf('\navi movie created in file: %s \n', avi_fName);
%
% Print image to file
% print -dpng -r300 KdVEqnPlot
```

Listing A.6.5. Code to be used following a KdV equation simulation run to produce a variety of 2D and 3D plots

An example of 3D plot is shown in Chapter 7.

ANIMATED GIF FILES

As an alternative to generating AVI movie files, it is also possible to generate *animated GIF files*. These files, which also produce a movie, are usually much smaller than AVI files and are particularly useful for displaying simulation results in Web pages. The process is simple and straightforward and involves saving image files at each simulation step and then assembling them into a composite animated GIF file. The one drawback is that GIF files cannot be generated directly from within Matlab using standard functions. However, many graphics packages provide this facility and a simple search will reveal a number of *freeware* and/or *shareware* utilities that can be downloaded from the Web and can perform the GIF animation task, for example, *VideoMach-standard*, a shareware product that can be downloaded from `http://www.gromada.com/videomach.html`.

EXAMPLE – 3D LAPLACE EQUATION MOVIE

The GIF animation procedure will be illustrated using Matlab code that can be used following the 3D Laplace equation code detailed in Chapter 11. This is an interesting graphical image problem as it relates to the solution of a 3D Laplace equation problem whereby the solution is obtained by converting it into a *pseudo time-dependent 3D problem*; that is, a 4D problem, and letting the simulation proceed to steady state that represents the final converged solution. Now, as mentioned in Chapter 11, it is not possible to view a 4D problem in the usual way, so we have to find a different approach.

The way we approach this difficult problem is to use the volume `slice` function in Matlab to produce a sequence of slices through the 3D volume at each time step and saving the corresponding image files. We then use these images to generate an animated sequence. This produces a clear idea of what is happening.

The following code generates a sequence of images from Matlab workspace data populated by a previous simulation run; that is, this code should be run following the Laplace 3D simulation code detailed in Chapter 11. A detailed description of the code will not be provided as ample comments are included (Listing A.6.6).

```
%
% Run after pde_1_main.m
% Generates a sequence of images from Matlab workspace data
% populated by a previous simulation run
  delta=1/(nx-1);
  [x,y,z]=meshgrid(0:delta:1,0:delta:1,0:delta:1);
  figure(1); clf; hold on; axis equal; axis on;
%
% Set suitable color map
  colormap(jet(16));
%
% Set Shading
  shading faceted
  for it=1:nout
%
%   Set 3D data at step 'it'
    w(:,:,:)=u(it,:,:,:);
%
%   Define slice planes to include in plot
    xslice=[0.1,.5,0.9]; yslice=[0]; zslice=[0,1];
%
%   Create slice image
    slice(x,y,z,w,xslice,yslice,zslice)
    xlabel('x');ylabel('y');zlabel('z');
    tStr=sprintf('Step No: %d',it);
    title(tStr);
    caxis([0,1]);
%
%   Include colorbar - scale
    colorbar
%
%   Set viewpoint
```

```
        view(-161,8);
%
%       Set file name with numeric increment
        file=sprintf('Movie_frame%d.png', 1000+it);
        disp(sprintf('Saving to %s', file));
%
%       Save image file
        print('-dpng','-zbuffer','-r100',file);
    end
```

Listing A.6.6. Code to be used following a 3D Laplace equation simulation run to produce a sequence of 3D `slice` plots

The code given in Listing A.6.6 generates in a set of image files that will form the basis of the animated GIF file. Shown in Figure A.6.1 is a composite image showing the results of steps 2–5 (where the simulation was previously run with a step of 0.04). It will be observed that at each step the color coding of each slice changes.

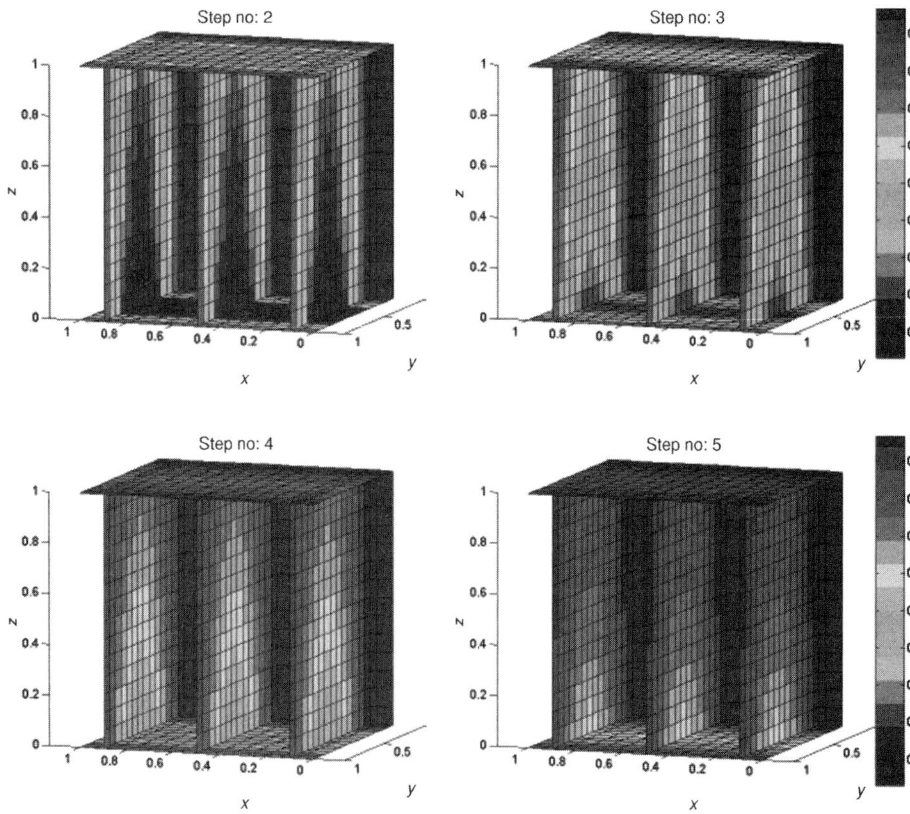

Figure A.6.1. Composite image showing results of steps 2 to 5 from Listing A.6.6 code run following `pde_1_main` of Chapter 11. (See also color plate in Color Plate Section)

Eventually, when fully converged to the solution (after approximately 12 steps), the images all become the same color corresponding to $u(x, y, z) = 1$.

The next step is to assemble into an animated GIF file using a suitable utility, such as the one mentioned earlier. This process is straightforward and most packages provide the facility to resize the images and set the number of frames to be displayed per second and other parameters that enable the user to achieve the desired overall aesthetic appearance of the animation.

EXAMPLE – SPHERICAL DIFFUSION EQUATION MOVIE

This example relates to the spherical coordinates diffusion problem described in Chapter 14. In order to display the results properly, we therefore need to perform a polar to Cartesian conversion on these data before creating the image. We will illustrate this by reference to the simulation performed by pde_2_main: first by the code to generate static plots of the inhomogeneous term $f(r, \theta, \phi)$, and second by the code to generate an animated GIF file of the simulation results $u(r, \theta, t)$.

The code given in Listing A.6.7 generates two static images side by side of the inhomogeneous function: one for the r–ϕ plane and one for the r–θ plane.

```
% Creates x-y and x-z plane polar plots of inhomogeneous 2D
% Gaussian source term 'f(r,theta,phi)' from pde_2_main.m
% code of Chapter 14
% Note: f is symmetrical about phi.
%
% Clear previous files
  clc; clear all
%
% Model parameters
  D=0.1;
  r0=5; th0=2*pi;
  std_0=1.0; std_pi2=2.0;
  tau=1.0;
  NN=101;
%
% Radial grid
  nr=NN; r=linspace(0,r0,nr);
%
% Angular grid
  nth=NN; th=linspace(0,th0,nth);
%
% Create polar mesh
  [R, Theta]=meshgrid(r, th);
  [X,Y]=pol2cart(Theta,R); % Convert to Cartesian coordinates
  [X,Z]=pol2cart(Theta,R); % Convert to Cartesian coordinates
%
```

```
% Calculate Inhomogeneous function data
  for i=1:nr, for j=1:nth
    F1(i,j)=exp(-(r(i)/std_0 )^2); % Symmetrical about phi
    F2(i,j)=exp(-(r(i)*sin(th(j))/std_pi2)^2 ...
              -(r(i)*cos(th(j))/std_0  )^2);
    end, end
%
% Convert function data to be suitable for use with
% the meshgrid using cubic interpolation
  Fxy=interp2(th,r,F1,Theta,R,'cubic');
  Fxz=interp2(th,r,F2,Theta,R,'cubic');
%
% Plot r-phi (x-y) plane of inhomogeneous source,
% f(r,theta,phi) over four quadrants (uses Cartesian
% coordinate data)
  figure()
  subplot(1,2,1)
  surf(X, Y, Fxy); shading interp;
  axis square; axis tight
  caxis([0,1]); colormap jet
  xlabel('X'); ylabel('Y'); zlabel('f(r,\theta)');
  title([{'f(r,\theta,\phi) - (r- \phi) Plane'}]);
  view([0,90]);
%
% Plot r-theta (x-z) plane of inhomogeneous source,
% f(r,theta,phi) over four quadrants  (uses Cartesian
% coordinate data)
  subplot(1,2,2)
  surf(X, Z, Fxz); shading interp;
  axis square; axis tight
  caxis([0,1]); colormap jet
  xlabel('X'); ylabel('Z'); zlabel('f(r,\theta,phi)');
  title([{'f(r,\theta,\phi) - (r- \theta) Plane'}]);
  view([0,90]);
% print -dpng  -r300 func_f1.png % Save image file
```

Listing A.6.7. Code to generate r–ϕ plane and r–θ plane plots of the inhomogeneous function $f(r, \theta, \phi)$

We can note the following points about this code section:

1. A polar coordinate mesh is created, based on grid variables `r` and `th`, which is then converted to Cartesian coordinates for subsequent use by the plot function `surf`.

```
%
% Create polar mesh
  [R, Theta]=meshgrid(r, th);
  [X,Y]=pol2cart(Theta,R); % Convert to Cartesian coordinates
  [X,Z]=pol2cart(Theta,R); % Convert to Cartesian coordinates
```

2. The inhomogeneous function is evaluated over the r–ϕ plane and the r–θ plane.

```
%
% Calculate Inhomogeneous function data
  for i=1:nr, for j=1:nth
    F1(i,j)=exp(-(r(i)/std_0 )^2); % Symmetrical about phi
    F2(i,j)=exp(-(r(i)*sin(th(j))/std_pi2)^2 ...
                -(r(i)*cos(th(j))/std_0  )^2);
    end, end
```

Note: The preceding code could have been implemented more efficiently using the following vectorized code, but we wish to emphasize the use of polar coordinates.

```
%
% Calculate Inhomogeneous function data
  F1=exp(-(Y/std_0).^2   -(X/std_0  ).^2);
  F2=exp(-(Z/std_pi2).^2 -(X/std_0  ).^2);
```

3. The function f data are now converted using the Matlab function `interp2` with *cubic interpolation* so as to be suitable for use with the meshes `[X,Y]` and `[X,Z]`. This step is required because the data `F1` and `F2` correspond to grid variables `r` and `th`, not the `meshgrid` variables `X`, `Y`, and `Z`.

```
%
% Convert function data to be suitable for use with the
% meshgrid using cubic interpolation
  Fxy=interp2(th,r,F1,Theta,R,'cubic');
  Fxz=interp2(th,r,F2,Theta,R,'cubic');
```

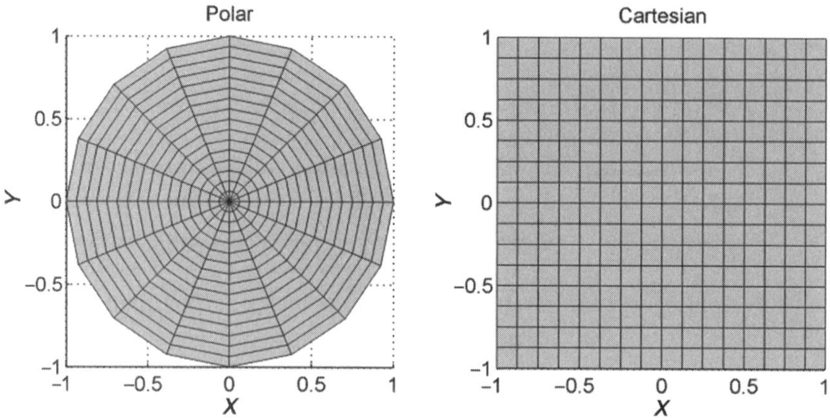

Figure A.6.2. Polar and Cartesian grid plots. (See also color plate in Color Plate Section)

That this last operation is necessary is seen from Figure A.6.2, which illustrates that conversion from a polar coordinate grid to a Cartesian coordinate grid requires interpolation; that is, there is not a simple mapping from each polar grid point to a corresponding Cartesian grid point or vice versa. Thus, to obtain $f(x, y)$ on the Cartesian grid at points (x_i, y_j), first we calculate the corresponding locations on the polar grid, $r_a = \sqrt{(x_i^2 + y_j^2)}$ and $\theta_b = \arctan(y_j/x_i)$, from which we obtain $f(x_i, y_j) = f(r_a, \theta_b)$ by interpolation of the function $f(r, \theta)$. The result is shown in Figure A.6.3.

The remainder of the code should be understandable from the embedded comments. Having demonstrated how polar coordinate data can be converted to Cartesian coordinate data and plotted using the Matlab `surf` function, we now proceed to the main task of this example, which is to create an animation of the spherical coordinate diffusion process described in Chapter 14.

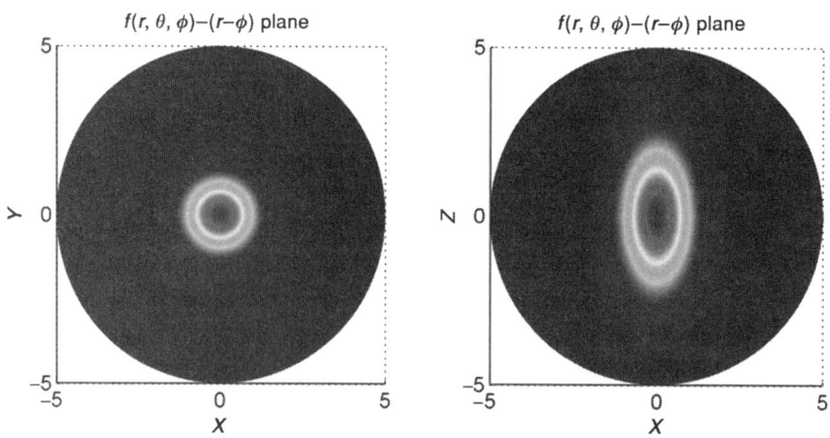

Figure A.6.3. Plots from Listing A.6.7 code showing Chapter 14 inhomogeneous function $f(r, \theta, \phi)$ data; *left* – r–ϕ plane, *right* – r–θ plane. (See also color plate in Color Plate Section)

The code given in Listing A.6.8 generates a sequence of images from Matlab workspace data populated by a previous simulation run; that is, this code should be run following the diffusion equation in spherical coordinates simulation code pde_2_main detailed in Chapter 14.

```
% Run after pde_2_main.m code from Chapter 14
% Generates a sequence of images from Matlab workspace data
% populated by a previous simulation run. Images can then be
% used to create an animated gif file.
%
% Set Up Plotting Vars
  r1=linspace(-r0,r0,2*nr);
  th1=linspace(0,2*pi,2*nr);
%
% Create mesh for single quadrant image
  [R1, Theta1]=meshgrid(r, th);
%
% Create mesh for four-quadrant mesh
  [R2, Theta2]=meshgrid(r1, th1);
%
% Convert meshes to Cartesian form
  [X1,Z1]=pol2cart(Theta1,R1);
  [X2,Z2]=pol2cart(Theta2,R2);
  plotType = 2; % Set plot type: 1=single-quadrant,
                % 2=four-quadrant
%
% Plot dependant variable, u(r,theta) over one quadrant
% at t=tf
  figure();
  for it=1:1:nout
%
% Create 2D matrix of simulation output data at t(it)
  U0(:,:)=u(it,:,:);
%
% Convert results data to be suitable for use with the
% meshgrid using cubic interpolation
  U1=interp2(th,r,U0,Theta1,R1,'cubic');
  if plotType == 1
%
%   Create plot from single quadrant data
    surf(Z1, X1, U1); hold on;
  else
%
%   Create four-quadrant data set from single quadrant data
%   by flipping (vertical/horizontal) and shifting operations
```

```
        U2=quadJoin(U1);
%
%       Create plot from four-quadrant data
        surf(Z2, X2, U2); hold on;
    end
    shading  interp
    colormap jet
    axis tight;  axis square
    colorbar(128); caxis([0,1]);  % Add color key
    xlabel('X'); ylabel('Z'); zlabel('u(r,\theta)');
    tmpStr=sprintf('at t=%2.1f',t(it));  % Prepare title text
    title([{'u(r,\theta), r=(X^2+Z^2)^{0.5},\theta=arctan(Z/X)'}
           {tmpStr}]);
    view([0,90]);  % Set azimuth and elevation
%
% Set file name with numeric increment
    file=sprintf('sphericalMov%d.png', 100+it);
    disp(sprintf('Saving image to file: %s', file));
%
% Save image file
    print( '-dpng', '-zbuffer', '-r100', file);
end
```

Listing A.6.8. Code to be used following the spherical coordinate diffusion process simulation described in Chapter 14. It produces a sequence of r–θ plane plots of the result $u(r, \theta, t)$

We can note the following points about this code section:

1. Because the problem is symmetrical about ϕ, the simulation only needs to generate results $u(r, \theta, \phi, t)$ for one quadrant of the r–θ plane. The remaining quadrants can be obtained by a simple transformation of this single quadrant. By setting `plotType=1`, a sequence of single-quadrant images is generated or by setting `plotType=2`, a sequence of four-quadrant images is generated, as shown in Figure A.6.4.

```
%
plotType=2; % Set plot type: 1=single-quadrant,2=four-quadrant
```

2. The following call to function `quadJoin`, discussed subsequently, performs the necessary transformations on the single-quadrant data `U1` to create the four-quadrant data `U2`.

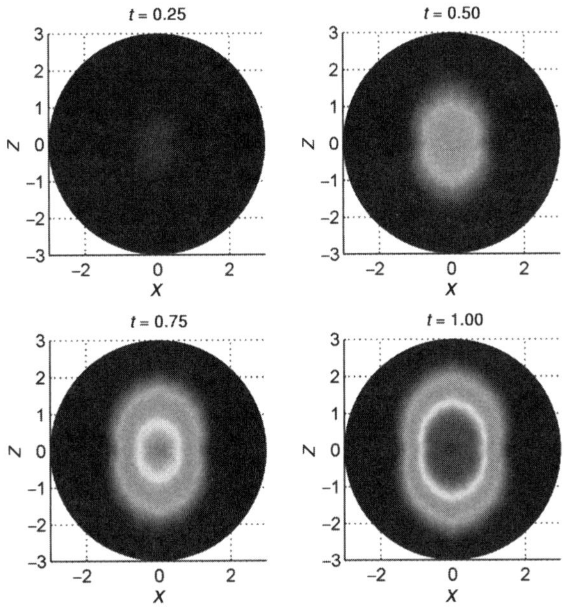

Figure A.6.4. Plots from Listing A.6.8. code showing time sequence of r–θ plane results from program pde_2_main of Chapter 14. (See also color plate in Color Plate Section)

```
%
%   Create four-quadrant data set from single quadrant data
%   by flipping (vertical/horizontal) and shifting operations
    U2=quadJoin(U1);
```

The function quadJoin given in Listing A.6.9 performs combinations of horizontal/vertical flips/shifts on single-quadrant data and combines the results to form a set of corresponding four-quadrant data.

```
  function B=quadJoin(A)
%
% Creates a four-quadrant image from a single quadrant image
% To be used with sphericalPlot.m
  [m,n]=size(A);
  B=zeros(2*n,2*n); % Pre-allocate array for four-quadrant
                    % data
  for i=1:n
    for j=1:n
      B(i      , n+1-j)= A(i,j);    % Top left quadrant
```

```
      B(i       , n+j  )= A(i,j);     % Top right quadrant
      B(2*n+1-i, n+1-j)= A(i,j);      % Bottom left quadrant
      B(2*n+1-i, n+j  )= A(i,j);      % Bottom right quadrant
    end
  end
```

Listing A.6.9. Code for function quadJoin

REFERENCES

[1] Biran, A. and M. Briener (1998), *Matlab 5 for Engineers*, Addison-Wesley, UK

[2] Hanselman, D. and B. Littlefield (2004), *Mastering MATLAB 7*, Prentice Hall, New Jersey

[3] Part-Enander, E. and A. Sjoberg (1999), *The Matlab 5 Handbook*, Addison-Wesley, UK

Index

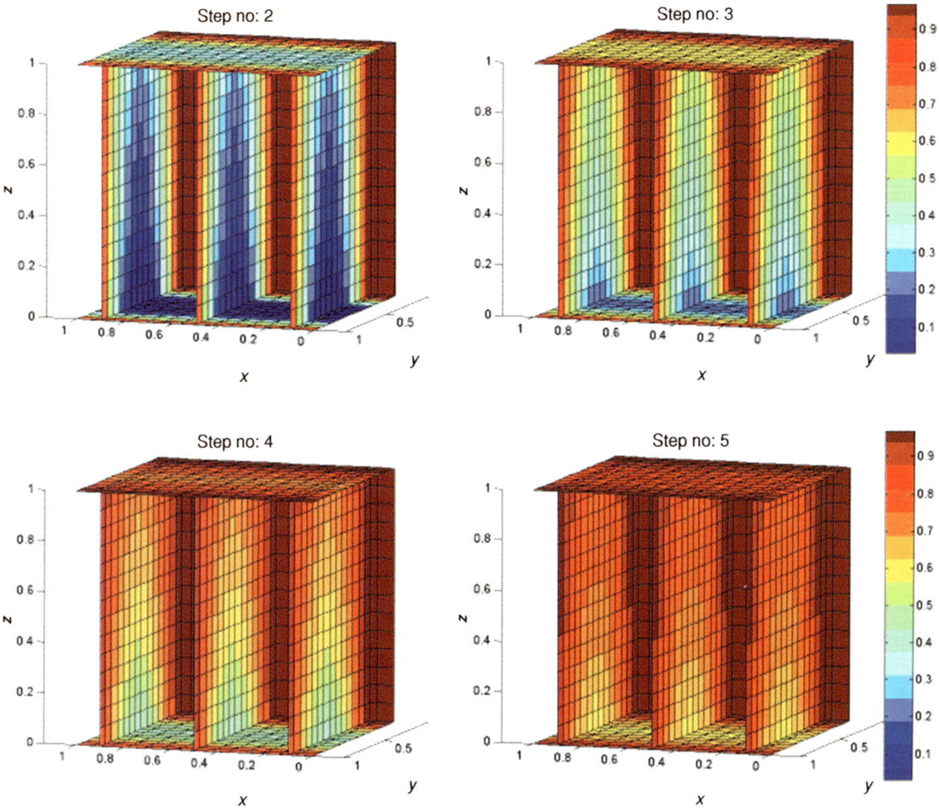

Plate A.6.1. Composite image showing results of steps 2 to 5 from Listing A.6.6 code run following `pde_1_main` of Chapter 11

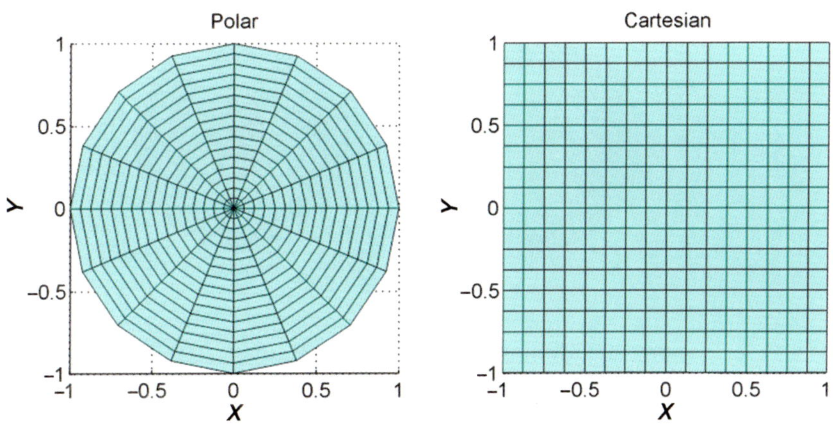

Plate A.6.2. Polar and Cartesian grid plots

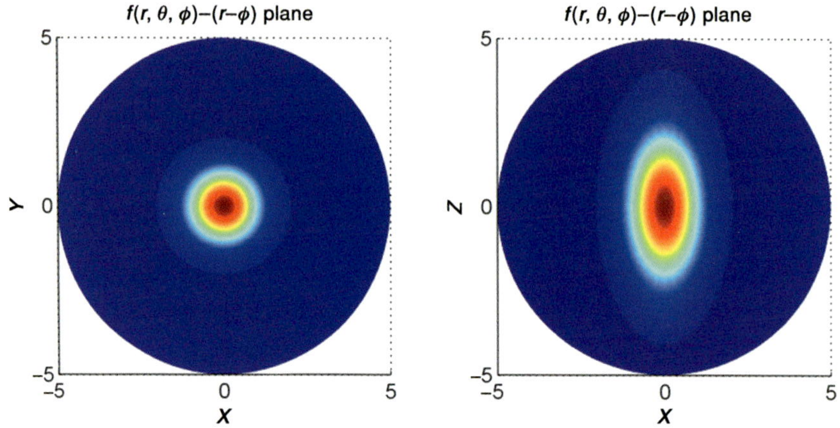

Plate A.6.3. Plots from Listing A.6.7 code showing Chapter 14 inhomogeneous function $f(r,\theta,\phi)$ data; *left* – r–ϕ plane, *right* – r–θ plane

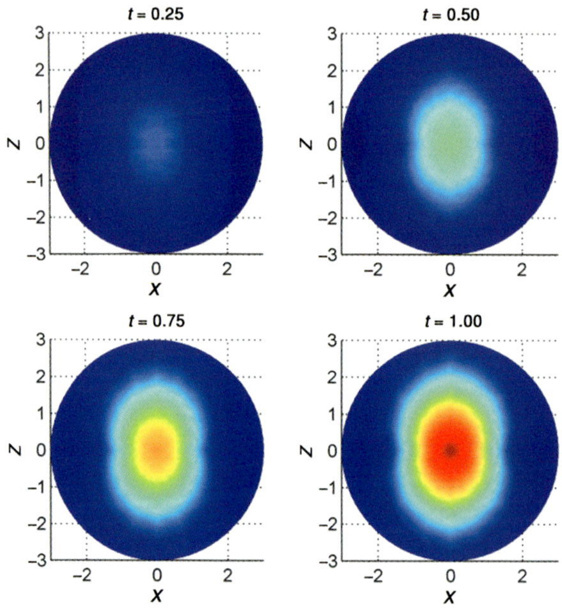

Plate A.6.4. Plots from Listing A.6.8. code showing time sequence of r–θ plane results from program `pde_2_main` of Chapter 14